LTE-Advanced Air Interface Technology

Xincheng Zhang
Xiaojin Zhou

CRC Press
Taylor & Francis Group
Boca Raton   London   New York

CRC Press is an imprint of the
Taylor & Francis Group, an **informa** business

CRC Press
Taylor & Francis Group
6000 Broken Sound Parkway NW, Suite 300
Boca Raton, FL 33487-2742

Version Date: 20120709

International Standard Book Number: 978-1-4665-0152-2 (Hardback)

Visit the Taylor & Francis Web site at
http://www.taylorandfrancis.com

and the CRC Press Web site at
http://www.crcpress.com

Contents

Preface

During the 400 days and nights before we finalized this book on January 1, 2012, we saw the sunset decorated by the raindrop, and were accompanied by the flowers blooming to withering. All those past memories are engraved on our hearts and will last eternally, cementing the most profound friendship in the world.

The number of mobile subscribers has been increasing tremendously since 2000 and reached five billion in the middle of 2010. The Third-Generation Partnership Project (3GPP) submitted Long Term Evolution Advanced (LTE-Advanced) in October 2009 to the International Telecommunications Union (ITU) as a proposed candidate International Mobile Telecommunications-Advanced (IMT-Advanced) technology. The specifications for LTE-Advanced became available in 2011 through Release 10. With this background, this book was planned since we had just finished our last coauthored LTE book *Technologies and Performance of LTE Air Interface.*

As a professional book on LTE-Advanced and related technical issues, this book thoroughly covers the performance targets and technology components studied by 3GPP for LTE-Advanced. It discusses in detail the highlighted evolutions in the first release of LTE-Advanced, Release 10, as well as LTE Release 9. It refers to numerous 3GPP drafts and various technical proceedings on LTE-Advanced. Therefore, besides being an explanatory text about LTE-Advanced air interface technologies, it tries to exploit the technical details in the 3GPP specification. It explains why the specification was written as it was, or what a description in the specification implies.

Chapter 1 provides a general description of the wireless cellular technology evolution and the performance targets of LTE-Advanced. Chapter 1 also gives a short discussion on the major technical features introduced in LTE-Advanced.

From Chapter 2 through Chapter 8, seven innovative technical features in LTE-Advanced are discussed in detail. Chapter 2 presents innovative concepts for carrier aggregation techniques. Chapter 3 explains collaborative multipoint (CoMP) theory and provides performance analysis. Enhanced multiantenna solutions or multiple-input, multiple-output (MIMO) technology, in particular multiuser and multilayer MIMO, are treated in Chapter 4. Relaying issues are discussed thoroughly in Chapter 5. The self-organizing network is addressed in Chapter 6, and heterogeneous networks are presented in Chapter 7. Finally, Chapter 8 discusses interference suppression and enhanced intercell interference coordination (eICIC) technology.

This book focuses on pure LTE-Advanced technical issues, so it is preferred that you already have technical knowledge of LTE when you read it.

Finally, although this is our first technical book in English on such a complicated topic and it may not be perfect, we will be gratified if even one point in the book can help you in your work on LTE-Advanced-related research or implementation.

We hope you will enjoy reading this book, and wish you good health and success in your work.

About the Authors

Xincheng Zhang graduated from the Beijing University of Posts and Telecommunications (Beijing, China) in 1992. He has been working for China Mobile for 20 years as a technical expert with a solid understanding of wireless communication technologies. He is working as a senior wireless network specialist in the fields of antenna arrays, analog/digital signal processing, radio resource management, propagation modeling, etc. He has participated in many large-scale wireless communication system designs for a variety of cellular systems using various radio access technologies, including the Global System for Mobile Communication (GSM),
Code Division Multiple Access (CDMA), Universal Mobile Telecommunications System (UMTS), and Long Term Evolution (LTE). He is the author or coauthor of eight books on Wideband Code Division Multiple Access (WCDMA), Worldwide Interoperability for Microwave Access (WiMAX), and LTE.

Xiaojin Zhou received his first master's degree in electronic and information engineering from the Beijing University of Posts and Telecommunications (Beijing, China), and his second master's degree in the engineering management program from the University of Ottawa (Ottawa, Canada). He has worked for China Mobile for many years, and has extensive knowledge of wireless communication systems and data communication networks. He now works for Ericsson, which he joined in 2008, and is actively participating in leading-edge LTE product design and implementation. He is the coauthor of several books on wireless technologies, including WiMAX and LTE.

Chapter 1

From LTE to LTE-A

Exponential growth in traffic volume calls for high data rates with mobility to meet the ever-increasing user demands. Multimedia traffic is increasing far more rapidly than speech traffic and will increasingly dominate traffic flows. Enhanced peak data rates are definitely necessary to support advanced services and applications. According to Nielsen's Law, network connection speeds for high-end home users will increase by 50% per year, or double every 21 months (as shown in Figure 1.1). This growth rate is slightly slower than Moore's Law of processor power growth (double in 18 months).

From 1980, with the bulky first-generation analog handsets, the wireless cellular communications industry traveled a long evolutionary path (as shown in Figure 1.2) and now enjoys a major presence in the world. When Global System for Mobile Communication (GSM) multiple access technology was selected 1987, Code Division Multiple Access (CDMA) was already proposed. When Wideband Code Division Multiple Access (WCDMA) multiple access technology was selected 1998, orthogonal frequency-division multiple access (OFDMA) was already proposed. The Long Term Evolution (LTE) study item was initiated in the Third-Generation Partnership Project (3GPP) back in 2004 to develop a framework for the evolution of the 3GPP radio access technology toward a high-data-rate, low-latency, and packet-optimized radio access technology. An LTE market opportunity has emerged during the past two years, and consequently most vendors now have relatively well-defined solutions that are scheduled to be ready for initial rollout. We believe that LTE will become the dominant mobile network technology and that most network operators will upgrade to it. But are there any other real options? Has the industry reached the *ultimate multiple access*?

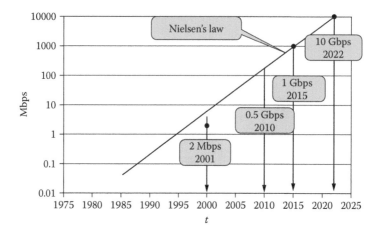

Figure 1.1 Nielsen's Law of Internet Speed.

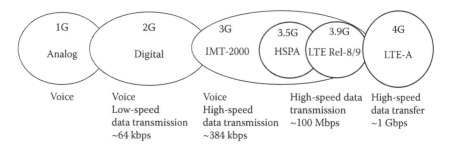

Figure 1.2 Mobile communication system evolution.

1.1 Review of LTE

1.1.1 Wireless Technologies

Wireless technology is playing a profound role in networking and communications because it provides two fundamental capabilities: mobility and access. Today's wireless market winners (shown in Figure 1.3) are medium-capacity mobile broadband networks, including Enhanced Data Rates for GSM Evolution (EDGE), WCDMA, High Speed Packet Access (HSPA), and the LTE of the International Telecommunications Union (ITU) family, wireless local-area networks (WLAN), and Worldwide Interoperability for Microwave Access (WiMAX) of the Institute of Electrical and Electronics Engineers (IEEE) family. Currently, IEEE 802.16-2004/802.16e (Portable and Mobile WiMAX) and 3GPP LTE are two major mobile broadband radio technologies.

In the ITU family, the Universal Mobile Telecommunications System (UMTS) employs the wideband CDMA radio-access technology to establish third-generation

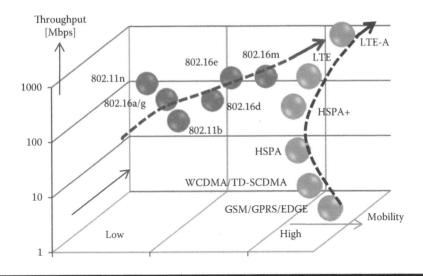

Figure 1.3 Popular wireless technologies. The dashed lines indicate the two evolution paths and directions of IEEE and 3GPP.

(3G) wireless networks. The primary benefits of UMTS include high spectral efficiency for voice and data, and simultaneous voice and data capability for users. Initial UMTS network deployments were based on 3GPP Release 99 specifications, in which the maximum theoretical downlink rate is just over 2 Mbps. Since then, Rel-5 defined High-Speed Downlink Packet Access (HSDPA), which delivers downlink (DL) peak theoretical rates of 14 Mbps, and Rel-6 defined High-Speed Uplink Packet Access (HSUPA). With HSPA-capable devices, the network uses HSPA (HSDPA/HSUPA) for data transmission. Now, HSPA+ with 42-Mbps capability on the downlink by higher-order modulation and multiple-input, multiple-output (MIMO) and 11.5 Mbps on the uplink (UL) have been widely deployed.

LTE is a standard for wireless data communications technology and an evolution of the GSM/UMTS standards. The LTE specification provides downlink peak rates of 300 Mbps (4x4 MIMO), uplink peak rates of 75 Mbps, and quality of service (QoS) provisions permitting round-trip times of less than 10 ms. LTE has the ability to manage fast-moving mobiles and provides support for multicast and broadcast streams. LTE supports scalable carrier bandwidths, from 1.4 MHz to 20 MHz, and supports both frequency-division duplexing (FDD) and time-division duplexing (TDD) system. The architecture of the network is simplified to a flat IP-based network architecture called the Evolved Packet Core (EPC), designed to replace the general packet radio service (GPRS) core network and support seamless handovers for both voice and data to cell towers with older network technology such as GSM, UMTS and CDMA2000. The possible peak data rates along the LTE evolution path are summarized in Figure 1.4.

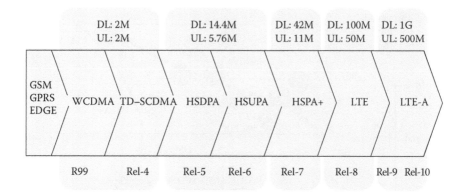

Figure 1.4   Evolution of the mobile network.

LTE Advanced (LTE-A) is expected to provide higher data rates while maintaining coverage proportional to LTE Release 8 (Rel-8). It aims to provide peak data rates of 1 Gbps in downlink and 500 Mbps in uplink, bandwidth scalability up to 100 MHz, increased spectral efficiency up to 15 bps/Hz in uplink (UL) and 30 bps/Hz in DL, improved cell-edge capacity, as well as decreased user and control plane latencies relative to LTE Rel-8.

1.1.2 LTE Performance

LTE Rel-8 is one of the primary broadband technologies based on Orthogonal Frequency-Division Multiplexing (OFDM), which is currently being commercialized. During 2005, LTE and System Architecture Evolution (SAE) study items were set up in 3GPP. Both operators and manufacturers were keen to push these study items because they were facing competition from other technologies such as WiMAX. Operators are looking for significant improvements compared to current UMTS releases that provide an architecture that is better suited to the shift from circuit-switched communications toward packet data and a radio technology that enables spectrum refarming. LTE Rel-8 provides high peak data rates of 150 Mbps (2x2 MIMO) on the downlink and 75 Mbps on the uplink for a 20-MHz bandwidth and allows flexible bandwidth operation of from 1.4 MHz up to 20 MHz. LTE Rel-8, which is mainly deployed in a macro/micro cell layout, provides improved system capacity and coverage, high peak data rates, low latency, reduced operating costs, multiantenna support, flexible bandwidth operation, and seamless integration with existing systems. Not only is OFDMA good for single cell, but also a better solution may be optimized for full intercell interference coordination. A comparison of the requirements and performance results of LTE is shown in Table 1.1.

The LTE network architecture is designed with the goal of supporting packet-switched traffic with seamless mobility, QoS, and minimal latency. LTE Rel-8 supports a cell average spectral efficiency gain of 2–3 times HSPA Rel-6, with

Table 1.1   LTE Requirements versus LTE Performance

Item		*LTE Requirement*	*LTE Performance*
Peak Data Rate	DL	> 100 Mbps	326.4 Mbps (4 layer)
		(5 bps/Hz)	172.8 Mbps (2 layer)
	UL	> 50Mbps	86.4 Mbps (64QAM)
		(2.5 bps/Hz)	57.6 Mbps (16QAM)
C-plane Latency	Idle to Active	< 100 ms	51.25 ms + 3 * S1 delay
	Dormant (DRX) to Active	< 50 ms	<< 51.25 ms
U-plane Latency		< 5 ms	4 ms

radio network user plane latency below 10 ms (round-trip time, or RTT). The packet-switched approach in LTE allows support for all services including voice through packet only connections. Therefore, a highly simplified flatter architecture is adopted by LTE with only two node types: evolved Node-B (eNB) and mobility management entity/gateway (MME/GW). Compared to 3G systems, LTE reduced the number of different types of radio access network (RAN) nodes and their complexity, reduced capital expenditure (CAPEX) and operating expenses (OPEX), and reduced the complexity of terminals. The LTE network would coexist with both the UMTS/HSPA terrestrial radio access network (UTRAN) and the GSM/EDGE radio access network (GERAN).

1.1.3 Challenges to the Next-Generation Network

Although LTE has much better performance than its predecessors, ongoing improved service requirements always put challenges ahead of wireless network development. The next-generation network requires lower radio latency, higher spectral efficiency, and more flexible and faster mobility, as well as some kind of cognitive radio that enables optimized spectrum usage over multiple operators. All these requirements need evolution at the air interface and new network topology development, as illustrated in Figure 1.5.

Furthermore, the IP-optimized mobile network in the future will provide various types of communications services through high bandwidth, low latency, and new packet-optimized broadband radio technologies. Based on the modern flat network structure, the next-generation network (as illustrated in Figure 1.6) will deploy many state-of-the-art technologies, including cooperative multipoint communication, cognitive radio, multilayer communication, heterogeneous networks and advanced MIMO, to implement a brand new mobile communication architecture. The flat network architecture will provide support to distributed antennas,

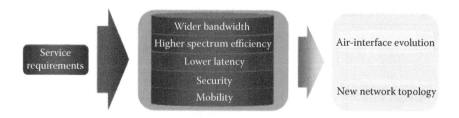

Figure 1.5   Challenges to mobile networks.

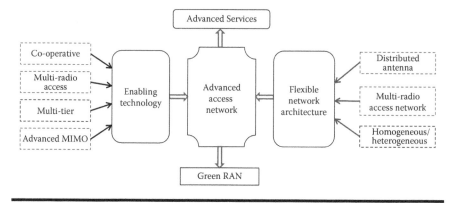

Figure 1.6   Next-generation mobile networks.

multiple radio access, and hybrid networking, while aiming to achieve "green" communication. With the existence of smart terminals and the popularity of mobile Internet, a more powerful mobile network should be developed to provide a better user experience.

1.2 Performance Targets of LTE-A

In order to exceed the performance requirements for International Mobile Telecommunications-Advanced (IMT-A), defined by ITU Radiocommunication Sector (ITU-R), the work on further evolution of LTE has started; it is called LTE-Advanced (LTE-A).

1.2.1 Background of LTE-A

When 3G wireless networks were established worldwide, the ITU-R organization specified the IMT-A requirements for 4G standards, setting peak speed requirements for fourth-generation (4G) service at 100 Mbps for high-mobility communication (such as from trains and cars) and 1 Gbps for low-mobility communication.

A 4G system is expected to provide a comprehensive and secure all-IP-based mobile broadband solution for laptop computer wireless modems, smartphones, and other mobile devices. Facilities such as ultrabroadband Internet access, IP telephony, gaming services, and streamed multimedia may be provided to users. 3GPP held two workshops on IMT-A, where the "Requirements for Evolved UTRA (E-UTRA) and Evolved Universal Terrestrial Radio Access Network (E-UTRAN)" were gathered. The resulting 3rd Generation Partnership Project Technical Report 36.913 was then published in June 2008 and submitted to the ITU-R; it defines the LTE-A system as their proposal for IMT-A.

1.2.2 Requirements of IMT-A

The requirements for IMT-A, set by the ITU, are performance oriented. The key considerations in IMT-A are shown in Figure 1.7. The system should have the flexibility to cost-effectively support a wide range of services and should be capable of interworking with other radio access systems and global roaming. From a performance perspective, the system should support 100 Mbps for high mobility and 1 Gbps for low mobility. The ITU recommends operation for at least 40-MHz and possible extensions up to 100-MHz radio channels and peak spectral efficiency of 15 bps/Hz in the downlink channel; the Voice over Internet Protocol (VoIP) capacity target is 40 users per cell (= sector) per MHz, with a 12.2-kbps AMR codec and 50% voice activity factor. Cell spectral efficiency targets are 2.60 bps/Hz for the DL and 1.80 bps/Hz for the UL. Target peak spectral efficiency is 15 bps/Hz in the DL and 6.75 bps/Hz in the UL. Expected cell-edge user spectral efficiency is 0.075 bps/Hz in the DL and 0.05 bps/Hz in the UL. The control-plane latency should be less than 100 ms, excluding paging delay and wireline network signaling delay. One-way user-plane delay, also called *transport delay*, is targeted to be less than 10 ms. It is an IP-to-IP delay (e.g., from the eNB's IP layer to the user equipment's [UE] IP layer). There are different classes of mobility ranging from stationary to high speeds from

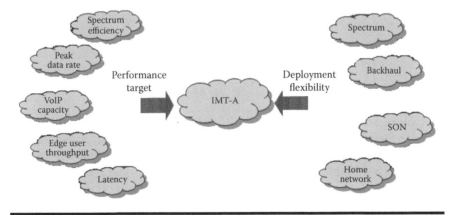

Figure 1.7 IMT-A key considerations.

Table 1.2   Some IMT-A Requirements

Test Environment	Average Spectral Efficiency (b/s/Hz/cell)		Cell-Edge User Spectral Efficiency (b/s/Hz/cell)		Minimum VoIP Capacity (UE/cell/MHz)
	Downlink	*Uplink*	*Downlink*	*Uplink*	
Indoor	3	2.25	0.1	0.07	50
Microcell	2.6	1.8	0.075	0.05	40
Urban	2.2	1.4	0.06	0.03	40
Highway	1.1	0.7	0.04	0.015	30

120 km/h to 350 km/h. The handover interruption time should be less than 27.5 ms in the case of intrafrequency handover and 40 ms in the case of interfrequency (but same frequency band) handover. Targets for interfrequency band handover and inter-RAT (radio access technology) handover are not specified. Such handover interruption times exclude the delay associated with interactions between the radio and the core network. Some requirements of IMT-A are summarized in Table 1.2.

Please note, cell spectral efficiency (η) is defined as the aggregate throughput of all users, that is, the number of correctly received bits divided by the channel bandwidth divided by the number of cells. The channel bandwidth for this purpose is defined as the effective bandwidth times the frequency reuse factor, where the effective bandwidth is the operating bandwidth normalized appropriately considering the UL/DL ratio.

The cell spectral efficiency is measured in b/s/Hz/cell. Let χi denote the number of correctly received bits by user i (DL) or from user i (UL) in a system comprising a user population of N users and M cells. Furthermore, let ω denote the channel bandwidth size and T the time over which the data bits are received. The cell spectral efficiency is then defined according to the following equation:

$$\eta = \frac{\sum_{i=1}^{N} \chi i}{T \cdot \omega \cdot M}$$

Cell-edge user spectral efficiency is defined as 5% point of the cumulative distribution function (CDF) of the normalized user throughput. With χi denoting the number of correctly received bits of user i, T_i the active session time for user i, and ω the channel bandwidth, the (normalized) user throughput of user i, γi, is defined by:

$$\gamma i = \frac{\chi i}{T_i \cdot \omega}$$

1.2.3 Requirements of LTE-A

In 2009, 3GPP worked toward identifying which LTE improvements were required to respect the IMT-A specifications. The 3GPP partners made a formal submission to the ITU proposing LTE Rel-10 and LTE-A. They suggested that this new release be evaluated as a candidate for IMT-A. The following paragraphs provide a summary of performance targets that LTE-A aims to achieve.

Data Rates and Spectral Efficiency: The target peak data rate is 1 Gbps in the DL and 500 Mbps in the UL. The peak spectral efficiency is 30 bps/Hz and 15 bps/Hz for the DL and the UL, respectively. Therefore, a 40-MHz spectrum can yield 1.2 Gbps in the DL, and the same amount of spectrum can yield 600 Mbps in the UL.

Mobility: Similar to LTE Rel-8, mobility up to 350 km/h (or 500 km/h depending on the frequency band) would be supported.

Latency: The transition from the idle to connected mode occurs faster than in Rel-8. Instead of the 100-ms target of Rel-8, a stricter 50-ms delay is targeted (as shown in Figure 1.8). When the user equipment (UE) performs discontinuous reception (DRX) in the active mode, the transition from the *dormant* to the *active* substate should occur within 10 ms.

Cost: The costs of the infrastructure and device should be lower. Power efficiency is also considered important. The backhaul does not need to be *wireline* (e.g., optical); it could use an LTE-based air interface. Self-organizing network (SON) features should be supported to reduce the amount of drive testing, leading to cost savings.

Spectrum: Consistent with IMT-A requirements, LTE-A will support several new bands in addition to LTE Rel-8 bands. More specifically, the following bands are initially targeted: 450–470 MHz, 698–862 Hz, 790–862 MHz, 2.3–2.4 GHz, 3.4–4.2 GHz, and 4.4–4.99 GHz (3GPP TR 36.913).

LTE-A is a smooth evolution of LTE and is backward compatible with Rel-8. LTE-A supports a variety of coverage scenarios, provides seamless coverage from macrocell to indoor, and focuses on resolving high-speed data transmission in low-speed mobile environments, including further reductions in technology costs and energy consumption. It offers high spectral efficiency (30 bit/s/Hz downlink, 15 bit/s Hz uplink), low latency, and high peak data rates (1-Gbps downlink, 500-Mbps uplink

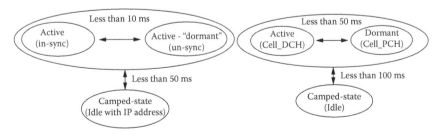

Figure 1.8 Transmission delay performance targets for LTE-A and LTE.

in 20 MHz). LTE-A can use up to 8 × 8 MIMO downlink and 4 × 4 MIMO uplink and introduces carrier aggregation of contiguous and noncontiguous spectrum allocations, relay technology, and collaborative multipoint (CoMP) transmission technology.

Table 1.3 includes a brief description of LTE-A requirements concerning spectral efficiency. The different scenarios listed in Table 1.3 correspond to variable ITU-defined test environments, as shown in Table 1.4.

Since the nominal target peak downlink data rate is ~1 Gbps, 8-layer spatial multiplexing is not required to achieve such a nominal peak rate for 100-MHz downlink transmission. With the current 4-layer Rel-8 setup, a nominal peak rate of ~1.5 Gbps can already be achieved with 100-MHz bandwidth. Consequently, supporting 8-layer spatial multiplexing seems to be intended to achieve an ~1-Gbps peak rate for lower downlink system bandwidth such as 50 MHz.

The performance improvements from LTE to LTE-A are summarized in Table 1.5.

1.2.4 LTE-A Overview

LTE-A (also known as LTE Rel-10, due to its first version number) significantly enhances the existing LTE Rel-8 and supports much higher peak rates, higher throughput and coverage, and lower latencies, resulting in a better user experience. Additionally, LTE-A will support heterogeneous deployments where low-power nodes composed of picocells, femtocells, relays, remote radio heads (RRH), and so on are placed in a macrocell layout. The LTE-A features enable one to meet or exceed IMT-A requirements. It may also be noted that LTE Rel-9 provides some minor enhancement to LTE Rel-8 with respect to the air interface and includes features such as dual-layer beamforming and time difference of arrival (TDOA)-based location techniques. In this book, an overview of the techniques being considered for LTE Rel-10 will be discussed. The topics include bandwidth extension via carrier aggregation to support deployment bandwidths up to 100 MHz and enhanced DL spatial multiplexing, including single-cell multiuser MIMO (MU-MIMO) transmission and CoMP transmission, UL spatial multiplexing including extension to 4-layer MIMO, and heterogeneous networks with emphasis on type-1 and type-2 relays.

To enhance the performance of LTE Rel-8 to meet the anticipated requirements of IMT-A, LTE-A adopted some advanced technologies to meet all the requirements from different aspects, which can be summarized as follows:

■ *Spectrum aggregation:* To maximize the flexibility in spectrum usage independent of regional regulations
■ *Hybrid OFDMA / single-carrier FDMA (SC-FDMA) UL multiple access:* Improvement in UL capacity
■ *UL single-user MIMO (SU-MIMO) 4x4:* Improvement on UL capacity and peak data rate
■ *DL MIMO 8x8:* Improvement on DL capacity and peak data rate

Table 1.3 LTE-A Requirements on Spectral Efficiency

Senarios	Downlink		Uplink	
	Cell Spectral Efficiency (bps/Hz/cell)	Cell-Edge User Spectral Efficiency (bps/Hz)	Cell Spectral Efficiency (bps/Hz/cell)	Cell-Edge User Spectral Efficiency (bps/Hz)
ITU Indoor Hot spot*	3	0.1	2.25	0.07
ITU Urban Micro*	2.6	0.075	1.8	0.05
ITU Urban Macro*	2.2	0.06	1.4	0.03
ITU Rural Macro*	1.1	0.04	0.7	0.015
3GPP Case 1	2.4 (2x2)	0.07 (2x2)	1.2 (1x2)	0.04 (1x2)
	2.6 (4x2)	0.09 (4x2)	2.0 (2x4)	0.07 (2x4)
	3.7 (4x4)	0.12 (4x4)		

Note: Assuming antenna configurations of DL 4x2 and UL 2x4 (up to 8 BS antennas allowed in M.2135).

Table 1.4 ITU Test Environment

	Indoor Hotspot (InH)	Urban Micro (UMi)	Urban Macro (UMa)	Rural Macro (RMa)
Environment	Office building	Urban area	Urban area	Rural area
User position	Indoor	50% indoor, 50% outdoor	Outdoors in car	Outdoors in car
Mobility	Low (3 km/h)	Low (3 km/h)	Medium (30 km/h)	High (120 km/h)
Application	Local area	Wide area	Wide area	Wide area
Cell radius	Small (ISD = 60 m)	Small (ISD = 200 m)	Medium (ISD = 500 m)	Large (ISD = 1732 m)
Antenna configuration	8×2	8×2, below roof	8×2, above roof	8×2, above roof
Radio condition	LOS, NLOS	LOS, NLOS, O-to-I	LOS, NLOS	LOS, NLOS
Interference	Low	High	High	High

Table 1.5 LTE-A Improvements

Performance	LTE-A (36.913v8.0.0-Case1)	LTE-FDD	Note
DL peak spectral efficiency (bps/Hz)	30 (8x8 MIMO)	7.5 (2x2 MIMO) 15 (4x4 MIMO)	
UL peak spectral efficiency (bps/Hz)	15 (4x4)	3.75 (1x2)	4x4 increase above 30%
DL average spectral efficiency (bps/Hz)	2.4 (2x2) 2.6 (4x2) 3.7 (4x4)	1.63 (2x2) 1.93 (4x2) 2.87 (4x4)	
UL average spectral efficiency (bps/Hz)	1.2 (1x2), 2 (2x4)	0.86 (1x2)	
DL cell-edge spectral efficiency (bps/Hz)	0.07 (2x2) 0.09 (4x2) 0.12 (4x4)	0.05 (2x2) 0.06 (4x2) 0.11 (4x4)	
UL cell-edge spectral efficiency (bps/Hz)	0.04 (1x2), 0.07 (2x4)	0.028 (1x2)	2-Tx antenna
Latency of connection setup	<50 (idle), 10 (sleep)	54.3–86	
Prescheduling RTT	<5 ms (not loaded)	10.2	
Non-prescheduling RTT		16.2	
HO interrupt time (ms)	(TR 25.913)	18.5	
VoIP capacity (UE/MHz)	60	44	Up 50%

- *MU-MIMO, network MIMO, distributed antenna MIMO, multimode adaptive MIMO:* Improvement on capacity and cell coverage
- *Superposition coding:* Improvement on DL capacity
- *Enhancement of MBSFN:* Improvement on MBSFN capacity
- *Improved beamforming (BF) techniques (e.g., adaptive BF):* To improve capacity and cell-edge coverage
- *Relay, remote radio equipment:* To improve capacity and cell coverage
- *Wireless network coding:* To improve capacity and cell coverage
- *Enhancements to codebooks and feedback mechanism for closed loop MIMO (CL-MIMO):* To improve CL-MIMO performance
- *Enhanced intercell interference management:* Improve capacity and cell-edge coverage
- *Wireless backhaul/sidehaul:* Simplify deployment constraints
- *Support of Home eNodeB/Femto/Picocells:* Extension of indoor coverage
- *SON:* Simplified network deployment and optimization
- Support both homogeneous and heterogeneous networks
- Support enhanced intercell interference coordination, with more advanced CoMP evolution

LTE standard evolution from Rel-8 to beyond Rel-10 is shown in Figure 1.9.

In summary, the technologies considered for LTE-A (listed in Figure 1.10) include extended spectrum flexibility to support up to 100-MHz bandwidth, enhanced multiantenna solutions with up to eight-layer transmission in the downlink and up to four layer transmission in the uplink, coordinated multipoint transmission/reception, and the use of advanced repeaters/relaying.

Each of the newly adopted technologies in LTE-A can generate performance gain to some extent. The gains attributable to each technology are listed in Table 1.6.

Moreover, from the UE perspective, network performance improvement raised requirements for new UE capacity definitions. Existing UE categories (categories 1–5, listed in Table 1.7) in LTE Rel-8/9 can be extended in LTE-A by additional signaling of the maximum number of supported UL layers, the number of supported component carriers (CCs) (depending on support for carrier aggregation (CA), and if supported, CA band combinations).

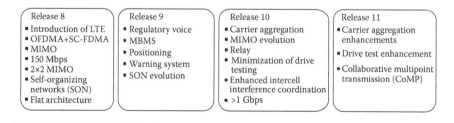

Figure 1.9　LTE standard evolution.

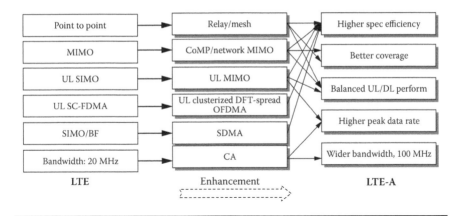

Figure 1.10 Technologies adopted in LTE-A.

New UE categories for LTE-A indicate L1 peak data rate and new features supported in LTE-A, such as higher-order DL MIMO, UL MIMO, carrier aggregation for downlink and uplink, and heterogeneous networks (HetNets). In order to avoid unnecessary mandatory UE features and simplify the interoperability test, there are 3 new categories introduced for LTE-A (as listed in Table 1.8). Additionally, UE categories 1–5 (Rel-8) can also be used for carrier aggregation, provided that the UE has CA capability.

As the maximum UE capability, the UE with 8 downlink layer reception and 4 uplink layer transmission would be required to identify the potential performance of LTE-A. The maximum DL/UL data rate mainly depends on the processing speed of the UE chips and the size of the memory/buffer. The number of layers for DL/UL MIMO relies on the number of receiving and transmitting antennas and the associated radio frequency (RF) chains at the UE. Whether CA can be supported and the number of CCs for CA in DL/UL also depends heavily on the RF capabilities of the UE.

1.3 LTE-A Technologies

1.3.1 Carrier Aggregation

In order to meet the increasing capacity requirement for next-generation systems, a high-bandwidth wireless transmission system for mobile communications has become a major trend, with increases in transmission bandwidth of mobile communication systems from 5 MHz (bandwidth of the initial design) of the UMTS to the 20 MHz of the LTE Rel-8 system and then the 100 MHz of the LTE-A system. Based on the existing RF power amplifier technology, its advantage is that it is easy to implement and is fully compatible with the LTE; the disadvantage is the relatively complex control channel structure.

Table 1.6 Technology Gain in LTE-A

Feature	Details	Gain
Adaptive Channel Estimation	• Reduced complexity minimum mean square error (MMSE) channel estimation for low Doppler cases, adapting to 2 x 1D Weiner filtering for high Doppler cases	Mitigates loss of over 10 dB for high mobility cases (120–350 km)
Interference Cancellation	• Least mean square (LMS) or gradient-based interference cancellation (IC) • Decision feedback interference cancellation • MMSE, successive interference cancellation (SIC), turbo SIC	Up to 9 dB gain
Power Control & Adaptive FFR	• Open and closed loop power control with fractional path loss optimized to interference over thermal (IoT) level • Tiered adaptive fractional frequency reuse (AFFR) • Network AFFR	Up to 10 dB gain in SINR
MIMO	• Support for STTD, SFBC, SM and CDD MIMO, Adaptive SIMO and MU-MIMO • Network MIMO (8x8 on DL, 4x4 on UL)	3–6 dB gain
CoMP	• Spatial partitioning and scheduling of fixed or dynamic transmissions and joint processing will reduce received interference	3–6 dB gain
Beamforming and Interference Suppression	• Beamforming (closed loop MIMO) and null steering to suppress high-gain interferers	Up to 10 dB gain
Relay Nodes	• Improve coverage, cell-edge throughput, and group mobility	3–6 dB gain
Adaptive Scheduling and Cognitive Radio	• Frequency selective scheduling (FSS) and frequency diverse scheduling (FDS) based on use of sub-band and wideband CQI feedback to select RBs with minimum SINR to adaptively avoid intercell interference	Up to 5 dB gain

Table 1.7 UE Categories in LTE

UE Category		DL				UL	
Peak Data Rate (DL/UL)		Maximum Number of Bits of a DL-SCH Transport Block Received within a TTI (DL TBS per TTI)	Maximum Number of Bits of a DL-SCH Transport Block Received within a TTI (DL TBS per CW)	Total Number of Soft Channel Bits (Soft Buffer)	Maximum Number of Supported Layers for Spatial Multiplexing in DL (DL Layer)	Maximum Number of Bits of a UL-SCH Transport Block Transmitted within a TTI (UL TBS per TTI)	Support for 64QAM in UL (UL modulation (MOD))
Cat. 1	10 Mbps / 5 Mbps	10296	10296	250368	1	5160	No
Cat. 2	50 Mbps / 25 Mbps	51024	51024	1237248	2	25456	No
Cat. 3	100 Mbps / 50 Mbps	102048	75376	1237248	2	51024	No
Cat. 4	150 Mbps / 50 Mbps	150752 (= $2N_{DR8}$)	75376 (= N_{DR8})	1827072	2	51024 (= N'_{UR8})	No
Cat. 5	300 Mbps / 75 Mbps	299552 (~ $4N_{DR8}$)	149776 (~ $2N_{DR8}$)	3667200	4	75376 (= N_{UR8})	Yes

Table 1.8 UE Categories in LTE-A

UE Category	Maximum Data Rate (DL / UL) (Mbps)	DL			UL		Support for 64QAM
		Maximum Number of DL-SCH TB Bits per TTI	Maximum Number of DL-SCH Bits per TB per TTI	Total Number of Soft Channel Bits	Maximum Number of UL-SCH TB Bits per TTI	Maximum Number of UL-SCH Bits per TB per TTI	
Category 6	DL 300 Mbps / UL 50 Mbps	301504	149776 (4 layers) 75376 (2 layers)	3667200	51024	51024	No
Category 7	DL 300 Mbps / UL 100 Mbps	301504	149776 (4 layers) 75376 (2 layers)	3667200	102048	51024	No
Category 8	DL 3000Mbps / UL 1500 Mbps	2998560	299856 (8 layers)	35982720	1497760	1497760	Yes

The multicarrier scheme uses a few carriers with no more than 20 MHz aggregated into 20 MHz–100 MHz transmission bandwidth (as shown in Figure 1.11). LTE-A allows 1 to 5 CCs of the same or different bandwidths to aggregate in DL and/or UL to support wider transmission bandwidth and peak data rates between eNB and the UE.

Flexible spectrum use is supported by using one or multiple component carriers, where each component carrier supports a scalable bandwidth up to 20 MHz (100 resource blocks). Multiple component carriers can be aggregated to achieve up to 100 MHz of transmission bandwidth. The aggregated component carriers can be either contiguous or noncontiguous in the frequency domain, including being located in separate spectrum (*spectrum aggregation*). The number of component carriers transmitted and/or received by a UE can vary over time depending on the UE capacity and instantaneous data rate. A Rel-8 terminal is assumed to be served by a single component carrier, while LTE-A terminals can be served simultaneously by multiple component carriers (as shown in Figure 1.12).

In the LTE-A specifications, two or more component carriers can be aggregated. Each component carrier can be configured in a backward-compatible way with LTE Rel-8, so each component carrier satisfies LTE Rel-8 bandwidth requirements (i.e., 5, 10, 15, or 20 MHz). Total aggregate bandwidth is up to 100 MHz. The number of component carriers on the UL and DL may be asymmetrical, and a configuration of more DL than UL CCs is the only permitted scenario. In typical TDD deployments, the number of CCs in UL and DL is typically the same.

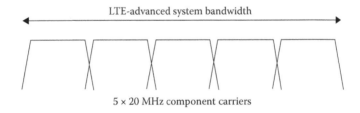

LTE-advanced system bandwidth

5 × 20 MHz component carriers

Figure 1.11 Component carriers to form LTE-Advanced system bandwidth.

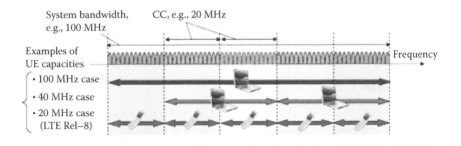

Figure 1.12 Multiple component carriers.

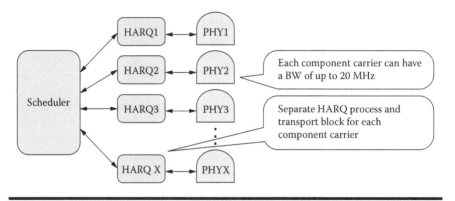

Figure 1.13 HARQ process in case of carrier aggregation. (PHY—physical; BW—bandwidth)

One transport block and one hybrid automatic repeat request (ARQ) entity is permitted per scheduled component carrier (as illustrated in Figure 1.13). N times (Nx) discrete Fourier transform spread OFDMA (Nx DFT-S-OFDMA) is the UL multiple access scheme (waveform) in LTE-A that supports wider bandwidth. It means that LTE-A UL continues on the SC-FDMA track. The reason for selecting Nx DFT-S-OFDMA is that it will provide full support for component carriers, with CC-specific modulation and coding scheme (MCS), hybrid automatic repeat request (HARQ), and power control. The selected approach also provides the best backward compatibility to LTE Rel-8 and enables the majority of existing control signaling to be directly borrowed from LTE Rel-8. In short, the UL multiple access scheme is DFT-precoded OFDM with one DFT per component carrier.

Furthermore, carrier aggregation is the best choice of network evolution for mobile operators. It is designed to obtain a large bandwidth by aggregating multiple carriers, which can be used by different wireless systems with different air interface technologies. For instance, a 20-MHz LTE single-carrier system and a 10-MHz UMTS double-carrier system can constitute a 30-MHz bandwidth collaborative communications system. Compared with providing the required transmission bandwidth by a totally new LTE-A system, creating greater bandwidth by multisystem coordination reduces investment in new systems, makes full use of existing system resources, is compatible with existing user terminals, and ensures the smooth evolution of the system.

1.3.2 CoMP

In wireless networks, multipath fading caused by the transmission environment seriously affects the performance of all wireless communications. If the wireless link changes slowly, it is possible that the channel will be in a state of deep decline for a long time, which makes communication among terminals impossible. In order to effectively overcome multipath fading, wireless terminals must have an antenna array. However, in many cases, limited physical size, manufacturing cost, and the hardware complexity of portable terminals make them impractical for achieving multiple transmit antennas. To address this shortcoming, space diversity technology based on collaborative communication has been proposed. In a cooperative communication system, antennas distributed in different locations cooperate and compose a distributed "virtual" multiantenna transmit diversity array (as illustrated in Figure 1.14).

3GPP standardization introduces a modern multipoint communication technology collaboration called multiuser *collaborative multipoint* (CoMP) transmission technology. CoMP technology refers to the collaboration of a few geographically separated nodes to provide service for multiple users through different collaboration approaches. These transmission nodes can be at a base station that has a complete resource management module, a baseband processing module, and radio unit, or multiple RF units and distributed antennas in different geographical positions, as well as relay nodes. The macro diversity used in soft handover of the CDMA mobile communication systems is the first use of CoMP technology in actual communication systems.

Past research into distributed antenna systems laid the foundation for today's CoMP technology. The generalized distributed antenna system (GDAS) was the first time the concept of CoMP technology was introduced in the distributed antenna system. Capacity theorems for the relay channel first introduced the concept of cooperative communication among cellular areas, and it is the prototype for modern multipoint collaboration communication technology, which marked the advent of multipoint collaboration communication for the entire cellular mobile communication system.

CoMP acts as a tool to improve high data rate coverage and cell-edge throughput, and also to increase system throughput. There are many advanced technologies

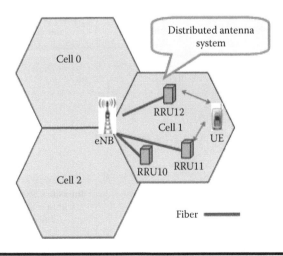

Figure 1.14 Distributed antenna system.

that can be deployed in CoMP implementation. Based on how the collaborated cells share data and transmit to users, CoMP technology can be divided into two categories: joint processing (JP) and coordinated scheduling/coordinated beam-forming (CS/CBF), as illustrated in Figure 1.15.

Joint processing means that data is available at each point in the CoMP cooperating set; it uses different eNB antennas to realize the spatial diversity of users and improve the cell-edge user performance. Joint processing can be further divided into joint transmission and dynamic cell selection. Both modes require detailed UE feedback on channel properties.

In joint transmission, multiple eNBs send data simultaneously to a single UE with the same time and frequency resource. In dynamic cell selection, an eNB is dynamically selected to send data to a UE. For dynamic cell selection, only one point serves at a time to maintain data transmission (TX) continuity when the UE moves from one cell to another cell. There is no handover procedure in this scheme and it can be applied to high-speed railway or highway scenarios.

In CS/CBF, data is only available at the serving cell (data transmission from that point), but user scheduling/beamforming decisions are made based on coordination among cells in the CoMP cooperating set. Although CS does not need intercell channel state information (CSI) measurement/feedback, CBF does need this measurement/feedback. Using interactive information from different cells and by means of resource (such as time, frequency, and space) scheduling as well as beamforming vector scheduling, CS/CBF is able to reduce intercell interference (ICI), thereby improving the cell-edge performance and the system throughput.

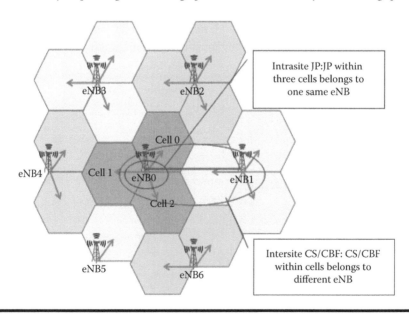

Figure 1.15   CoMP categories.

In the JP scheme, data is available at multiple transmission points and are coherently/noncoherently transmitted to the UE according to the scheduling decision. Data availability at multipoints and physical DL shared channel (PDSCH) transmission to UEs from multiple points (part of or an entire CoMP cooperating set) at a time, the types of JP include: SU JP, MU-JP, intrasite JP, intersite JP, coherent JP, and noncoherent JP.

In the CBF scheme, data is available and transmitted to the UE from the serving cell while some cooperation occurs among multiple cells for interference and/or beam coordination.

The networkwide coordination requires that CSI and scheduling information be exchanged among all the involved coordinating points over the backhaul interface, and it also requires all the cells in the network to complete the scheduling process before data transmission. It is a challenge for backhaul design as well as eNB processing capability. In the CBF scheme, the UE needs to feed back channel quality indicator (CQI) information and CSI corresponding to the cells that are significantly interfering in the networkwide coordination case.

First, each UE feeds back the preferred TX precoder of the serving cell and channel information concerning significant interferers (interfering cells, some predefined thresholds), then the scheduler of each cell coordinates its beam to reduce interference to the already scheduled cells; that is to say, we schedule the UEs in cell $n + 1$ based on the scheduling results in cells $[1, 2, ..., n]$. So, CBF requires CQI/CSI feedback for multiple cells. Network CBF requires CSI and scheduling information exchanged over the backhaul in the short- to midterm interval; intrasite CBF has no requirement on the backhaul.

For example (Figure 1.16), the scheduler starts from cell 1; the scheduling for cell 2 is based on the scheduling results and transmitting weight of cell 1. The scheduling for the current cell is based on the scheduling and transmitting weight of the previous cells. Data is transmitted until all the cells complete their scheduling.

In the CBF scheme, each cell determines its own beam cyclic period/pattern and schedules the UE to its preferred beam(s) in the predefined beam position. This is an effective method to alleviate the random and severe interferences from the neighboring cells, especially in a heavily loaded system. Each cell independently selects the beam cycling pattern based on its user load and user spatial distribution, while maintaining the common cycling period and switching the beam synchronously. This can also be applied to a bursty traffic system with a simple scheduling adjustment. The beam patterns are semistatically changed and exchanged among the coordinated cells. There is no need to exchange any CSI or data through the backhaul. The UE will feed back a specific CQI corresponding to the predefined beam pattern in the serving cell. The CBF scheme requires similar feedback overhead as Rel-8 and fetches very limited impact on backhaul.

Of all the CoMP schemes, intrasite JP is the most competitive. It outperforms the other CoMP schemes significantly, even if it only implements coordination within one site. The intrasite CoMP JP is feasible in the practical system, without

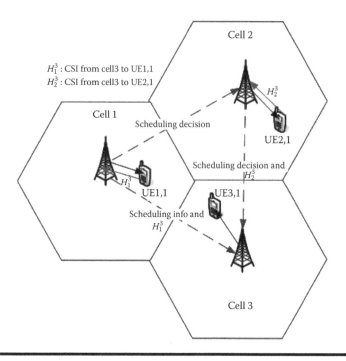

Figure 1.16 Cell coordinates for CBF.

impact on the backhaul link. CS is one of the simplest CoMP schemes. CBF is a moderate CoMP scheme and can increase cell average and cell-edge throughput, respectively, with the coordination.

Many kinds of physical layer transmission technologies are required to support CoMP, such as MIMO technology adaptive to multiple-cell joint transmission, precoding technology, network coding technology, efficient channel estimation, and joint detection technology. Meanwhile, advanced efficient radio resource management schemes are the main factors that affect the performance of CoMP, such as cell resource allocation strategy, load balancing, cooperative cell selection mechanisms, and the effective switching strategy of joint transmission.

1.3.3 Advanced MIMO

Antenna array processing techniques are key components of the emerging 4G wireless standards. There are a wide variety of antenna array techniques available, including spatial division multiple access (SDMA), adaptive antenna systems (AASs), MIMO, and fixed and adaptive beamforming. Each technique offers potentially significant coverage and/or capacity advantages.

MIMO technology is based on multiple antennas at both the transmitter and receiver, including the single-input multiple-output (SIMO) system, the multiple-input single-output (MISO) system, and multiple-input multiple-output (MIMO) system. After space–time processing, the symbol stream to be transmitted is mapped to the transmitting antenna and transmitted to the receiver through the wireless channel. The receiving node detects the data stream of parallel spatial subchannels by corresponding MIMO processing. From the point of view of information theory, multiantenna technology increases channel capacity exponentially as compared with a single antenna system (single-input, single-output, or SISO) by introducing space resources and is one of the most powerful technologies to support the high speed and high capacity of next-generation mobile communication systems.

LTE-A should target a downlink peak data rate of 1 Gbps and an uplink peak data rate of 500 Mbps. Application of spatial multiplexing is important to enable such a throughput increase. In Rel-8, downlink MIMO transmission already supports up to 4 transmit antennas at eNB and either 2 or 4 receive (RX) antennas at the UE; LTE-A will support UL single-user MIMO and DL 8x8 MIMO. The development of MIMO features from LTE to LTE-A is summarized in Table 1.9.

According to the different way of multiantenna processing, MIMO technology can be divided into several types. Different MIMO technologies use various space–time process approaches and have their own characteristics. Spatial multiplexing MIMO technology, such as Vertical-Bell Laboratories Layered Space-Time (V-BLAST) and Diagonal-Bell Laboratories Layered Space-Time (D-BLAST), multiplex data to each antenna and get the full multiplexing gain. MIMO technology is suitable for use in an environment with much scatter and low spatial correlation among antennas. Its capacity is close to theoretical channel capacity in a rich scattering channel environment. Its performance is largely constrained by spatial correlation between antennas; as the antenna correlation increases, the performance deteriorates sharply. Spatial diversity MIMO technologies, such as space–time trellis coding (STTC), space–time block coding (STBC), and space–frequency block coding (SFBC), transmit data in different space–time (frequency) units and get high spatial diversity gain. Diversity transmission is more effective against the spatial correlation between antennas and is robust in variable channel environments. However, because of diversity transmission, its spectral efficiency is relatively low.

Beamforming aims to improve the signal to interference plus noise ratio (SINR) to a single user; that is, parallelism is not exploited in the channel, and therefore, the capacity benefit is less than spatial multiplexing (SM). Beamforming typically uses an antenna array with 4 to 8 closely spaced (usually 1/2 wavelength spaced) columns providing some benefit to uplink budget and enhanced SINR on the downlink. It encompasses a range of techniques including switched fixed beams or adaptive beams. Beamforming technology implements transmitting or receiving beam shaping, and gets beamforming gain by using channel angle information. Its performance is better when the spatial correlation is large or the signal direction is

Table 1.9 LTE MIMO Features Roadmap

	Rel-8	Rel-9	LTE-A and Beyond
DL Added Features	SISO/SIMO Transmit diversity: – SFBC/SFBC+FSTD – Large-delay CDD SU-MIMO: – Codebook-based – Up to 4x4 MIMO MU-MIMO: – Codebook-based – Rank-1 per UE Beamforming: – Non-codebook-based – Rank-1 per UE	Dual-layer Beamforming: – Non-codebook-based, – Up to rank-2 per UE – Up to 2-UEs with orthogonal DM-RSs MU-BF	SU-MIMO: – Codebook design for 8 TX, – Up to 8x8 MIMO MU-Beamforming: – Dual codebook PMI feedback, – Non-codebook-based TX – Up to rank-4 per UE, – Up to 4-UEs with orthogonal DM-RSs MU-BF CSI-RS for inter-cell CSI measurement for CoMP/HetNet
UL Added Features	Single-carrier TX with SISO/SIMO MU-MIMO: – Overlapped BW of paired UE		PUCCH transmit diversity SU-MIMO: – UL codebook – Up to 4x4 MIMO MU-MIMO: – Flexible UE paring SRS enhancement

obvious. However, it does not apply in an environment with low spatial correlation and the transmitter has high requirements for channel information.

1.3.4 Self-Organizing Network

Future networks will be increasingly complex and heterogeneous, and the task of ensuring end-to-end performances will be even more challenging than today because these networks will be composed of combinations of a large number of different wireless and wired nodes and devices, and a wide variety of applications and protocols. The management of these future networks poses challenges that are not limited to individual point-to-point applications; they involve the network as a whole as with CoMP and MIMO, and radio network control will become quite complicated (handover, interference coordination, etc.). A SON has a larger potential for improving network performance, simplifying management, and reducing cost.

The SON is a concept that originated from the Next-Generation Mobile Network (NGMN)* Alliance. A SON network can automatically extend, change, configure, and optimize its topology, coverage, capacity, cell size, and channel allocation, based on changes in location, traffic pattern, interference, and the situation/environment. The SON is an exchange-to-exchange (e2e) self-aware and self-optimizing system with a series of features and solutions. The SON aims at reducing the cost of installation and management by simplifying operational tasks through automated mechanisms such as self-configuration and self-optimization. The network should be running at its best and rapidly adjusting to any changes in the environment and conditions. For the processes that are too fast, too granular, and/or too complex for manual intervention, the automation is very important. SON algorithms represent a continuation of the natural evolution of wireless networks, where automated processes are simply extending their scope (e.g., Radio Resource Management, or RRM) deeper into the network. From an operator's perspective, the main drivers for SON in LTE are shown in Figure 1.17.

The SON is designed to automatically configure and optimize the LTE network (as illustrated in Figure 1.18) by self-configuration, self-optimization, and self-healing. *Self-configuration* involves plug-and-play behavior when installing network elements, which reduces costs and simplifies the installation procedure. *Self-optimization* means automatic optimization based on network monitoring and measurement data obtained from various network nodes and terminals. *Self-healing* means that the system detects problems itself and mitigates or solves these problems

---

* NGMN is an initiative by a group of mobile operators to provide a coherent vision for technology evolution beyond 3G for the competitive delivery of broadband wireless services. Their objective is to establish clear performance targets, fundamental recommendations, and deployment scenarios for a future wide area mobile broadband network.

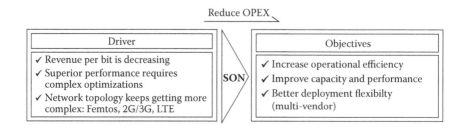

Figure 1.17 Main drivers for SON.

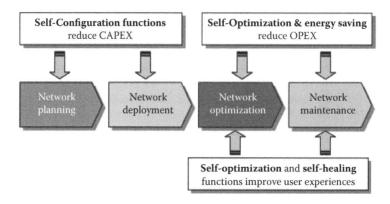

Figure 1.18 Main objectives of SON.

to avoid unnecessary user intervention and to significantly reduce maintenance costs. The main objectives of the SON is to reduce CAPEX and OPEX.*

The SON will bring us operational benefits as described in Table 1.10. Support for the SON is a part of LTE. Several use cases that could benefit from SON have been introduced in the Rel-8 and the work is continuing for LTE-A. LTE Rel-8 currently supports the following SON functions: dynamic configuration of S1 and X2, physical cell ID (PCI) selection, and automatic neighbor discovery, among others. LTE-A will have new requirements and new features for the system that will require

---

* Survey data show that investment by telecommunications network operators follows the Pareto law: pre-equipment investment is only 20% of the total network investment, and operating costs (including maintenance costs, marketing costs, labor costs) account for 80%. Reducing operating costs is the key for network construction operators. For example, minimization of drive tests (MDT) aims to achieve driving test functions in the terminal to minimize the drive-test cost of the network operator, which is significant to CAPEX and OPEX savings. Terminals supporting minimum drive tests can automatically record the measurement index of special situations (such as random access failure and catch failure of broadcast information) and send it to the network at the right time. The operator can adjust or optimize the network parameters according to these indicators.

Table 1.10 Operational Benefits by SON

Self-Configuration	– Flexibility in logistics (eNB not site specific)
	– Reduced site / parameter planning
	– Simplified installation; less prone to errors
	– No / minimum drive tests
	– Faster rollout
Self-Optimization	– Increased network quality and performance
	– Parameter optimization Reduced maintenance, site visits
Self-Healing	– Error self-detection and mitigation
	– Speed up maintenance
	– Reduce outage time

new SON use cases needs to be designed. LTE-A will also address the SON functions needed to support LTE-A-specific architectures and functions, for example, relay. In LTE-A, power efficiency in the infrastructure and terminal is an essential element. LTE-A is going to cover further enhancements in SON as well, for instance:

■ Special care will be taken for special deployment scenarios such as network sharing.
■ Special care will be taken for mass deployment scenarios as in the case of home eNBs, i.e., inbound and outbound mobility for home eNBs and problems caused by incorrect behavior of home eNBs.
■ Drive tests need to be avoided.
■ The impact on UE complexity and power consumption needs to be taken into account.

Operation and maintenance tasks will be minimized to the best possible extent. In addition, all the interfaces specified will be open for multivendor equipment interoperability. From these requirements, initial use cases can be derived (as shown in Figure 1.19), such as optimization of energy consumption, SON in heterogeneous deployments, plug-and-play self-configuration, mobility-related use cases, and avoidance of drive tests, cell outages, coverage hole management, and so on.

The SON use cases were identified by 3GPP* using coverage and capacity optimization, energy savings, interference reduction, automated configuration of physical cell identity, mobility robustness optimization, mobility load-balancing

* TR 36.902 v9.0.0 includes self-configuring and self-optimizing network use cases and solutions. See bibliography.

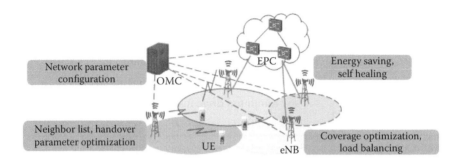

Figure 1.19 SON function.

Table 1.11 Self-Configuration

	Rel-8	*Rel-9*
Self-Configuration	– Automated configuration of PCI	– Automated configuration of PCI
	– ANR	– ANR
	– Self-configuration of eNBs	– Self-configuration of eNBs Automatic Software Management
	– Automatic Software Management	– Inter-RAT ANR
		– Automatic Radio Configuration Function

optimization, random access channel optimization, automatic neighbor relation function, and intercell interference coordination.

In LTE Rel-8, the following SON functions were specified, including eNB self-configuration, automatic neighbor relations (ANR), automatic physical cell ID (PCI) allocation, and self-healing. In Rel-9, the specified SON functions include PCI optimization, mobility robustness optimization, random access channel (RACH) optimization, interference control, mobility load balancing, capacity and coverage optimization, energy savings, and so on. In LTE-A, the SON functions that were specified or extended, include minimization of drive test, SON procedure coordination, mobility load balancing, capacity and coverage optimization, and energy saving. All of these SON features are summarized in Tables 1.11, 1.12, and 1.13, respectively.

In conclusion, SON features provide strong capabilities for automation of management tasks, from automatic configuration management at the network element level to large-scale optimization tasks at the network level. These SON features improve the quality and performance of the network while reducing the OPEX of the network operators.

Table 1.12 Self-Optimization

	Rel-9	*Rel-10*
Self-Optimization	– Coverage and capacity optimization – Mobility load balancing – Mobility robustness optimization – Avoidance of drive tests – SON evaluation scenario – RACH optimization	– Coverage and capacity optimization (spillover, new features like relays) – Mobility load balancing – Mobility robustness optimization (spillover, new features like relays) – Avoidance of drive tests – SON evaluation scenario – RACH optimization – Interference reduction – Intercell interference coordination – Energy savings – Control and resource optimization of relays

Table 1.13 Self-Healing

	Rel-9	*Rel-10*
Self-Healing	Cell outage compensation/ mitigation	Self-healing

1.3.5 Relay

Relays are mobile network base stations that connect to the network via a wireless backhaul link instead of using a dedicated wired or wireless (e.g., microwave) backhaul link. The purpose of using relays is to provide coverage extension to high-shadowing regions or locations where dedicated backhaul links are not deployed. Relays can be used to enhance capacity while maintaining good cost versus performance trade-offs.

Relays have been standardized in WiMAX by the IEEE 802.16 Relay Task Group, in the IEEE 802.16j-2009 standard. Relays for LTE were standardized by 3GPP in Rel-9 and specified in more detail in Rel-10 of LTE-A. Since relaying can be used for improving the SINR conditions at the cell edge with low additional infrastructure cost, for LTE-A it is considered a tool to improve cell-edge performance and fill coverage holes.

Relay node deployment can cover a wide range (as shown in Figure 1.20), including indoor relays, fixed outdoor relays, and even mobile relays that are attached to a vehicle and provide coverage for passengers. LTE-A focuses on fixed relays because they are expected to have the widest deployment. Mobile relays in particular are not in the current interest of standardization.

Relays can be considered an advanced form of a repeater. A typical repeater works at the analog RF level, amplifying the received signal from the eNB and forwarding it to the UEs in the DL. Similarly, it receives signals from the UEs in the UL and forwards them to the eNB after amplification. Thus, the repeater increases the coverage as the link budget needs to be satisfied between the eNB and the repeater and between the repeater and the UE, rather than between the eNB and the UE. This kind of amplifying and forwarding repeater adds noise to the signal in the process. In general, a relay is a smarter repeater that regenerates and amplifies only the relevant parts of the received signal based on the subset of users targeted by the relay. Therefore, there is no noise or interference enhancement due to relay, compared with the repeater.

Relays can have some intelligence built into them. For example, the transmit power of the relay could be controlled, and a relay could be activated only when users were in the coverage area of it. A relay doing just physical layer processing is a Layer 1 relay. Another type of relay is a decode-and-forward relay; such a decoder does not add noise as does its analog counterpart, since the signal is actually decoded and retransmitted like the transmission done by the original source/

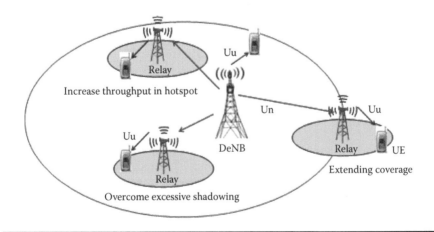

Figure 1.20 Deployment of relay nodes. The radio interface between the DeNB and the RN is called the Un interface. Uu is the air interface between the UE and the RN.

transmitter of the signal. Decode-and-forward relay increases the overall delay and could be considered an example of Layer 2 forwarding. New interfaces and protocols are required between the original transceiver (e.g., eNB) and the relay transceiver. A Layer 3 relay receives and forwards IP packets (Packet Data Convergence Protocol [PDCP] service data units [SDUs]). Therefore, the user packet at the IP layer is viewable at Layer 3 relays. A Layer 3 relay has all functions that an eNB has, and it conventionally communicates with its donor eNB through an X2-like interface. The functions of different types of relay nodes are illustrated in Figure 1.21.

In LTE-A specifications, inband Layer 3 (L3) relays (Type-1) have been selected, outband relays (Type-1a) and inband relays with adequate antenna isolation between backhaul and access links (Type-1b) are also considered, since they are expected to have little impact on specifications. Multimedia broadcast single-frequency network (MBSFN) subframes are used to allow the relay node (RN) to "mute" access transmission during backhaul reception.

From the UE's point of view, an RN is a cell of its own. An RN has a cell identity of its own, transmits its own synchronous channels and reference signals, and broadcasts system information. The UE receives and sends HARQ feedback directly from and to the RN. A relay node appears as Rel-8 eNB to Rel-8 UEs, so that it can provide coverage and some capacity enhancement, and is expected to work smoothly.

Inband backhauling on the eNB–RN link is realized using MBSFN subframes. The RN transmits an empty MBSFN subframe (i.e., a subframe that contains only control signals) to the UEs when it wants to receive data in the backhaul link. Control/data transmission between the eNB and the RN needs to be redesigned in the MBSFN subframe.

Outband backhauling (Type-1a) and inband backhauling with antenna isolation (Type-1b) are also LTE-A options. With these two relay types, the backhaul

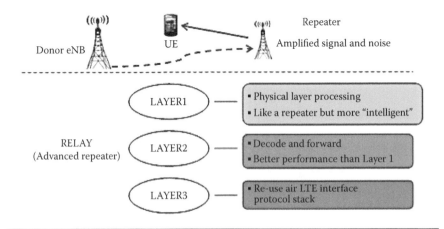

Figure 1.21 Types of relay nodes.

link and access link are on different frequencies (bands), or are isolated and operating completely independently.

Because each relay node provides its own cell identity, including physical cell ID and reference signals, for Rel-8 UEs, whether they connect via an RN or directly via an eNB, they will be transparent. Later releases may add support in the UE to distinguish between RN and eNB connections, but this is for a future specification.

Resource partitioning between the RN and its donor eNB is performed in the donor eNB on a subframe or Transmission Time Interval (TTI) basis. It is not possible to split resources between directly connected UEs and UEs that are connected via an RN.

In summary, relay networks are expected to fulfill the demanding coverage and capacity requirements in a cost-efficient way. Currently, Layer 3 inband relaying is being standardized by 3GPP LTE-A. The relay link transmission is time-division multiplexed with the access link transmission, whereas macro users share the same resources with the relays. Therefore, system performance depends strongly on the resource sharing strategy among and within the links, as well as how well the competition for resources is managed at the eNB.

Chapter 2

Carrier Aggregation

In a Long Term Evolution Advanced (LTE-A) system, wider bandwidths (up to 100 MHz) are needed to support high peak data rates (1 Gbps in the downlink and 500 Mbps in the uplink) and average throughout per cell. In order to support bandwidths larger than 20 MHz, carrier aggregation (CA) with two or more component carriers (CCs) is considered in LTE-A. An LTE-A terminal with transmission and/ or reception capability beyond 20 MHz can simultaneously transmit and receive data on multiple component carriers. The number of component carriers transmitted and/or received to/from a mobile terminal can vary over time depending on the instantaneous data rate. Up to five component carriers can be aggregated, allowing for transmission bandwidths up to 100 MHz. Backward compatibility to an LTE Rel-8 system is required in the carrier aggregation design to allow LTE Rel-8 terminals to operate in an LTE-A network. However, not all component carriers are necessarily Rel-8 compatible.

2.1 Basic Concepts of CA

2.1.1 CA Types and Scenarios

In addition to the peak data rate, another motivation for carrier aggregation is to facilitate efficient use of fragmented spectrum. In LTE-A carrier aggregation, each component carrier can take any of the channel bandwidths of 1.4, 3, 5, 10, 15, and 20 MHz that are supported by LTE Rel-8. Component carriers do not have to be of the same frequency. Operators with a fragmented spectrum can also provide high-data-rate services through CA technology.

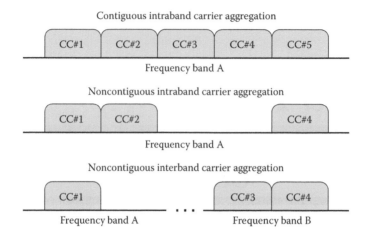

Figure 2.1   Carrier aggregation types.

Depending on the spectrum usage, three possible CA types are considered in LTE-A: intraband contiguous CA, intraband noncontiguous CA, and interband CA, as illustrated in Figure 2.1.

For intraband contiguous CA, the spacing between center frequencies of contiguously aggregated CCs is a multiple of 300 kHz, in order to be compatible with the 100-kHz frequency raster of LTE Rel-9 while preserving orthogonality of the subcarriers with 15-kHz spacing. The nominal channel spacing between two adjacent aggregated CCs is defined as follows (3GPP TR 36.808 V1.7.0):

Nominal channel spacing

$$= \left\lfloor \frac{BW_{Channel(1)} + BW_{Channel(2)} - 0.1 \left| BW_{Channel(1)} - BW_{Channel(2)} \right|}{0.6} \right\rfloor 0.3 \ [\text{MHz}]$$

where $BW_{Channel(1)}$ and $BW_{Channel(2)}$ are the channel bandwidths of the two respective CCs with values in MHz. The channel spacing for intraband contiguous CA can be adjusted to any multiple of 300 kHz less than the nominal channel spacing to optimize performance in a particular deployment scenario.

Intraband noncontiguous CA is likely to be very useful in North America where operators have scattered spectrum even within a given band (e.g., Sprint in the PCS-1900 band).

A typical example of interband noncontiguous CA is the aggregation of the 2.6-GHz band and the 800-MHz European Digital Dividend band in Europe.

Noncontiguous CA has the advantage of having spectral diversity gain, with different types of fading channels on different frequencies. On the other hand,

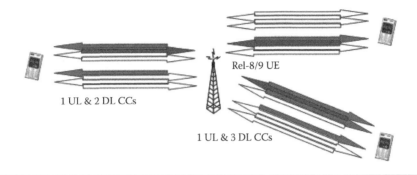

Figure 2.2 Asymmetric DL and UL CC configurations.

contiguous CA can save much spectrum because many subcarriers that are used as guard bands can be employed for data and control signal transmission.

LTE-A supports aggregation of up to 5 downlink (DL) CCs and 5 uplink (UL) CCs, irrespective of intraband or interband CA. It is possible to configure user equipment (UE) connecting to the same evolved Node-B (eNB) to aggregate a different number of CCs. For frequency-division duplexing (FDD) operation, the number of aggregated carriers in the uplink and downlink may be different (as shown in Figure 2.2), while LTE-A supports only the case where the number of downlink CCs is not less than the number of uplink CCs. However, in time-division duplexing (TDD) operation, the number of CCs and the bandwidth of each CC in the UL and DL will be the same.

The same frame structure is used in all aggregated CCs. For TDD carrier aggregation, the uplink–downlink configuration across all CCs should be the same.

Five different deployment scenarios are considered for CA in LTE-A, assuming cells with two different CC frequencies F1 and F2, and F2 is larger than F1. In Scenario 1, the F1 and F2 CCs are colocated and overlaid. If F1 and F2 are at the same band or the frequency separation is small, it would lead to nearly the same coverage for all CCs. In Scenario 2, the F1 and F2 CCs are colocated and overlaid, but the coverage of them is different due to large frequency separation between CCs, which leads to the different path losses. Only F1 provides sufficient coverage and F2 is used to provide throughput. Mobility is performed based on F1 coverage. Scenarios 1 and 2 are illustrated in Figure 2.3.

In Scenario 3 (shown in Figure 2.4), the F1 and F2 CCs are colocated and F1 and F2 are typically on different bands. F2 antennas are directed to the cell boundaries of F1 so that cell-edge throughput is increased. F1 provides sufficient coverage but F2 potentially has holes, for example, due to larger path loss. Mobility is based on F1 coverage.

In Scenario 4, one CC provides macrocell coverage and remote radio head (RRH) cells are placed at traffic hotspots for additional throughput by the other CC. Mobility is accomplished based on macrocell coverage. F1 and F2 are generally

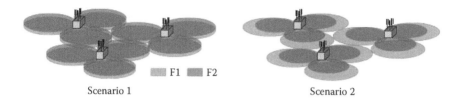

Scenario 1   Scenario 2

Figure 2.3   CCs with overlapping radio coverage.

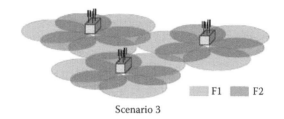

Scenario 3

Figure 2.4   CCs with different radio coverage.

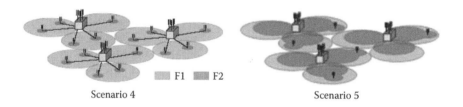

Scenario 4   Scenario 5

Figure 2.5   CCs cooperating with RRHs and repeaters.

on different bands in Scenario 4. Scenario 5 is similar to Scenario 2, but frequency selective repeaters are deployed so that coverage is extended for one of the carrier frequencies. It is expected that the F1 and F2 cells of the same eNB can be aggregated where coverage overlaps. Scenarios 4 and 5 are illustrated in Figure 2.5.

Scenarios 4 and 5 are not supported in LTE-A on uplink operation. Generally, uplink CA is supported for intraband CC configurations with Scenarios 1–3 only. However, all CA scenarios should be supported for the downlink in LTE-A.

2.1.2 Component Carriers

Although all CCs in LTE-A (Rel-10) are designed to be backward compatible, some other kinds of component carrier implementations were discussed in 3GPP TSG RAN WG1 (RAN1) that were at least technically valuable or might be considered in the future evolution.

2.1.2.1 Backward-Compatible Carrier

A backward-compatible carrier (Figure 2.6) is accessible to UE of all existing LTE releases. LTE-A UE must be able to share spectrum with LTE Rel-8 UE. A backward-compatible carrier can be operated as a single carrier (stand-alone) or as a part of carrier aggregation. For an FDD system, backward-compatible carriers always occur in pairs, that is, DL and UL.

2.1.2.2 Non-Backward-Compatible Carrier

A non-backward-compatible carrier is not accessible to UE of earlier LTE releases, but is accessible to UE of the release defining such a carrier. A non-backward-compatible carrier can be operated as a single carrier (stand-alone, thus synchronization channel/physical broadcast channel [SCH/PBCH] exists as Carrier 2 in Figure 2.7) if the non-backward-compatibility originates from the duplex distance or otherwise as a part of carrier aggregation (SCH/PBCH may not be included on this CC as Carrier 1 in Figure 2.7).

Non-backward-compatible CCs are considered to allow more efficient operation of LTE-A UE. In non-backward-compatible DL CCs, it is useful to allow an operation without a physical DL control channel (PDCCH) region and possible to transmit the physical DL shared channel (PDSCH) from the first orthogonal frequency-division multiplexing (OFDM) symbol for improving DL data throughput. This is beneficial especially for an operation scenario with a small number of UEs in the system (e.g., home eNB, hotspot), where the PDCCH region on the

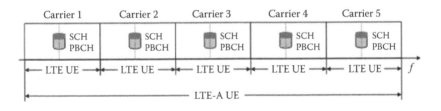

Figure 2.6 Backward-compatible CCs.

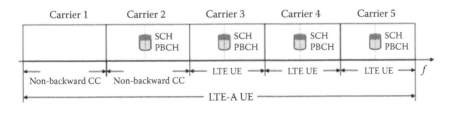

Figure 2.7 Non-backward-compatible CCs.

remaining DL CCs is sufficient to accommodate the required PDCCHs. A DL CC without PDCCH can be indicated by special signaling, such as using the 4th physical control format indicator channel (PCFICH) state, or by a semistatic configuration via higher layers.

2.1.2.3 Carrier Segments

Carrier segments are defined as the bandwidth extensions of a Rel-8-compatible component carrier (no larger than 110 resource blocks [RBs] in total) and constitute a mechanism to utilize frequency resources in case new transmission bandwidths are needed in a backward-compatible way that complements carrier aggregation means. The advantage is to reduce additional PDCCH transmission that would be required in a carrier aggregation setting as well as the use of small transport block (TB) sizes for the part corresponding to the segment. The notion of a carrier segment allows for aggregating additional resource blocks to a component carrier, while still retaining the backward compatibility of the original carrier bandwidth. Carrier segments are always adjacent and linked to one carrier and cannot be stand-alone segments. They do not provide synchronization signals, system information, or paging, and therefore cannot be used for random access or UE camping. They support the same hybrid automatic repeat request (HARQ) process, PDCCH indication, and transmission mode as their linked CC. Figure 2.8 illustrates the notion of carrier segments.

2.1.2.4 Extension Carrier

The *extension carrier* is a carrier that cannot be operated as a single carrier (standalone), but must be a part of a component carrier set where at least one of the carriers in the set is a stand-alone-capable carrier. For example, no SCH is an option for an extension carrier; a component carrier without SCH would have to be located close to a component carrier with SCH for reliable synchronization and is probably not suitable for interband noncontiguous carrier aggregation. No control

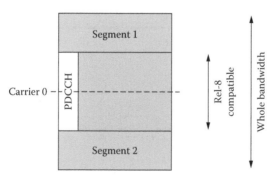

Figure 2.8   Carrier segments.

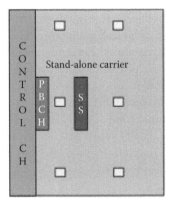

Figure 2.9   Stand-alone carrier and extension carrier.

channels (e.g., PDCCH, physical HARQ indicator channel [PHICH], or PCFICH) is another option for an extension carrier in order to prevent access from Rel-8/9 UE and to reduce the overhead. The extension carrier (as shown in Figure 2.9) supports a separate HARQ process, PDCCH indication, and transmission mode, which is configured from its linked CC.

2.2 CA Configurations

2.2.1 *PCell and SCell*

In LTE-A, each aggregated component carrier appears as a separate cell with its own cell ID. A UE configured for CA connects to one primary serving cell (PCell) and up to four secondary serving cells (SCells). The CCs corresponding to the PCell are referred to as the downlink and uplink primary component carriers (PCCs), while the CCs corresponding to an SCell are referred to as downlink and uplink secondary component carriers (SCCs).

When UE is configured for CA, it has only one radio resource control (RRC) connection with the network via the PCell. At RRC connection establishment/reestablishment/handover, the PCell provides the non-access stratum (NAS) mobility information (e.g., tracking area identity [TAI]), and at RRC connection reestablishment/handover, the PCell provides the security input. The association between the DL PCC and the corresponding UL PCC is cell specific and signaled as part of the system information (in system information block [SIB]2). This is similar to the case without carrier aggregation.

A UE operates in its PCell in a manner similar to a Rel-8/9 serving cell. The PCell is configured per UE, that is to say, different UE may have different PCell within one eNB. The PCell can only be changed with a handover procedure (i.e., with a security key change and random access channel [RACH] procedure). Unlike

SCells, a PCell cannot be deactivated and cross-scheduled. Radio link failure (RLF) is triggered based on PCell monitoring only, so radio link monitoring is not necessary for SCells. According to Rel-8 procedures, upon RLF, the UE will fall back to non-CA mode. UL PCC is used for carrying physical uplink control channel (PUCCH) acknowledgment/negative acknowledgment (ACK/NACK), scheduling request (SR), and periodic channel state information (CSI) from a UE.

SCells may be configured once an RRC connection is established and may be used to provide additional radio resources. An SCell consists of a DL resource and optional UL resource (i.e., there can be more DL resources than UL resources, but not vice versa). When serving cells are mentioned in LTE-A specs, this can refer to either PCells or SCells.

Depending on UE capabilities, SCells can be configured to combine with the PCell to form a set of serving cells. The functions specific to PCells and SCells are compared in Table 2.1.

The configured set of serving cells for UE must be Rel-8/Rel-9 backward compatible and always consist of one PCell and one or more SCells. The usage of UL resources by the UE in addition to the DL resources is configurable for each SCell. The number of DL SCCs configured is therefore always larger or equal to the number of UL SCCs, and no SCell can be configured for usage of UL resources only. A serving cell can be configured for use only as an SCell (i.e., prevent UEs from camping on it) if, for example, master information block (MIB) and SIBs are not broadcast in that cell. The serving cells can be contiguous or noncontiguous (in frequency) and have different bandwidths, which do not need to provide the same coverage.

2.2.2 SCell Activation/Deactivation

SCells are configured based on UE capability. Activation and deactivation of the respective carrier component will follow the traffic need and the UE reported CQI. SCells can be viewed as having three *states*:

Table 2.1 Functions of the PCell versus the SCell

PCell	SCell
PDCCH/PDSCH/PUSCH/PUCCH transmits on the PCell	PDCCH/PDSCH/physical uplink shared channel (PUSCH) transmits on the SCell
Measurement and mobility procedure are based on the PCell Random-access (RA) procedure is performed over the PCell	Medium access control (MAC)-layer-based activation/deactivation is supported for SCells for UE battery saving
Same discontinuous reception (DRX) cycle applied for both PCell and SCell	Same DRX cycle applied for both PCell and SCell

- **Non-configured:** UE performs radio resource management (RRM) measurements when applicable interfrequency measurement is configured for the UE (same as LTE Rel-8).
- **Configured but deactivated:** UE does not receive PDCCH or PDSCH and does not perform channel quality indicator (CQI) measurements.
- **Configured and activated:** UE performs normal monitoring of PDCCH and PDSCH, performs CQI measurements, and maintains the path loss estimation.

Thanks to the backward-compatible design of CA, the primary synchronization signal (PSS) and secondary synchronization signal (SSS) are transmitted on all CCs for UEs to facilitate cell search. Once a UE makes a successful cell search, it considers the current cell as the PCell and performs the random access procedure on the uplink CC associated with the PCell. Even in case of CA, no more than one random access procedure is ongoing at any time. When CA is configured, the first three steps of the contention-based random access procedure occur on the PCell. After contention resolution, the UE will then establish an RRC connection to its PCell by following the usual Rel-8/9 procedures, as shown in Figure 2.10.

During the RRC UE capability enquiry procedure, the eNB that owns the UE's PCell may inquire whether the UE has the capability to support CA. The BandCombinationParameters field (as shown in Figure 2.11) in the UE Evolved Universal Terrestrial Radio Access (EUTRA) capability information element (IE) within the RRC message "UECapabilityInformation" defines the carrier aggregation (and multiple-input, multiple-output [MIMO]) capabilities supported by the UE for configurations with interband noncontiguous, intraband noncontiguous,

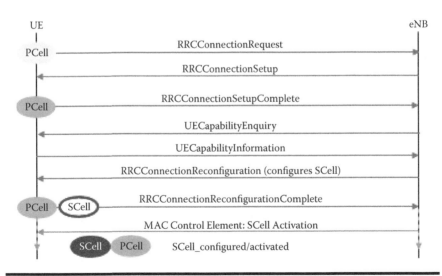

Figure 2.10   Initial access and entering CA mode.

```
SupportedBandCombination-r10 ::= SEQUENCE (SIZE (1..maxBandComb-r10)) OF BandCombinationParameters-r10

BandCombinationParameters-r10 ::= SEQUENCE (SIZE (1..maxSimultaneousBands-r10)) OF BandParameters-r10

BandParameters-r10 ::= SEQUENCE {
    bandEUTRA-r10              INTEGER (1..64),
    bandParametersUL-r10       BandParametersUL-r10       OPTIONAL,
    bandParametersDL-r10       BandParametersDL-r10       OPTIONAL
}

BandParametersUL-r10 ::= SEQUENCE (SIZE (1..maxBandwidthClass-r10)) OF CA-MIMO-ParametersUL-r10

CA-MIMO-ParametersUL-r10 ::= SEQUENCE {
    ca-BandwidthClassUL-r10          CA-BandwidthClass-r10,
    supportedMIMO-CapabilityUL-r10   MIMO-CapabilityUL-r10       OPTIONAL
}

BandParametersDL-r10 ::= SEQUENCE (SIZE (1..maxBandwidthClass-r10)) OF CA-MIMO-ParametersDL-r10

CA-MIMO-ParametersDL-r10 ::= SEQUENCE {
    ca-BandwidthClassDL-r10          CA-BandwidthClass-r10,
    supportedMIMO-CapabilityDL-r10   MIMO-CapabilityDL-r10       OPTIONAL
```

Figure 2.11 UE capacity signaling.

or intraband contiguous carrier aggregation. For each band in a band combination, the UE provides, via the BandParametersUL/DL, the supported CA bandwidth classes (and the corresponding MIMO capabilities).

The eNB may then configure a UE that supports CA with one or more SCells in addition to the PCell that is initially configured, through the RRC connection reconfiguration procedure after the initial security activation procedure.

One or more SCells can be configured in one RRCConnectionReconfigutratrion message (as shown in Figure 2.12). Each SCell is assigned an SCell identity, *SCellIndex*, with a range of 1 to 7. The corresponding physical cell ID and carrier frequency are configured as well.

The default state for a newly configured SCell is *deactivated*. Deactivation of an SCell means that PDCCH reception on this SCell (for both DL and UL grants) is stopped, as well as the PDSCH reception and physical uplink shared channel (PUSCH) (including retransmissions), sounding reference signal (SRS), and CQI transmissions. When DL SCC is activated or deactivated, the linked UL SCC is also activated or deactivated. DL SCC activation/deactivation can be explicit or implicit.

Explicit activation/deactivation of the DL SCC is via Medium Access Control (MAC) signaling. The LTE-A newly defined activation/deactivation MAC element (Figure 2.13) is identified by a MAC protocol data unit (PDU) subheader with logical channel ID (LCID) 27 carrying an 8-bit bitmap for the downlink activation and deactivation of SCells. With the bitmap, a single activation/deactivation command can activate or deactivate a subset of the SCells.

The "Ci" field indicates the activation/deactivation status of the SCell with SCellIndex i. The Ci field is set to "1" to indicate that the SCell with SCellIndex i is to be activated. The Ci field is set to "0" to indicate that the SCell with SCellIndex i is to be be deactivated.

Implicit deactivation of DL SCC can be triggered via a timer called the *sCell-DeactivationTimer*. One deactivation timer is maintained per SCell, but one common value is configured per UE by RRC.

Once configured, for simplicity reasons, the UE operates in CA mode until it remains connected to the same eNB, except when the radio connection is lost or for handover and the UE tries to reconnect. In this case, the UE will autonomously release any previously configured SCells before attempting the RRC connection reestablishment. The eNB will then have to newly configure the CA mode with a different PCell and SCell performed by RRC. At intra-LTE handover, RRC can also add, remove, or reconfigure SCells for usage with the target PCell. When adding a new SCell, dedicated RRC signaling is used for sending all required SCell system information; that is, while in connected mode, UEs need not acquire broadcasted system information from the SCells.

Because CA can be configured to a UE only during the attach procedure, RRC connection reconfiguration procedure, or handover (HO) procedure, a UE can be configured to CA only in RRC-CONNECTED state. A UE in RRC_IDLE state always behaves as a single-carrier UE.

```
RRCConnectionReconfiguration-v1020-IEs ::= SEQUENCE {
     sCellToReleaseList-r10          SCellToReleaseList-r10        OPTIONAL,   -- Need ON
     sCellToAddModList-r10           SCellToAddModList-r10         OPTIONAL,   -- Need ON
     nonCriticalExtension            SEQUENCE {}                   OPTIONAL    -- Need OP
}

SCellToAddModList-r10 ::=           SEQUENCE (SIZE (1..maxSCell-r10)) OF SCellToAddMod-r10

SCellToAddMod-r10 ::=               SEQUENCE {
     sCellIndex-r10                 SCellIndex-r10,
     cellIdentification-r10         SEQUENCE {
           physCellId-r10                PhysCellId,
           dl-CarrierFreq-r10            ARFCN-ValueEUTRA
     }                                                            OPTIONAL,   -- Cond SCellAdd
     radioResourceConfigCommonSCell-r10   RadioResourceConfigCommonSCell-r10 OPTIONAL,   -- Cond
SCellAdd
     radioResourceConfigDedicatedSCell-r10 RadioResourceConfigDedicatedSCell-r10 OPTIONAL, --
Cond SCellAdd2
     ...
}
```

Figure 2.12 SCell add/release signaling.

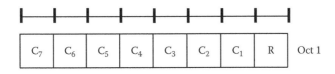

| C_7 | C_6 | C_5 | C_4 | C_3 | C_2 | C_1 | R | Oct 1 |

Figure 2.13   Activation/deactivation MAC element.

2.3 CA Layer 2 Structure

In LTE, the MAC layer mainly performs HARQ and scheduling, the radio link control (RLC) layer mainly performs segmentation and concatenation of packets and automatic repeat request (ARQ). In case of CA in LTE-A, the multicarrier nature of the physical layer is exposed only to the MAC layer for which one HARQ entity is required per serving cell. The multiple CCs of carrier aggregation are not visible to the packet data convergence protocol (PDCP) and RLC layers.

It is notable that minimal standardization and implementation impact on LTE Rel-8 are required to support CA. The protocol separation to different carriers is done inside the MAC layer. The MAC layer makes multi-codeword (MCW) transmission on a per-CC basis, and selects adaptive modulation and coding (AMC), HARQ, and MIMO schemes on a per-CC basis. Multiple HARQ processes need to be supported for a single user with multiple ACK/NACK. The eNB must monitor multiple CQI reports from carriers being aggregated.

The advantage of maintaining the physical layer design for each component carrier allows greater reuse of existing LTE Rel-8 structures (hardware, software; e.g., same transport-block sizes, soft buffer sizes, no/limited impact on MAC and RLC design). Furthermore, with the CA technology, it will be possible to schedule a user on multiple component carriers simultaneously, but some new issues should be considered on resource scheduling. In the noncontiguous interband aggregation scenario, where the aggregated carriers belong to different frequency bands, the fading characters, such as the path loss and Doppler shift are different between carriers (as illustrated in Figure 2.14).

As a result, allowing for link adaptation (including rank adaptation) in the frequency domain (i.e., adaptation per component carrier) becomes increasingly important at higher bandwidths, especially if the component carriers are located in vastly different parts of the spectrum.

In LTE-A, for both the uplink and downlink, each component carrier has an independent HARQ entity; therefore, there is one independent HARQ process for each DL and UL CC. This allows separate link adaptation and MIMO support for each carrier, which should improve throughput since each data transmission can be independently matched to channel conditions on each carrier. As a result, data processing at the physical layer can be thought of as being independent of the carrier.

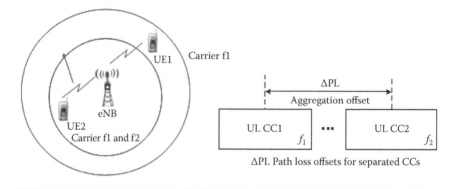

Figure 2.14   Different path loss between two CCs.

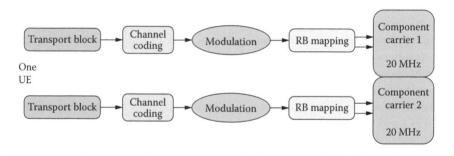

Figure 2.15   Multiple TBs on multiple CCs.

From a UE's perspective, there is one transport block (in absence of spatial multiplexing) and one HARQ entity per scheduled component carrier. Each transport block is mapped to a single component carrier. A UE may be scheduled over multiple component carriers simultaneously (as illustrated in Figure 2.15).

There is one PDCP entity and one RLC entity per radio bearer and it is not visible from RLC on however many CCs the physical (PHY) layer transmission is conducted. RLC can handle data rates up to 1 Gbps. Dynamic Layer 2 packet scheduling across multiple CCs is supported. Due to independent HARQ per CC, HARQ retransmissions are transmitted on the same CC as the corresponding original transmission. The Layer 2 user plane structure for the downlink with CA configured is shown in Figure 2.16.

The user plane structure of UL uses the same general principle as for DL, with separate transport channels per CC and independent synchronous HARQ per CC. If UE is scheduled on multiple CCs, the UE decides how to multiplex data from different radio bearers on CCs (based on logical channel prioritization rules). The scheduler can choose the appropriate carrier for data transmission if the interference on other carriers is above a set threshold. The Layer 2 user plane structure for the uplink with CA configured is shown in Figure 2.17.

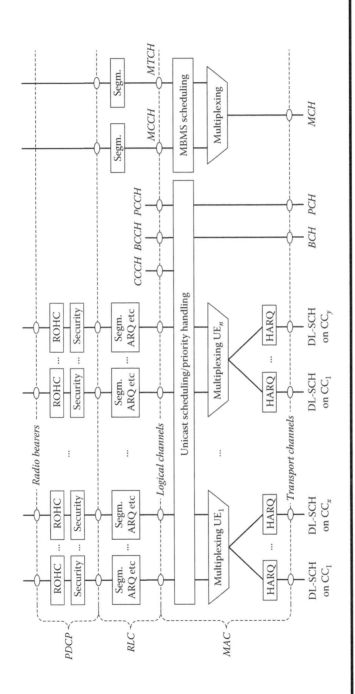

Figure 2.16 Layer 2 structure for DL CA.

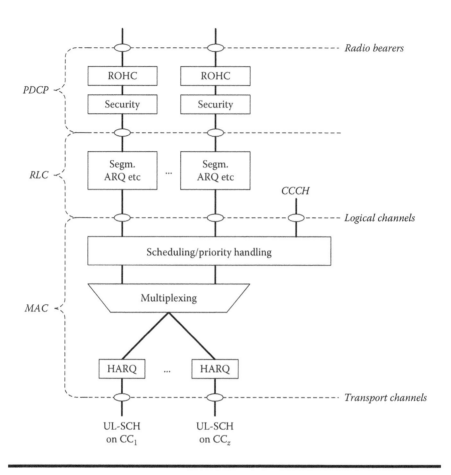

Figure 2.17   Layer 2 structure for UL CA.

2.4 DL CA Operations

2.4.1 System Information Broadcasting

With CA, each serving cell has its own broadcast control channel (BCCH). BCCH acquisition and monitoring is the same as Rel-8. UE is required to monitor the BCCH on the PCell only for system information (SI) acquisition. Dedicated signaling is used to convey SI of SCells.

An SCell is configured to UE via the RRCConnectionReconfiguration message, which includes the SI necessary for transmission and reception of the added SCell (similar to Rel-8 HO). Changes in system information on a configured SCell are handled by removal or addition of the SCell. Consequently, there is no need for the UE to monitor the BCCH on SCCs. The scenarios for acquiring system information in CA cases are listed in Table 2.2.

Table 2.2 System Information for Carrier Aggregation

	Broadcast Signaling	*Dedicated Signaling*	*Broadcast Signaling*
Rel-8, Rel-9	For normal SI acquisition	At HO Step 1: Acquiring SI from HO command	At SI Change Step 1: Checking paging of value tag
PCell	Step1: Reading MIB/ SIB1/SIB2/.../SIB*n* (System Information block 1/2/*n*)		Step 2: Reading MIB/ SIB1/SIB2/.../SIB*n*
SCell	For normal SI acquisition Step1: Acquiring SI from RRC message for SCell addition		At SI Change Step 1: Acquiring SI from RRC message for SCell addition

The frame timing and system frame number will be aligned across CCs that are candidates for CA; if they are not aligned, additional complexity is seen in handling the DRX, CQI, and SRS configuration of SCCs. Additionally, for time-division duplexing (TDD), TDD-Config is the same for all candidate CCs.

2.4.2 PDSCH Transmission

The procedure for data transmission on PDSCH is illustrated in Figure 2.18. In LTE, multiple UEs may be assigned to the DL-SCH by the eNB because it is a shared channel. The UE observes the cell-specific reference signal (CRS) and additional CSI RSs to estimate prevailing downlink channel conditions and quantify the CSI. The CSI, in general, involves the determination of quantities such as CQI, precoding matrix indicator (PMI), and rank indicator (RI). Periodic CSI reporting for a given UE is configured on one UL CC to support as many as five DL CCs.

The scheduler is executed at the eNB to determine which user's data should be transmitted. In LTE-A, the scheduler complexity increases further as additional decisions (e.g., CC selection in case of CA) are made. When the scheduler selects a user for downlink data transmission, it uses the CQI value reported by the UE and the data buffer that is waiting for transmission to decide the resource blocks, data rate, and modulation scheme for the transmission. Information on how the data is transmitted is sent on the PDCCH. Cross-carrier scheduling was introduced in LTE-A to support CA.

The UE receives the data on one or more CCs, verifies the checksum, and transmits an ACK or NACK to the eNB according to the verification result. Downlink HARQ processing is deployed in CA scenarios with one HARQ entity per CC.

The multiple steps of the DL-SCH physical layer processing in LTE Rel-8 are also applicable with carrier aggregation. For transmission on multiple CCs in parallel to

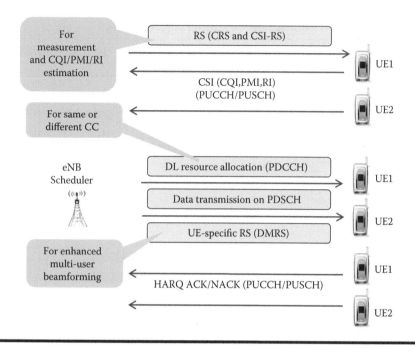

Figure 2.18   Data transmission on PDSCH.

the same UE, the transmissions on the different carriers correspond to separate transport channels with separate and more or less independent physical layer processing.

2.4.3 DL Control Channels

Control signaling serves several main purposes in LTE: it provides signaling related to scheduling assignments, it provides an acknowledgment in response to data transmission, it provides channel state feedback information, and it provides power control information. To support carrier aggregation, these functions must be extended to support multiple component carriers. It should be noted that in general the principle is to extend the LTE design to multiple carriers when possible.

In the downlink, three control channels are presented: the PDCCH is used for scheduling assignments and power control, the PCFICH is used to indicate the size of the DL control region, and the PHICH is used to provide an acknowledgment in response to UL data transmission.

2.4.3.1 PDCCH

Currently in Rel-8 FDD operation, the UL and DL carriers are paired. This means that a UL grant transmitted using a PDCCH on a given DL carrier implicitly indicates a particular UL carrier. In the case of carrier aggregation, LTE-A PDCCH

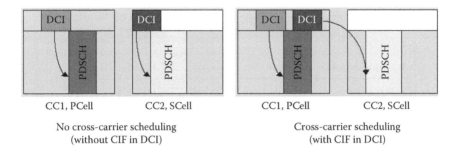

CC1, PCell   CC2, SCell   CC1, PCell   CC2, SCell

No cross-carrier scheduling   Cross-carrier scheduling
(without CIF in DCI)   (with CIF in DCI)

Figure 2.19   Cross-carrier scheduling.

can be separately coded for each CC; one PDCCH indicates one scheduled CC (i.e., downlink control information (DCI) is separately encoded for each CC), while multiple PDCCHs should be separately encoded and transmitted to schedule PDSCH/PUSCH in multiple CCs for a UE in a subframe. In this case, LTE-A will reuse the Rel-8 PDCCH structure (same coding, same control channel element [CCE]-based resource mapping) and DCI formats on each component carrier. In other words, PDCCH on a component carrier assigns PDSCH resources on the same component carrier and PUSCH resources on a single linked UL component carrier. A separate grant is made for each component carrier based on the DCI format for a single carrier. PDCCH on a component carrier can also assign PDSCH or PUSCH resources in one of multiple component carriers using the carrier indicator field (CIF) allowing for cross-carrier scheduling (as illustrated in Figure 2.19); such a DCI may be used to address any supported component carrier. Rel-8 DCI formats can be extended with a fixed 3-bit CIF configured upon higher-layer signaling.

CIF can eliminate the need to define an explicit relationship between DL and UL carriers for DCI purposes, and it also allows uplink heavy allocation to be efficiently supported. The main reasons for cross-carrier scheduling are load balancing and interference management in heterogeneous cell deployment. It is possible to use the CIF to schedule PUSCH transmission in multiple component carriers from a single DL carrier. Characteristics of the CIF are described as follows:

▪ Configuration for the presence of CIF is UE specific (i.e., not system specific or cell specific).
▪ CIF is a fixed 3-bit field. The CIF location is fixed irrespective of the DCI format size.
▪ Cross-carrier scheduling for DCI format 0, 1, 1A, 1B, 1D, 2, 2A, 2B in a UE-specific search space should always be supported by an explicit CIF.
▪ The CIF is not included in DCI format 0 or 1A in the common search space when the cyclic redundancy check (CRC) is scrambled by cell radio network temporary identifier (C-RNTI) or semipersistent scheduling RNTI (SPS C-RNTI).

Table 2.3 Transmission Mode and DCI Format to be Monitored

Transmission Mode	DCI Format to be Monitored
1. Single-antenna port; port 0	DCI 0/1A, DCI 1
2. Transmit diversity	DCI 0/1A, DCI 1
3. Open loop spatial multiplexing	DCI 0/1A, DCI 2A
4. Closed loop spatial multiplexing	DCI 0/1A, DCI 2
5. MU-MIMO	DCI 0/1A, DCI 1D
6. Closed loop Rank = 1 precoding	DCI 0/1A, DCI 1B
7. Single-antenna port; port 5	DCI 0/1A, DCI 1

■ DCI formats do not have CIF when CRC is scrambled by SI-RNTI, paging RNTI (P-RNTI), random access RNTI (RA-RNTI), or temporary C-RNTI (TC-RNTI).

In LTE Rel-8, the set of PDCCH candidates to monitor are defined in terms of search spaces. The PDCCH search space is split into the UE-specific search space and the common search space. In the UE-specific search space, the UE searches aggregation levels 1, 2, 4, and 8. In the common search space, the UE searches aggregation levels 4 and 8. At each aggregation level, the UE performs blind decoding for two different DCI* lengths on the specified number of PDCCH candidates, which results in 44 blind decodes (32 on the UE-specific search space and 12 on the common search space). For the UE-specific search space, UE monitors DCI 0/1A and DCI that is semistatically configured via RRC signaling depending on the transmission mode as shown in Table 2.3. For the common search space, UE monitors DCI 0/1A/3/3A and DCI 1C.

In the case of CA, there is one control region per component carrier. PDCCH supports one carrier with separate coding. The UE decodes a set of PDCCH according to all the monitored DCI format candidates on one or more activated serving cells as configured by higher-layer signaling for control information in every non-DRX subframe.

Cross-carrier scheduling for all DCI formats currently allowed in the UE-specific search space is supported by explicit CIF, but not allowed in the common search space for format 0/1A. For any DL carrier without CIF where the UE monitors

---

* In Rel-8, two different DCI sizes are blindly decoded. One DCI size is used for the downlink assignment, and this DCI size depends on the downlink transmission mode configured by the RRC signaling. The other DCI size is used for the uplink assignment and downlink compact assignment, and this DCI size is fixed since only one transmission mode is supported in the Rel-8 uplink.

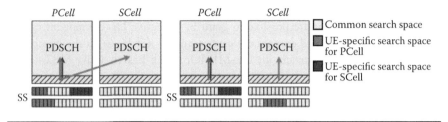

Figure 2.20   Search spaces with and without cross-carrier scheduling.

PDCCH, the search space is the same as in Rel-8. The UE monitors one common search space at aggregation levels 4 and 8 on the PCell (as illustrated in Figure 2.20). The UE does not need to monitor the common search space in the SCell.

For a given UE, search spaces located on each PDCCH CC are individually defined per aggregation level and per CC. A UE's search spaces on a PDCCH CC are shared by DCIs with the same size. The same hashing function is used to generate search spaces of multiple CCs where the offset between search spaces for different CCs is a function of the CIF. No standard solution is needed to handle ambiguities related to CIF configuration. If a PDCCH candidate of the common search space DCI format without CIF and a PDCCH candidate of the UE's search space DCI format with CIF both have the same payload size, scrambled by C-RNTI, and have the same starting CCE, the UE will assume that only the PDCCH candidate of the common search space DCI format may be transmitted.

In LTE-A, due to the introduction of additional DCI formats and due to carrier aggregation, the number of blind decoding operations will increase; the supported maximum number of blind decodes is in line with the number of aggregated CCs. However, the UE does not have to monitor the common search space on any secondary component carrier.

In LTE-A, the UE needs to monitor three DCI formats: DCI 0/1A, DCI for configured downlink transmission mode, and DCI for configured uplink transmission mode (UL-MIMO is supported by introducing a transmission mode in the UL with a new DCI format that is only transmitted in the UE-specific search space on PDCCH and is not bit aligned to any DCI format). Therefore, the total number of blind decoding in a CC is increased to 60 (48 on the UE-specific search space and 12 on the common search space). The total number of blind decoding may become $60 \times N$ (N denotes the number of component carriers). In a live network, we should decrease the blind decoding times for complexity and power reduction reasons.

In LTE-A specification, the UE shall monitor only one common search space on the primary cell, so the total number of blind decoding can be decreased significantly. Note that in the case of CA with one PCC and multiple SCCs configured, the actual number of blind decodes is up to $44 + 32 \times N\_DL\_SCC + 16 \times N\_UL\_SCC + 16 \times N\_ULM\_CC$, where N_DL_SCC is the number of active

DL secondary component carriers, *N_UL_SCC* is the number of secondary UL component carriers that are possible to grant by an active DL component carrier that is not the SIB2-linked component carrier, and *N_ULM_CC* is the number of configured component carriers for UL-MIMO, which has an active SIB2-linked DL component carrier (it is possible to grant by an active DL component carrier that is not the SIB2-linked component carrier). The actual blind decode number is recalculated with the consideration of CC activation and deactivation.

Furthermore, some evolutional technology on PDCCH is under study and may be adopted in the future. Precoding of the PDCCH may be used to significantly improve cell-edge performance and coverage. PDCCH beamforming may be considered as a potential PDCCH performance enhancement technique for LTE-A. PDCCH beamforming can be implemented using a precoding vector that is preselected by eNB and fed back by the UE.

2.4.3.2 PCFICH

In LTE, the physical control format indicator channel (PCFICH) carries information about the number of OFDM symbols used for transmission of PDCCHs in a subframe. LTE-A uses the Rel-8 design (modulation, coding, mapping to resource elements) for PCFICH.

In the case of carrier aggregation in LTE-A, the PCFICH indicates the independent control region size per component carrier, which enables individually adjusted control regions per carrier. A UE will therefore independently decode the PCFICH and determine the data boundary as in LTE Rel-8. In the case of a PDCCH-less carrier (e.g., an extension carrier in heterogeneous network), this would be known to the user and no special handling would be necessary. Since the control region may be of a different size on different component carriers, a UE needs to receive the PCFICH on each of the component carriers on which it is scheduled.

When cross-carrier scheduling is used, the PDCCH related to a certain PDSCH transmission is transmitted on a component carrier other than the PDSCH itself. Therefore, the UE needs to know the starting position for the data region on the carrier on which the PDSCH is transmitted. Although it can be done by decoding the PCFICH on each carrier on which a PDSCH is scheduled, a serious situation must be considered that is caused by incorrect decoding of the PCFICH.

In LTE Rel-8, if the PCFICH is incorrectly decoded, it is very likely that the PDCCH in the control region that schedules the PDSCH will be lost. In CA with cross-carrier scheduling, if the PCFICH error occurs on a carrier with PDSCH transmission indicated by PDCCH on another carrier, it may lead to the UE's downlink HARQ buffer corruption, as illustrated in Figure 2.21.

Therefore, for cross-carrier scheduled transmissions, a standardized solution is introduced to handle the PCFICH error problem. For UE that is cross-carrier scheduled using CIF, a single value of the PDSCH starting position on the CC that carries the PDSCH will be indicated to the UE via RRC signaling, rather

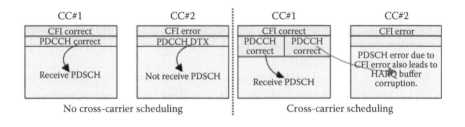

Figure 2.21   CFI error in cross-carrier scheduling.

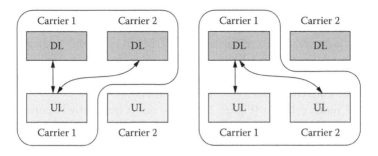

Figure 2.22   Carrier aggregation configuration.

than obtaining it from the PCFICH on the corresponding CC. The semistatically configured data region starting position may differ from the value signaled on the PCFICH on the CC that carries the PDSCH transmission.

2.4.3.3 PHICH

LTE-A will reuse PHICH physical transmission aspects from Rel-8, including orthogonal code design for each PHICH group, modulation scheme, scrambling sequence, the time and frequency resource mapping to resource elements (REs), and so on.

In CA, PHICH is transmitted only on the same DL carrier that is used to transmit the UL grant. Since separate HARQ processing is used in LTE-A, associated control signaling will be required for each of the component carriers. In addition, LTE-A UE can be configured with UE-specific UL/DL carrier aggregation configurations that are a subset of the system configuration as shown in Figure 2.22.

PHICH resource mapping rules depend on the CA configuration. For 1:1 or n:1 mapping between DL CCs and UL CCs, or scheduling without CIF, the PHICH resource mapping reuses the Rel-8 mapping scheme. For the 1:n mapping case for an uplink-heavy aggregation scenario, or scheduling with CIF, a single set of PHICH resources are shared by all UEs (Rel-8 UEs and LTE-A UEs).

The PHICH acknowledgment is transmitted on the same DL carrier as the UL scheduling assignment, which allows many simultaneous UE-specific configurations to be supported. This feature solves the issue with uplink-heavy aggregation and also does not require an explicit DL–UL pairing relationship to be defined. Although this requires the eNB to manage the PHICH, it is similar to the PDCCH management that will also be required.

When cross-carrier scheduling is used, one downlink CC may have to carry PHICH transmissions for multiple uplink CCs. There is an increased probability of PHICH collisions occurring because the PHICH index is determined from the lowest physical resource block (PRB) of the corresponding PUSCH transmission, which may be the same on multiple uplink CCs.

The potential PHICH resource collision already exists in LTE Rel-8 in the case of uplink multiuser MIMO (MU-MIMO) or when the cell is configured with a small number of PHICH groups. The problem can be resolved by an eNB scheduler using the demodulation reference signal (DMRS) cyclic shift indication in the corresponding UL grant. In LTE, higher layers provide the number of configured PHICH resources for a carrier, which is proportional to the DL bandwidth and can be set to four different values ($Ng = \{1/6, 1/2, 1, 2\}$).

PHICH index collision (Figure 2.23) may occur frequently because the starting UL PRB index allocated for multiple UL CCs is the same. If there are more UL PRBs than PHICH channels, there is a high probability that a PHICH index

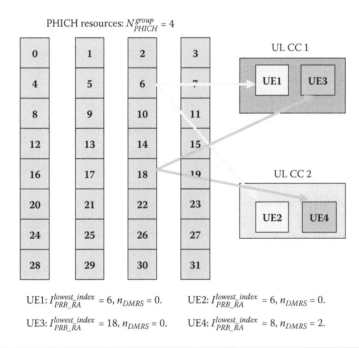

PHICH resources: $N_{PHICH}^{group} = 4$

UE1: $I_{PRB\_RA}^{lowest\_index} = 6, n_{DMRS} = 0.$ UE2: $I_{PRB\_RA}^{lowest\_index} = 6, n_{DMRS} = 0.$

UE3: $I_{PRB\_RA}^{lowest\_index} = 18, n_{DMRS} = 0.$ UE4: $I_{PRB\_RA}^{lowest\_index} = 8, n_{DMRS} = 2.$

Figure 2.23 PHICH index collision (one DL CC reserves PHICH resources for two UL CCs).

collision will occur. Although the eNB can avoid some PHICH index collisions by adjusting n-DMRS, this would increase the complexity for the eNB scheduler due to joint coordination PHICH resources over multiple UL CCs. It was proposed that the UL CC-specific PHICH resource offset, which is equivalent to adjusting the UL PRB index, could be applied to avoid possible PHICH index collision.

2.5 UL CA Operations

First, a typical uplink data transmission procedure is summarized (in Figure 2.24) as follows:

1. If new data has arrived in the data buffer of a UE, and there are no PUSCH resources available in this transmission time interval (TTI) (for the UE), the UE will signal a scheduling request (SR) on the PUCCH. If no SR resource is available, the UE will start a random access procedure on the physical RACH (PRACH) to signal the SR.
2. The eNB sends a UL grant for the UE on the PDCCH.
3. The UE uses the received UL grant to send its buffer status report (BSR) so the eNB can make a decision on additional UL resources to be allocated. If there is an additional resource, data will also be transmitted.
4. Based on the UE BSR, the eNB sends the UE an additional UL grant, if needed.
5. The UE transmits UL data on the PUSCH.
6. The eNB sends back a HARQ ACK/NACK for the received UL data.

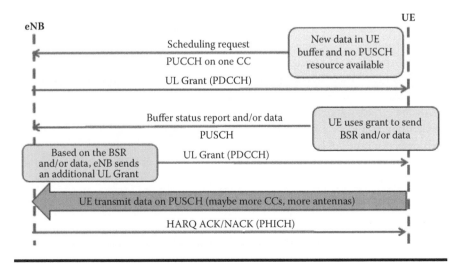

Figure 2.24 Uplink data transmission procedure.

Thanks to the evolution in LTE-A, PUSCH transmission may occur using one or more CCs and one or more antennas. Recall that up to (4x4) single-user MIMO (SU-MIMO) is supported in LTE-A. The same precoding is used for the PUSCH and the DMRS. Besides, with the introduction of CA, all the above-mentioned physical channels or signals need to be redefined or further clarified.

2.5.1 Cell Search and Random Access

In a carrier aggregation system, wider bandwidth and system throughput will result in better opportunities for system access. Assuming asymmetric carrier aggregation, the following cell search and broadcast channel (BCH) reception procedure can be applied for both Rel-8 UEs and LTE-A UEs. Especially for LTE-A UEs, it would be beneficial to receive the system information about DL and UL carrier configuration in the accessing cell so that the UE cell-search complexity does not increase according to system bandwidth when a single-component carrier is used for initial access.

- The UE performs a cell search on a 100-KHz frequency raster.
- UE performs an initial cell search based on the synchronization channel (SCH) for a coarse timing and frequency synchronization, if the UE detects a SCH signal on one of the aggregated DL CCs (DL carrier #0), and then the UE receives the PBCH in that DL CC. After receiving the PBCH, DL bandwidth, and the number of transmission antennas, PHICH configurations are obtained.
- The UE receives SI-2 through the DL CC #0. Uplink E-UTRA Absolute Radio Frequency Channel Number (EARFCN), UL bandwidth, and several physical channel configurations are acquired.
- After the UE receives the PBCH and SI-2, the information about paired UL CC (UL CC #0) corresponding to the DL CC (DL CC #0) is obtained.

After the initial cell search and BCH reception, some important control signaling, including the PRACH configuration for random access opportunities, is obtained by the UE, then the UE performs an initial random access procedure to establish a connection with the eNB according to the configured PRACH parameters.

In the random access procedure, the UE transmits a random access preamble by choosing one from a maximum of 64 possible candidates on a PRACH. If the eNB detects a random access preamble, it sends a random access response allocating a temporary UE identity and radio resources for the uplink transmission of the initial RRC message.

LTE-A's random access procedure is the same as Rel-8. The RA procedure is always performed on the PCell. With asymmetric downlink-heavy carrier aggregation, several downlink CCs may be associated with one uplink CC. The ambiguity in the random access procedure arises from the inability of the eNB to determine which downlink CC the UE has selected for initial access because downlink CC identification information is not conveyed in the preamble transmission. This ambiguity

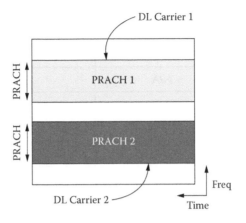

Figure 2.25 PRACH for asymmetric CA.

must be resolved by the eNB so that it can transmit downlink data to the UE in the appropriate downlink carrier. An example of the solution is shown in Figure 2.25, in which each downlink CC is associated with a dedicated PRACH in time and frequency. With this approach, the eNB can uniquely determine the appropriate downlink carrier based on which PRACH or preamble set the UE selects.

2.5.2 UL Control Channels

2.5.2.1 PUCCH in CA

The physical uplink control channel, PUCCH, carries uplink control information. UL control signaling carries acknowledgments for the DL data transmission and channel state feedback (e.g., CQI, PMI, or RI). In the uplink with carrier aggregation, the simplest possible approach is to support an individual UL control channel per carrier based on the Rel-8 structure. Thus, each UL control channel may be configured independently to support one DL–UL carrier pair. In the case of asymmetric carrier aggregation, additional UL control signaling (ACK/NACK and CQI) may be configured to support multiple DL carriers on a single UL carrier. One concern for the UL is the limited transmission power available to the UE. As a result, it may not be possible to transmit feedback or acknowledgments from many component carriers simultaneously. In this case, it is possible that some form of control information multiplexing, such as ACK/NACK bundling, may be used. In addition to carrier aggregation, within the same carrier, a UE has the ability to simultaneously transmit both the PUSCH and PUCCH. The total cumulative power that the UE is allowed to transmit across all component carriers should not exceed the maximum UE power. Under a transmit power limitation, power reduction based on per-channel scaling is employed where scaling factors are used so that power reduction is first performed on the PUSCH.

Table 2.4 Supported PUCCH Formats

PUCCH Format	Modulation Scheme	Number of Bits per Subframe, M_{bit}
1	N/A	N/A
1a	BPSK	1
1b	QPSK	2
2	QPSK	20
2a	QPSK + BPSK	21
2b	QPSK + QPSK	22
3	QPSK	48

Simultaneous transmission of PUCCH and PUSCH from the same UE is supported if enabled by higher layers. For a TDD system, the PUCCH is not transmitted in the UpPTS field. The PUCCH supports multiple formats, as listed in Table 2.4. Formats 2a and 2b are supported for normal cyclic prefixes only. All PUCCH formats use a cell-specific cyclic shift.

The LTE-A PUCCH design supports up to five DL component carriers. PUCCH transmission of ACK/NACK, SR, and periodic CSI is possible only on the UL PCC. If the UE has a PUSCH transmission on the UL PCC, then any uplink control information (UCI) on the PUSCH is carried on the PCC. In transmissions of one or multiple PUSCHs and no PUSCH on the PCC, any UCI on the PUSCH is carried on a single PUSCH on an SCC.

A single CC for PUCCH transmission of ACK/NACK, SR, and periodic CSI will be semistatically assigned. The ACK/NACK transmission method for carrier aggregation in LTE-A UE that supports up to 4 ACK/NACK bits PUCCH format 1b with channel selection is adopted. While for LTE-A UE that supports more than 4 ACK/NACK bits, both PUCCH format 1b with channel selection and PUCCH format 3 are supported and are configured by higher-layer signaling. Periodic CSI reporting for a given UE is configured on one UL CC to support as many as five DL CCs. The Rel-8 PUCCH 1a/1b resource is used when the PDSCH is received on the PCell only. Explicit signaling is used for indicating the ACK/NACK resource for channel selection for non-cross-carrier scheduling or for cross-carrier scheduling from the SCell.

2.5.2.2 ACK/NACK Transmission

In LTE Rel-8 TDD, both PUSCH and PUCCH can carry ACK/NACK(s) corresponding to multiple DL subframes. One of the supported feedback modes is ACK/NACK bundling; in such a mode, the AND operation is performed per codeword across multiple ACK/NACK bits within the subframes in the bundling window and will generate 1 or 2 bundled ACK/NACK bits for feedback. Such a mode is useful for coverage-limited UE. The other is ACK/NACK multiplexing; in such a mode, the AND operation is performed within a DL subframe across spatial code words (i.e., ACK/NACK spatial bundling), and ACK/NACK multiplexing is achieved via the channel selection method, which boosts DL throughput compared to ACK/NACK bundling. ACK/NACK feedback mechanisms specified in the Rel-8 TDD need to be reused as much as possible in LTE-A.

In LTE-A, a single UE-specific UL CC is configured semistatically for carrying PUCCH ACK/NACK. Simultaneous ACK/NACK transmission from one UE on PUCCH in multiple UL CCs is not supported. Multiple simultaneous PUCCH transmission for ACK/NACK in multiple nonadjacent PRBs is not supported either. For a CA-capable UE that is configured for single UL/DL carrier-pair operation, single antenna PUCCH resource assignment is done as in Rel-8.

All HARQ ACK/NACKs for a UE can be transmitted on the PUCCH in absence of PUSCH transmission. The PUCCH format for ACK/NACK transmission is decided according to the number of ACK/NACK bits. PUCCH format 1a/1b in Rel-8 can support up to 2 ACK/NACK bits. PUCCH format 1b with channel selection can support up to 4 ACK/NACK bits. Just a reminder: format 1b with channel selection is applied in Rel-8 TDD to avoid throughput performance degradation due to subframe bundling. It can support up to 4 ACK/NACK bits if simply reusing the mapping rule specified in Rel-8.

The new PUCCH format (PUCCH format 3) is introduced to support larger numbers of ACK/NACK bits, which is based on DFT-S-OFDM.

For LTE-A UE that supports up to 4 ACK/NACK bits, PUCCH format 1b with channel selection is used for ACK/NACK transmission. For LTE-A UE that supports more than 4 ACK/NACK bits, both PUCCH format 1b with channel selection and PUCCH format 3 are supported. PUCCH format 1b with channel selection is supported up to 4 ACK/NACK bits, and PUCCH format 3 for the full range of ACK/NACK bits. The UE is configured by higher layers to adopt PUCCH format 3 or PUCCH format 1b with channel selection.

The ACK/NACK resource for DL feedback transmission may be retrieved implicitly or explicitly, as illustrated in Figure 2.26.

If the UE is configured for PUCCH format 1b with channel selection and PDSCH transmission on the PCell, implicit ACK/NACK resource allocation is deployed for dynamic scheduling as in Rel-8.

If the UE is configured for PUCCH format 1b with channel selection and PDSCH transmissions on SCells, for non-cross-carrier scheduling or for cross-carrier scheduling from the SCell, explicit ACK/NACK resource allocation configured by RRC is deployed. The PDCCH corresponding to the PDSCH on the SCell indicates a resource (ACK/NACK resource indicator, ARI) derived from the RRC configured resources. For cross-carrier scheduling from the PCell, implicit ACK/NACK resource allocation is used.

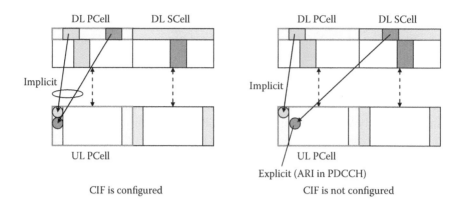

Figure 2.26 ACK/NACK resource allocation with channel selection.

If the UE is configured to use PUCCH format 3, explicit ACK/NACK resource allocation will be configured by RRC. The PDCCH corresponding to the PDSCH on the SCell indicates a resource ARI derived from the RRC configured resource(s). If no PDCCH corresponding to the PDSCH on the SCells is received, and the PDSCH is received on the PCell, the Rel-8 PUCCH 1a/1b resource is used. The UE assumes the same ARI for all PDCCHs corresponding to the PDSCH on the SCells.

For resource allocation of PUCCH format 3 (DFT-S-OFDM), the transmit power control (TPC) field in the PDCCH corresponding to the PDSCH on the PCell is used as the TPC command, while the TPC field (2 bits) in the PDCCH corresponding to the PDSCH on the SCell is used as the ARI.

2.5.2.3 PUCCH Format 3

In LTE-A, carrier aggregation needs more ACK/NACK bits (up to 20 bits with 5 component carriers) to be supported during one UL subframe. In case the UE receives the PDSCH transmissions from multiple DL carriers in a subframe, it has to feed back multiple ACK/NACKs associated with the different TBs in one UL subframe. For each CC configured with dual codeword transmission, the total number of HARQ feedback states is 5 per DL CC, which are (ACK,ACK), (ACK,NACK), (NACK,ACK), (NACK,NACK), and discontinuous transmission (DTX). So the total number of feedback states for n activated DL component carriers can be represented by 5^n. For the maximum number of 5 activated DL component carriers, each scheduled with dual codeword transmission, the number of bits required to feed back all states becomes 12 bits for an FDD system, therefore a new PUCCH format 3 for carrier aggregation was introduced to meet this requirement.

The new PUCCH format (PUCCH format 3), which can convey up to 20 ACK/NACK bits, is introduced in LTE-A. UE capable of operating on more than two downlink CCs needs to support PUCCH format 3, since more than 4 bits of HARQ acknowledgments are required. For such UE, PUCCH format 3 can also

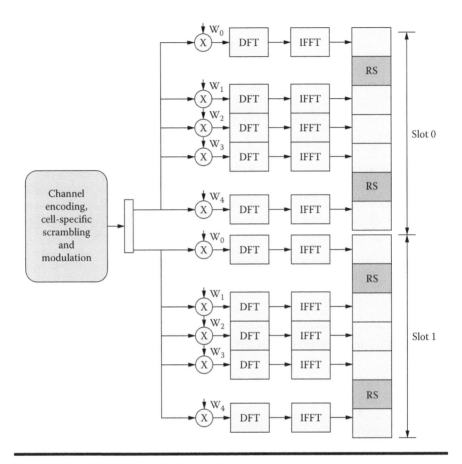

Figure 2.27   PUCCH format 3 (normal CP).

be used for less than 4 bits ACK/NACK relating to simultaneous transmission on multiple CCs when it is configured by higher-layer signaling not to use PUCCH format 1 with resource selection.

PUCCH format 3 is based on DFT-S-OFDM, which is the same transmission scheme as that used for PUSCH. The processing procedure for PUCCH format 3 is illustrated in Figure 2.27. Block coding is applied followed by scrambling using a cell-specific scrambling sequence to randomize intercell interference. The resulting 48 bits are quadrature phase-shift keying (QPSK)-modulated and divided into two groups, one per slot, of 12 QPSK symbols each.

For a normal cyclic prefix (CP) (7 OFDM symbols per slot), the RS is in symbols 1 and 5, leaving 5 symbols for data transmission. For extended CP, the RS is in symbol 3. PUCCH format 3 adopts SF = 5 (multiplexing capability, orthogonal cover code [OCC] on the RS is not used). The length-5 orthogonal cover sequences are obtained as 5 DFT sequences.

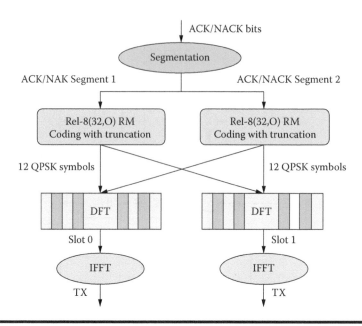

Figure 2.28 Coding design for PUCCH format 3.

LTE-A adopts UE-specific and cell-specific scrambling applied to coded channel bits for randomization of PUCCH format 3, similar to PUCCH format 2 scrambling. The scrambling operation is identical across single-carrier frequency-division multiple access (SC-FDMA) symbols in a slot, because symbol index–dependent scrambling would increase intersymbol interference (ISI) within SC-FDMA symbols in low-Doppler, frequency-selective channels.

Cell-specific cyclic shift hopping is also adopted, targeting randomization of intercell interference. The random cyclic shift sequence is identical to Rel-8 and cyclic shift is applied before DFT.

For PUCCH format 3 with an ACK/NACK payload size less than or equal to 11 bits, the (32, O) rate matching (RM) code from Rel-8 with circular buffer rate matching (single RM code) is reused. ACK/NACK bit mapping is the same as in Rel-8.

In case of an ACK/NACK payload size larger than 11 bits, dual RM coding is used, as illustrated in Figure 2.28. The ACK/NACK bits are equally segmented into two ACK/NACK blocks of length ceil (N/2) and N–ceil (N/2), where N is the ACK/NACK feedback payload size. For each ACK/NACK block that contains eleven or fewer bits, each block is encoded with the Rel-8 RM (32, O) coding with the last 8 rows punctured. Each ACK/NACK block is then modulated into 12 QPSK symbols, and the 24 QPSK symbols collected alternately from the two ACK/NACK blocks that are transmitted on two slots.

When ACK/NACK is transmitted on PUSCH with payload size larger than 11 bits, same as operations for PUCCH format 3, the ACK/NACK bits are

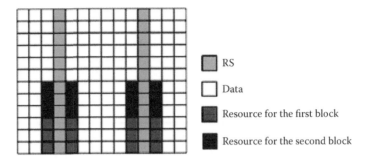

Figure 2.29   ACK/NACK resource mapping scheme.

segmented into two blocks with length ceil (N/2) and N-ceil (N/2) respectively, where N is the ACK/NACK feedback payload size. Each ACK/NACK block is encoded using the RM (32, O) code with circular buffer rate matching. The output bit sequence of the channel coding block is obtained by concatenation of the two encoded blocks, which will map the encoded ACK/NACK bits from the two ACK/NACK blocks to the resource elements on PUSCH as shown in Figure 2.29.

PUCCH format 3 supports full ACK/NACK feedback for FDD and ACK/NACK feedback for TDD with spatial bundling. For PUCCH format 3, similar to PUCCH format 1/2, a resource can be represented by a single index from which the orthogonal sequence and the resource-block number can be derived. A UE can be configured with four different resources for PUCCH format 3.

2.5.2.4 Scheduling Request

In CA, a scheduling request is per-UE based, that is, one request for all UL component carriers. The scheduling request is transmitted on PUCCH and is semistatically mapped onto one UE-specific UL component carrier.

2.5.2.5 CSI Reporting

To support link adaptation and precoding at the eNB, UE needs to report CQI, PMI, and RI for each component carrier. LTE Rel-8 supports two types of CSI feedback channels: periodic CSI and aperiodic CSI. Periodic CSI is transmitted using PUCCH Format 2/2a/2b, with a payload size up to 11 bits. Aperiodic CSI is transmitted on the PUSCH, with a payload size up to 64 bits. Aperiodic CSI can be transmitted with or without simultaneous UL data and triggered with an indication in the UL grant.

The LTE-A methodology of CSI transmission will reuse PUCCH format 2 for CSI reporting on the PUCCH. The periodic CSI reporting (PUCCH) is semistatically mapped to one UE-specific UL CC (PCell). The CSI reporting mode could

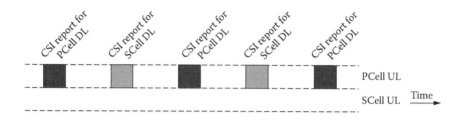

Figure 2.30   Periodic CSI reporting for CA.

be independently configured for each DL CC. The following reporting types are supported for periodic CSI:

- The Type 1 report supports CQI feedback for the UE selected sub-bands.
- The Type 1a report supports sub-band CQI and secondary PMI feedback.
- The Type 2, Type 2b, and Type 2c reports support wideband CQI and PMI feedback.
- The Type 2a report supports wideband PMI feedback.
- The Type 3 report supports RI feedback.
- The Type 4 report supports wideband CQI.
- The Type 5 report supports RI and wideband PMI feedback.
- The Type 6 report supports RI and PTI feedback.

Since the periodic CSI reports (as shown in Figure 2.30) for different downlink CCs may still collide (in same subframes), some additional priorities will be defined for different CCs and drop periodic CSI reports according to the priorities.

If simultaneous PUCCH and PUSCH is not configured and there is at least one PUSCH transmission, all UCI is piggybacked on a PUSCH. If there is a PUSCH on the PCell while there is no aperiodic CSI triggered, all UCI is piggybacked on the PCell; otherwise an SCell is selected according to SCell indices. When an aperiodic CSI is triggered, the UCI is mapped on the PUSCH that contains the aperiodic CSI.

Simultaneous PUCCH and PUSCH transmission is enabled by UE-specific higher-layer signaling. In case of periodic CQI/PMI/RI alone or ACK/NACK alone, the UCI is transmitted on the PUCCH. For any collision between periodic CSI and aperiodic CSI for the same or different DL CCs, periodic CSI is dropped as in Rel-8. Type 1a, 2a, 2b, and 2c CSI reports should be coded as for CQI/PMI on the PUSCH; Type 5 and 6 CSI reports should be coded as for the RI in the PUSCH transmission. The RI report per cell is up to 3 bits. The tail-biting convolutional code (TBCC) is applied with joint coding of CQI/PMI and 8-bit CRC.

Some simultaneous UCI scenarios are considered in LTE-A: In case of SR+ACK/NACK with PUCCH format 3, the positive/negative SR bit is appended at the end of the ACK/NACK information bits. For SR+ACK/NACK with channel selection with format 1b, spatial bundling is used for FDD, full bundling is used for TDD, and 1 or 2 bits bundled ACK/NACK is transmitted on the SR resource. When simultaneous PUCCH and PUSCH is not configured and when CQI and ACK/NACK are to be transmitted in the same subframe, CQI is dropped unless simultaneous ACK/NACK and CQI is enabled and the UE only received the PDSCH on the PCell. When simultaneous PUCCH and PUSCH transmission is configured, ACK/NACK is always transmitted on the PUCCH and CQI is transmitted on the PUSCH if available.

2.5.3 UL Multiple Access

LTE Rel-8 utilizes SC-FDMA for uplink transmission. Uplink multiple access schemes need to be standardized for LTE-A to extend the bandwidth beyond 20 MHz, meanwhile backward compatibility shall be maintained to LTE Rel-8 as for cubic metric and cell coverage. Therefore the following aspects were evaluated in UL multiple access (UL MA) scheme design in LTE-A:

- Performance, especially with UL-MIMO
- Receiver complexity, system design flexibility, and flexibility in resource allocation depending on spectrum division
- Peak-to-average power ratio (PAPR) should be low at the cell edge to maintain coverage equivalent to SC-FDMA.
- Discontinuous resource allocation is required to enhance average throughput.
- Rel-8 DFT-S-OFDMA will remain due to the backward compatibility requirement.

It is necessary to adopt UL multiple access in LTE-A not only for the multiple-carrier (CA) scenario, but also for the single-carrier scenario. In the current Third-Generation Partnership Project (3GPP) agreements, clustered DFT-S-OFDM is deployed for the single-carrier (intra-CC) scenario, and Nx-DFT-S-OFDM is deployed for the multiple-carrier (inter-CC) scenario.

2.5.3.1 Clustered DFT-S-OFDM

In LTE Rel-8 with SC-FDMA (as shown in Figure 2.31), the uplink resource allocation is always contiguous in frequency domain to maintain beneficial low cubic metric properties. However, more flexible resource allocation to better utilize frequency domain packet scheduling gains is seen as necessary for LTE-A, under which condition the cubic metric (CM) requirements need to be relaxed. In this case, clustered DFT-S-OFDM can be used to provide more freedom for resource allocation, while keeping the modifications compared to Rel-8 at a minimum. The new UL access scheme has some advantages including more flexible scheduling, assignment of only the preferred frequency for the UE, and a degraded PAPR of UE transmission.

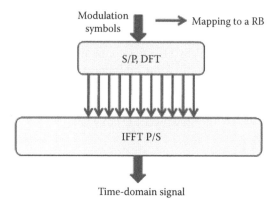

Figure 2.31   SC-FDMA transmitter.

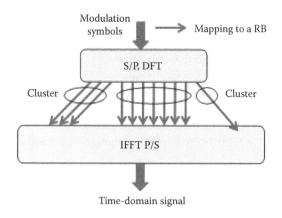

Figure 2.32   Clustered DFT-S-OFDM transmitter. (S/P—serial/parallel)

In the so-called clustered DFT-S-OFDM transmission scheme, a cluster is defined as a chunk of consecutive PRBs and resource block groups (RBGs) and is separate from any other clusters in the frequency domain. Spectrum elements after DFT precoding are divided into two or more parts called *clusters*, and each cluster is mapped to allocated resource blocks (see Figure 2.32. This option can provide more uplink scheduling flexibility for the UE with better geometry while allowing single-carrier transmission for power-limited UE by scheduling localized RBs only.

Clustered DFT-S-OFDM transmission supports noncontiguous resource allocation, and supports PUSCH transmission from the UE with a sufficient amount of power headroom with and without spatial multiplexing. With clustered DFT-S-OFDM, simultaneous PUCCH and PUSCH transmission will be supported, as

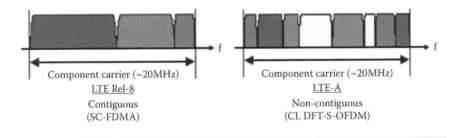

Figure 2.33   PUSCH transmission with clustered DFT-S-OFDM.

well as noncontiguous PUSCH transmission with a single DFT per component carrier. In the current LTE-A specification (Rel-10), at most two PRB clusters are supported for PUSCH transmission (as shown in Figure 2.33).

2.5.3.2 Nx DFT-S-OFDM

SC-FDMA allows only continuous resource allocation that does not make full use of multiuser diversity in the broadband channel. Therefore, it is not suitable for CA scenarios. Multiple UL transmission schemes including OFDMA, clustered DFT-S-OFDMA, and Nx DFT-S-OFDMA have been considered to support carrier aggregation.

With clustered DFT-S-OFDMA, the data packet undergoes the coding and modulation process prior to DFT precoding and is then mapped to multiple clusters. Each cluster is contiguous to minimize PAPR/CM, although multiple clusters may be mapped to the same carrier to allow noncontiguous resource allocation. This is suitable when physical layer segmentation and aggregation of data is used (i.e., one transport block in total component carriers). Therefore, if clustered DFT-S-OFDMA is adopted to support CA, redesign of the physical layer will be required and there will be no independent link adaptation and HARQ across different component carriers. An illustrative block diagram of clustered DFT-S-OFDMA is shown in Figure 2.34. The key advantage of this method is the slightly lower CM than Nx SC-FDMA. However, such a CM advantage will be somewhat negated by slightly inferior link performance. Moreover, when the UE implementation has multiple transmit chains, the signal phase between the chains before analog combination may be difficult to align precisely. In the case of random phase among clusters, the CM of the composite signal gets worse by 0.6dB (QPSK) and 0.4 dB (16/64-QAM) in the 4-cluster example, as opposed to the phase-aligned case with, for example, a single transceiver implementation.

Nx DFT-S-OFDMA can achieve wider bandwidth by adopting parallel multi-CC transmission. This scheme can satisfy requirements for peak data rate while maintaining backward compatibility, low cost, and fast development by reusing the Rel-8 specification.

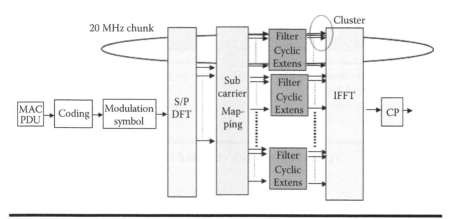

Figure 2.34 Clustered DFT-S-OFDMA.

With Nx DFT-S-OFDMA, the data packet is segmented into several segments or codewords. Each segment undergoes separate DFT precoding and is then mapped into a component carrier. This is equivalent to using N parallel DFT-S-OFDM transmitters, which allows for per-carrier AMC and HARQ. There can be multiple DFT precoding blocks and the output of each DFT precoding block is mapped to consecutive RBs only, but the outputs of different DFT precoding blocks can be mapped to separate frequency bands. Disadvantages of this method include additional control overhead because individual control signaling per carrier may be required along with additional segmentation overhead. This scheme is suitable when MAC layer segmentation and aggregation of data is used (since the OFDMA method has no DFT precoding, it can support both MAC data aggregation and physical layer data aggregation). That is, each component carrier provides a separate data stream that is aggregated or segmented at the MAC layer. This requires separate HARQ processing and associated control signaling and a bit more overhead. A key advantage of this approach is separate link adaptation and MIMO support for each carrier, which should improve throughput. An illustrative block diagram of Nx DFT-S-OFDMA is shown in Figure 2.35.

In general, the CM increases with N and is comparable to OFDMA with a large number of codewords. The key advantage of this method is that it can be implemented with few changes required in the physical layer specifications where the transmission (TX) architecture consists of parallel Rel-8 transmitters and receive (RX) architecture consists of parallel Rel-8 receivers, which can be realized with one single inverse fast Fourier transform (IFFT) if the subcarrier rasters match. This option can provide more uplink scheduling flexibility for the UE with better geometry while allowing single-carrier transmission for power-limited UE by scheduling a single DFT block only.

From the perspective of CM performance, the size of each cluster (or each DFT block) does not affect the CM performance of clustered DFT-S-OFDMA or Nx DFT-S-OFDMA. Generally, the CM property of clustered SC-FDMA

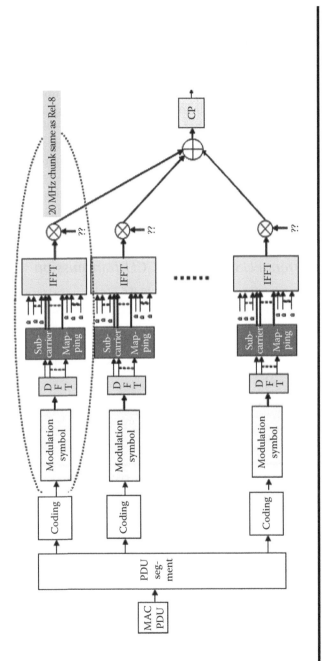

Figure 2.35 Nx DFT-S-OFDMA.

depends on the number of clusters, cluster size, and cluster placement. For a given combination of clusters (DFT blocks) and RBs per cluster (DFT blocks), the CM value of clustered DFT-S-OFDMA is smaller than that of Nx DFT-S-OFDMA. The gap decreases as the modulation order increases. As for the block error rate (BLER) comparison between the two schemes, for all cases, Nx DFT-S-OFDMA slightly outperforms clustered DFT-S-OFDMA and the gain is around or less than 0.3 dB. Finally, the Nx DFT-S-OFDMA scheme was adopted by 3GPP for uplink transmission in CA scenarios, together with clustered DFT-S-OFDMA adopted for intra-CC uplink transmission (as illustrated in Figure 2.36). Consequently both frequency-contiguous and frequency-noncontiguous resource allocation is supported on each component carrier. The Nx DFT-S-OFDMA scheme, with one DFT and one transport block per component carrier, can provide lower CM and comparable performance with OFDMA within the considered signal-to-noise ratio (SNR) region. Independent PUSCH power control will be provided per CC.

2.5.3.3 Concurrent PUSCH and PUCCH Transmission

LTE-A supports simultaneous PUCCH and PUSCH transmission within a CC or over different CCs (as illustrated in Figure 2.37). The UE will transmit both

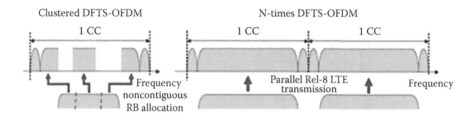

Figure 2.36   Clusterized DFT-S-OFDMA and Nx DFT-S-OFDMA.

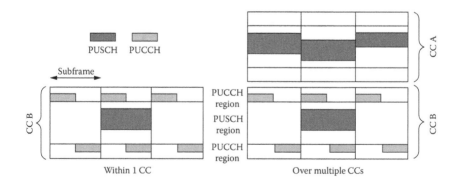

Figure 2.37   Simultaneous PUCCH and PUSCH transmission.

the PUCCH and one or more PUSCHs simultaneously in a UL CC without UCI piggyback. PUCCH is transmitted only on the PCC.

LTE-A supports concurrent PUSCH and PUCCH transmissions based on the benefit of utilizing the PUCCH resources that remain empty when the UCI is included in the PUSCH, and at the same time preserving resources in the PUSCH to be used for data transmission. Despite the benefit of freeing up PUSCH resources for data transmission, concurrent PUSCH and PUCCH transmission has several limitations that are discussed in the following text.

First of all, concurrent transmission can only apply for UE that is not power limited. Due to the CM benefit of the single carrier, the UE can transmit ΔCM dB (CM difference depends on the PUSCH modulation) more power for time-division multiplexing (TDM) of UCI and data in the PUSCH than in concurrent PUCCH and PUSCH transmissions. Therefore, concurrent PUCCH and PUSCH transmission does not have advantages for low signal-to-interference plus noise ratio (SINR) or coverage-limited UE.

As we know, the PUCCH is interference limited and it is not clear whether the trade-off for transmitting UCI in the PUCCH versus in the PUSCH is always positive. Sometimes UCI transmission in the frequency diversity PUSCH will not require a lot of resources and is likely to result in better performance than frequency-hopping (FH) transmission in the PUCCH. Especially for relatively large PRB allocations, the inherent frequency diversity in the PUSCH is likely to be larger than the one achievable by PUCCH transmission.

Furthermore, because the power control loops are different between the PUSCH and the PUCCH, there is an obvious limitation when the combined PUSCH and PUCCH signal power reaches P_{max}. Because the UCI does not benefit from HARQ and has higher reception reliability requirements, the power of the PUCCH should not be affected. The eNB scheduler can account for the reduction in power of the PUSCH, including the CM increase, through the appropriate selection of the data modulation and coding scheme (MCS).

Finally, concurrent multichannel transmissions in one CC should also be considered in case of PUCCH-only transmissions. For example, due to the single-carrier requirement in LTE, whenever the transmission of CQI and SR coincide in the same subframe, CQI is dropped. However, in LTE-A, single carrier is no longer always the case, and when the UE has only CQI and SR transmissions in a subframe, it is reasonable to support concurrent transmissions of CQI and SR to avoid dropping CQI reports.

The TDM of UCI and data should be the default mode in LTE-A, and concurrent PUCCH and PUSCH transmissions from a UE should be configurable.

2.5.3.4 UL Noncontiguous Transmission

LTE-A supports noncontiguous resource assignment with a single DFT per component carrier using clustered DFT-S-OFDM. Discontinuous resource allocation can achieve higher multiuser diversity gain than continuous resource allocation. With noncontiguous resource allocation, PUSCH resource fragmentation can be eliminated or greatly reduced due to multiple carrier transmission. On the other hand, one drawback of supporting noncontiguous resource allocation is the increase in the cubic metric.

In theory, increasing the number of clusters for PUSCH transmission can improve the throughput gains from frequency diversity. In practice, noncontiguous resource allocation is suitable for a highly frequency-selective channel. For less frequency-selective channels, which are more likely in femtocells, noncontiguous resource allocation will have minimal gains. In addition to the throughput performance, UE complexity, testing complexity, and PDCCH overhead need to also be taken into account when considering noncontiguous PUSCH transmissions. Figure 2.38 shows link-level performance for QPSK and 16 quadrature amplitude modulation (16-QAM) under the ideal TU-3 channel estimation. For single-input multiple-output (SIMO) 1Tx-2Rx configuration, the gain is approximately 0.15dB for QPSK and 0.08dB for 16-QAM at the 10% BLER operating point. Due to the increase in the cubic metric, a net performance loss is seen. We also see that current Rel-8 intrasubframe hopping of the PUSCH can already provide sufficient frequency diversity, so the frequency hopping is not supported simultaneously with noncontiguous PUSCH resource allocation in LTE-A.

LTE-A supports dynamic switching between Rel-8 single-cluster transmission and LTE-A multicluster PUSCH transmission. From the simulation result shown in Figure 2.39, we can see that CM increases when the number of clusters increases.

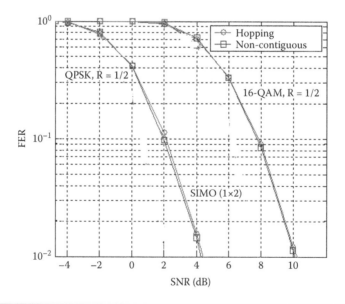

Figure 2.38 Link frequency diversity gain with noncontinuous allocation.

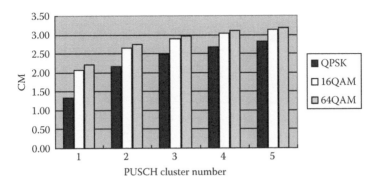

Figure 2.39   CM of clustered PUSCH.

In the LTE-A specification, the maximum number of UL clusters is two for the consideration of relative low CM property. The CM increase translates to lower maximum transmit power, and therefore lower coverage to cell-edge users, especially in noise-limited environments, resource allocation signaling complexity, and UL scheduling efficiency. In the simulation, the size of each cluster has a value from the following set: $N \times 1RB$, $N \times 2RBs$, $N \times 3RBs$, $N \times 4RBs$, $N \times 5RBs$, where N is an integer number.

2.5.3.5 OFDMA versus DFT-S-OFDMA

UL single-user MIMO is being introduced to increase the data rate and improve coverage. UL SU-MIMO was considered in LTE Rel-8, but compared to the added benefit, it was found to be too expensive for the terminals due to the need for multiple power amplifiers. Now, up to four TX antenna transmissions will be specified for LTE-A UL.

Since UL-MIMO is introduced in the UL, the MIMO performance of OFDMA- versus DFT-S-OFDMA-based schemes is currently being debated. It is well known that with a minimum mean squared error (MMSE) receiver and certain antenna configurations, the MIMO performance of OFDMA is better than that of DFT-S-OFDMA. But with 2x4 spatial multiplexing, the performance is almost equal. However, MMSE is not considered a state-of-the-art receiver for modern base stations. Better performance is obtained with turbo equalization. Therefore, MMSE combined with successive interference cancellation (SIC)* using turbo decoder soft outputs has been considered as a practical receiver.

With state-of-the-art receivers, the MIMO performance differences of the studied UL MA schemes are negligible.

---

* Receiver-side techniques are also already quite optimal: MMSE receiver combined with successive interference cancellation (SIC) is the baseline receiver for MIMO in 3GPP studies.

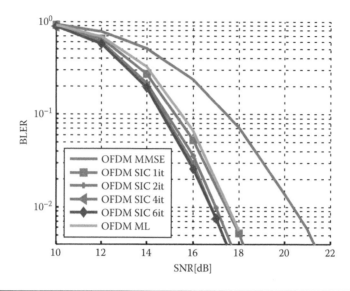

Figure 2.40   Performance comparison among MMSE, ML, and Turbo SIC receivers for OFDMA.

It has been claimed that OFDMA is superior to SC-FDMA since it allows for maximum likelihood (ML) receivers. However, as shown in Figure 2.40, simple and practical turbo SIC receivers provide better performance than ML receivers also with OFDMA. This is due to the fact the turbo SIC receivers can utilize the channel coding, while an ML receiver that exploits channel coding would not be feasible from an implementation point of view. The simulations are done in the spatial channel model (SCM) with 2x2 spatial multiplexing using 16-QAM modulation and a coding rate of 2/3. The performance of the ML receivers operating on the symbol level is worse than what can be achieved with a simple SIC receiver. The performance of MMSE is not adequate even with OFDMA. Therefore, turbo SIC is the preferred receiver for SU-MIMO uplink in LTE-A regardless of the multiple access scheme.

Figure 2.41 shows the performance of OFDMA and DFT-S-OFDMA with turbo SIC and MMSE receivers in SCM channel considering MCSs from QPSK to 16/64 QAM. It can be seen that there are no differences in performance between the two UL multiple access schemes with the 2x4 antenna configuration; with 2x2 antennas, OFDMA performs marginally better with some coding rates of 16/64 QAM. With QPSK, the performance of OFDMA and DFT-S-OFDMA is the same.

2.5.4 LCP and BSR

In LTE Rel-8, quality of service (QoS) is supported in the MAC layer by prioritizing data transmission of logical channels based on their configured priority and prioritized bit rate (PBR). In practice, the MAC layer performs logical channel

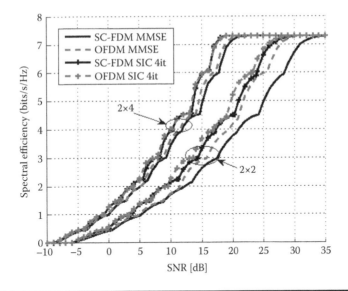

Figure 2.41   Performance of OFDMA and DFT-S-OFDMA with Turbo-SIC and MMSE receivers.

prioritization (LCP) with which logical channels are served in order of their priority to meet their PBR target on average.

In CA, different CCs may provide similar QoS. It is left to the UE implementation how to distribute UL data if the UE was granted UL resources spanning multiple UL CCs. UE can apply LCP to each CC or to an aggregate UL grant.

The buffer status report (BSR) procedure is used to provide the serving eNB with information about the amount of data available for transmission in the UL buffers of the UE.

In CA, the BSR may be transmitted on any UL CC, that is, BSR may be transmitted on any PCell or SCell. Same as in Rel-8, only zero or one regular/periodic BSR plus zero or more padding BSR(s) may exist in a MAC PDU and still max one BSR per TB. If multiple BSRs are included in one TTI, for any logical channel group (LCG) the same value shall be indicated.

Because the higher data rates are supported in CA, one additional BSR table will be specified (range from 0–3000 kbytes), with values that follow the Rel-8/Rel-9 exponential distribution. RRC will tell the UE which table to use. The UE can be assigned extended BSR reporting irrespective of whether it is configured with CA/MIMO or not.

2.5.5 UL Power Control

The uplink power control in LTE is based on both signal-strength measurements done by the terminal itself (for open loop power control) and measurements by the base station (closed loop power control). The latter measurements are used to generate power control commands that are subsequently fed back to the terminals as part of the downlink control signaling.

The uplink power control in LTE-A is similar to Rel-8, that is, it will compensate for distance-dependent path loss and shadowing, while reducing the interference generated toward neighboring cells, and will support component carrier-specific UL power control. However, since in LTE-A multiple PUSCHs may be transmitted in parallel on different CCs and simultaneous PUSCH/PUCCH may be transmitted on the same or different CCs, uplink power control in a CA scenario becomes much more complicated.

Generally, LTE-A supports larger bandwidths through CC-specific UL power control for both contiguous and noncontiguous channel aggregation. CC-specific power control is expected to be needed for noncontiguous channel aggregation due to potentially quite different propagation conditions on different CCs. On the other hand, CC-specific power control might also be needed with contiguous channel aggregation due to different interference conditions and requirements on different CCs.

In LTE-A, power scaling can be very complicated since multiple PUSCHs and PUCCHs are allowed to transmit simultaneously on a single CC while uplink transmissions may be on multiple CCs at the same time. Two types of power headroom reports (PHRs) that consider the PUSCH transmission and simultaneous PUCCH and PUSCH transmission will be provided.

2.5.5.1 Path Loss Derivation

The UL path loss (PL) estimation is a key factor for UL power control and is highly dependent on the configuration of the component carriers. It should be possible for the network to completely compensate for the path loss to ensure reliable UL power control and fair scheduling among CCs.

In LTE-A, since a DL CC may be used by different UE to estimate the path loss of different UL CCs, when two UL CCs are aggregated but located in different bands, the path loss gap can be significant (as illustrated in Figure 2.42).

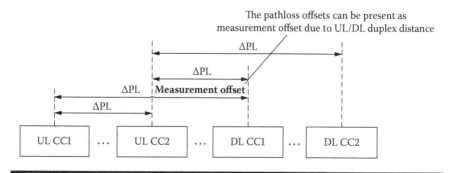

Figure 2.42   Path loss offsets for separated CCs.

For an LTE-A user in carrier aggregation mode, the aggregation offset should be able to be completely compensated for in order to ensure reliable UL PL compensation and fair joint scheduling among CCs. For accurate path loss estimation, a reference downlink CC is defined in LTE-A for each uplink CC. The path loss reference for an uplink PCC is the downlink PCC. The path loss reference for an uplink SCC can be either the SIB2-linked downlink CC or the downlink PCC, according to the RRC configuration.

2.5.5.2 Power Scaling

In CA, maximum UE power per CC is denoted by $P_{CMAX,c}$. Because it is not the usual case that eNB schedules the UE uplink transmission on all configured CCs, and even the UE will transmit on all configured CCs, it is less likely that the UE is using maximum power $P_{CMAX,c}$ of all the CCs to transmit on all CCs. So usually, or at least it is not forbidden, the sum of $P_{CMAX,c}$ for all configured CCs may exceed the maximum UE output power.

Since UL power control is performed per CC, the power control algorithm of each CC will explicitly ensure that the total transmit power for a CC does not exceed $P_{CMAX,c}$ for that CC. However, the separate power-control algorithms on different CCs do not ensure that the total transmit power for all CCs does not exceed P_{TMAX}. If the UE total transmit power exceeds P_{TMAX}, power scaling should be adopted to reduce the total transmit power within the range of P_{TMAX}.

Power scaling is per-CC based, with prioritization of power scaling in cases of power limitation. PUCCH power is first prioritized, and remaining power may be used by the PUSCH. Furthermore, the PUSCH with UCI is prioritized over the PUSCH without UCI. The overall priority order is: PUCCH > PUSCH with UCI > PUSCH without UCI. In other words, in case the UE does not have enough power, power is scaled down first on the PUSCH without UCI, then on the PUSCH with UCI, and last on the PUCCH.

Prioritization occurs regardless of the same or different CCs. UCI cannot be carried on more than one PUSCH in a given subframe. The UE scales the power of all PUSCHs without UCI equally, with the assumption that in the power-limited case, all UL CCs will have roughly the same UL QoS. The PUSCH power scaling can be expressed as follows:

$$\sum_c w_c \cdot \hat{P}_{PUSCH,c} \leq \left(\hat{P}_{TMAX} - \hat{P}_{PUCCH} \right)$$

where w_c is a scaling factor for the PUSCH on carrier c. All power values in the equation are linear values. Equal power scaling is applied to all PUSCH clusters on a given CC. For any PUSCH with UCI, the scaling factor w_c should be set to 1. For the remaining PUSCH, the scaling factors w_c are set to the same value less than or equal to 1.

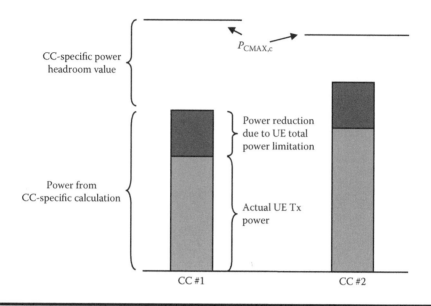

Figure 2.43 CC-specific PHR.

2.5.5.3 *Power Headroom Reporting*

Power headroom in Rel-8 represents the difference between the allowed UE maximum transmit power P_{CMAX} and the current PUSCH transmit power. UE will report PHR to its serving eNB for scheduling decision and link adaptation purposes.

In LTE-A, the PHR is per-CC based. PHR is computed for all activated CCs (as illustrated in Figure 2.43) and may be transmitted on any PCell or SCell. Since simultaneous PUSCH and PUCCH transmission on the same CC is supported, two types of PHR are supported for each CC separately:

■ Type 1: $P_{CMAX,c}$ – PUSCH power
■ Type 2: $P_{CMAX,c}$ – PUCCH power – PUSCH power

The SCell always reports Type 1, and the PCell, with possible simultaneous PUCCH + PUSCH configured, may report Type 1 or Type 2. The PHR report always contains reports for all activated CCs, for which LTE-A introduced an extended PHR format using the extended PHR MAC control element (CE).

The extended PHR MAC CE is identified by a MAC PDU subheader with logical channel ID (LCID) 17. It has a variable size and a structure as shown in Figure 2.44. The contents of the extended PHR MAC CE include a bitmap of reported CCs, field V indicating whether the PH value is based on a real transmission or a reference format, and $P_{CMAX,c}$ used in reporting the PHR value for that CC. Whether the UE uses the extended PHR format is configured by RRC. The support of extended PHR is mandatory for UE supporting uplink CA and for UE supporting simultaneous PUSCH and PUCCH transmission.

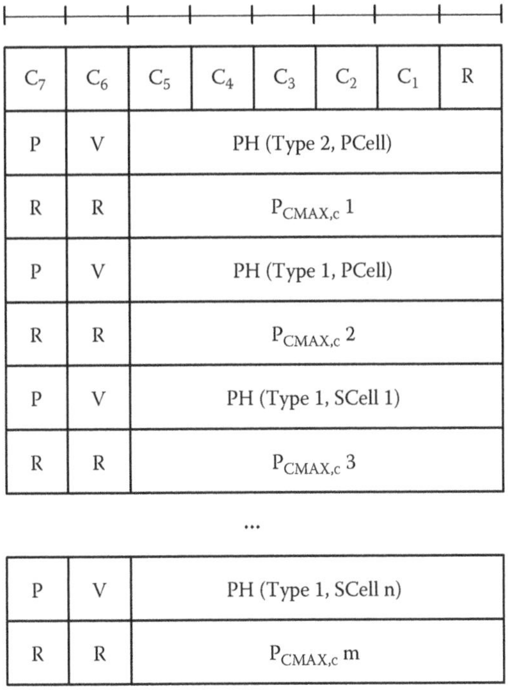

C_7	C_6	C_5	C_4	C_3	C_2	C_1	R
P	V	PH (Type 2, PCell)					
R	R	$P_{CMAX,c}$ 1					
P	V	PH (Type 1, PCell)					
R	R	$P_{CMAX,c}$ 2					
P	V	PH (Type 1, SCell 1)					
R	R	$P_{CMAX,c}$ 3					

...

P	V	PH (Type 1, SCell n)					
R	R	$P_{CMAX,c}$ m					

Figure 2.44   Extended power headroom MAC CE.

2.5.5.4 TPC Command Transmission

TPC in the UL grant is applied to the UL CC for which the grant applies; TPC in the DL grant is applied to the UL CC on which the ACK/NACK is transmitted. DCI format 3/3A is supported only for the same CC on which the TPC commands are transmitted for multiple UE power control. There is no consensus to adopt cross-carrier group UL power control in LTE-A.

The PUCCH is power controlled by the PCell. Please recall that the TPC command in downlink assignment relating to a SCC is reinterpreted as an acknowledgment resource indicator (ARI).

2.6 CA Scheduling

2.6.1 Scheduling Rules

In LTE-A, there may exist multiple DL CCs and multiple UL CCs in a cell, and the eNB should be able to schedule multiple DL/UL CCs to a UE in a subframe. As we know, there is at most one transport block (or two in case of spatial multiplexing) and one independent HARQ entity per UL/DL CC.

Each transport block and its potential HARQ retransmissions are mapped to a single CC (same CC in the case of retransmissions). The UE may be scheduled over multiple CCs simultaneously so additional complexity is introduced. LTE-A supports a higher number of parallel transmission chains to one UE (up to 10 with 5 CCs and dual-stream MIMO on all CCs); assuming 8-ms HARQ round-trip time (RTT) in FDD as for Rel-8, CC-specific HARQ means up to 80 HARQ channels per UE. Based on similar considerations, the eNB/UE might need to transmit up to 10 ACK/NACK per TTI; in TD-LTE this number can increase to 90 ACK/NACK when considering a frame format with 9 DLs and 1 UL. ACK/NACK multiplexing clearly becomes an issue, especially in a coverage-limited TDD uplink, and ACK/NACK bundling might have a significant impact on RRM algorithms.

Moreover, cross-carrier scheduling is supported in LTE-A through the CIF in the PDCCH grant/assignment. Cross-carrier scheduling can be useful in deployment scenarios when the PDCCH cannot be received reliably on all CCs (e.g., the heterogeneous network [HetNet] case) or when a narrow-bandwidth carrier is aggregated with a wide bandwidth carrier. Cross-carrier scheduling allows the PDCCH of one serving cell to schedule resources on another serving cell, with the following restrictions:

- The PCell is always scheduled from the PCell's PDCCH.
- The SCell configured with the PDCCH is scheduled from its own PDCCH.
- The SCell without a configured PDCCH is scheduled from the PCell or other SCells; DL and UL are scheduled from the same PDCCH.

For the DL/UL PDCCH assignment/grant without the CIF:

- The DL assignment received in the PCell without the CIF corresponds to the DL transmission in the PCell.
- The DL assignment received in SCell *n* without the CIF corresponds to the DL transmission in SCell *n*.
- The UL grant received in the PCell without the CIF corresponds to the UL transmission in the PCell.
- This corresponds to the UL transmission on the UL CC of the SCell where the UL grant is received. If the UL CC is not configured in that SCell, the grant is ignored by the UE.

Semipersistent DL resources can be configured only for the PCell and only PDCCH allocations for the PCell can override the semipersistent allocation (appropriate since semipersistent scheduling [SPS] is used mainly for voice over IP [VoIP], which does not need aggregation of resources through CA). Combination of CA and TTI bundling cannot be configured for a UE.

■ A5: The PCell becomes worse than threshold 1 and the neighbor becomes better than threshold 2. The measurement object for A3/A5 can be any frequency including the SCell (SCell in this case is included in the comparison).

■ A3 and A5 implicitly are linked to the PCells as the serving cell. The measurement object linked to an A3 or A5 event can be any frequency, and if an SCC is the target object, the corresponding SCell is included in the comparison.

■ A new measurement event A6 (illustrated in Figure 2.46) is introduced for CA: an intrafrequency neighbor becomes offset better than the SCell for which the neighbor cells on an SCC are compared to the SCell of that SCC.

With an A6 event, the deconfiguration of the actual SCell will likely happen together with the configuration of a new SCell. Event A6 occurs when an intrafrequency neighbor is offset better than an SCell. Event A6 compares the neighbor cell and the SCell that are on the same carrier frequency. A stronger neighbor on an SCC may point to the need for handover (HO), as shown in Figure 2.47.

The eNB needs to configure the UE so that the UE can make measurements on the CCs. Interfrequency neighbor measurements are needed for all the carrier

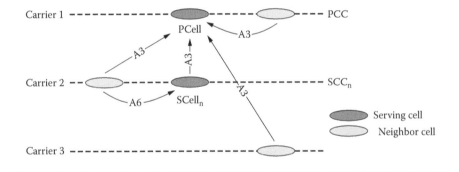

Figure 2.46   Measurement events A3 and A6.

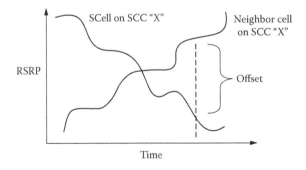

Figure 2.47   A better candidate serving cell is reported by the UE via event A6.

frequencies that do not have a serving cell (PCell or SCell) for a UE. The measurements on activated SCells can be done without measurement gaps. RRM measurements on deactivated cells will need gaps. The UE will signal as part of its capability information if it needs gaps for some band combinations when configured with CA.

2.7.2 Intra-LTE Handover

During intra-LTE handover, the source eNB selects a target PCell and indicates the selection in the relevant X2/S1 handover request message (same as Rel-8). The target eNB decides which SCells (if any) to configure at handover. The source eNB may include a list of the best cells on each frequency for which measurements are available, in order of decreasing RSRP. Available measurement information for cells in the list may also be included by source eNB. That measurement information is used by the target eNB to determine which (if any) SCells to configure for use after HO. The source eNB does not need to be aware of the CA capability of the target eNB (i.e., the list could include cells not under target eNB or cells that the target eNB cannot aggregate).

Upon handover, SCells are deactivated. SCells are not automatically released by the UE upon handover.

2.8 RF Consideration for CA

An LTE-A terminal with reception and/or transmission capabilities for carrier aggregation can simultaneously receive and/or transmit on multiple component carriers. Operation on wider and possibly aggregated bandwidth targeted for LTE-A implies higher complexity for both UE and the eNB. An LTE Rel-8 terminal can receive and transmit on a single component carrier only, provided that the structure of the component carrier follows the Rel-8 specifications. It will be possible to configure all component carriers as LTE Rel-8 compatible, at least when the aggregated numbers of component carriers in the UL and the DL are the same.

2.8.1 eNB RF Consideration

Current linear power amplifier (LPA) technologies are capable of supporting 20–30 MHz modulation bandwidth at efficiencies required for the Rel-8 eNB. The design of the eNB LPA is critical for higher bandwidth (expected to exceed 100 MHz for allocations above 2 GHz) support for LTE-A. Design issues impacting the LPA's modulation and tuning bandwidth are LPA device impedances and matching techniques, LPA linearization techniques, and LPA efficiency techniques.

Presently, limitations in linearization and efficiency technology will limit practical LPA modulation bandwidth to the range of 30–40 MHz, so it is feasible to support 40 MHz with only a single analog-to-digital converter (ADC). Therefore,

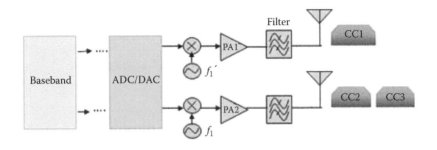

Figure 2.48 DL signal processing in CA. (DAC—digital to analog converter)

eNB transmission bandwidth greater than 40 MHz (as well as noncontiguous CA) requires multiple RX chains and ADC (Figure 2.48).

LPA combining techniques will need to be considered. Techniques for combining LPA resources include hybrid combining, cavity combining, coherent combining, and combining using the Fourier transform matrix (FTM), and so on.

Cavity combining is implemented by adding cavity filters at the output of the LPAs to be combined. Cavity combining has wideband capabilities, adding approximately 0.5 dB–1.5 dB of output loss. Its effectiveness is lost for contiguous bands due to loss of isolation between ports as the cavity filters approach each other's passband.

Coherent combining is implemented by using input splitters and output combiners. For coherent combining to be effective, the signals must remain coherent (matched gain, phase, delay) through each of the LPAs combined. Coherent combining is effective for contiguous bands.

FTM combining is implemented by using an input FTM and an output FTM. As with coherent combining, the signals must remain coherent through each of the LPAs combined.

The choice of combiner techniques depends on trade-offs among design criteria such as cost, complexity, LPA bandwidth, total eNB transmission bandwidth, and whether the bands to be combined are contiguous or noncontiguous. Obviously, this increases power consumption (pretty much linearly with the number of RX chains), so limiting reception on multiple CCs as much as possible should result in power savings.

Continuous bandwidths larger than 20 MHz will increase the frequency domain size of the channel estimate and increase the size of the frequency domain equalizer (FDE). That will lead to a linear increase in computational complexity, a maximum fivefold increase in computational complexity compared to one 20-MHz LTE allocation.

Furthermore, since DL power is shared among all resource blocks (RBs), more bandwidth means less power per resource block (as illustrated in Figure 2.49). To maintain coverage, higher bandwidth requires higher total power, and multi-layer transmission schemes also require higher total power.

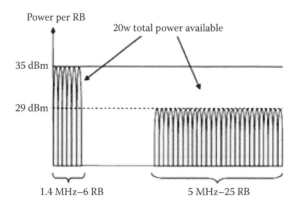

Figure 2.49 DL power distribution.

LTE-A CA will be designed considering these limitations, and obviously there will be different UE categories for different bandwidth capabilities. This is necessary for efficient use of the available eNB hardware; for instance, if eight transmit antennas are used, the corresponding eight power amplifiers need to be exploited efficiently.

2.8.2 UE RF Consideration

In LTE-A and future LTE versions, the following new features will be introduced for UL transmission:

- **UL-MIMO:** Signals are transmitted from multiple antennas, as compared to DL-MIMO, requiring the implementation of multiple transceivers, multiple power amplifiers (PAs), and multiantenna design, all in a small UE form factor.
- **UL noncontiguous Tx:** Signals are transmitted from multiple PRB clusters.
- **UL collaborative multipoint (CoMP):** Signals are cooperatively received by multiple cell receivers.
- **UL CA:** Signals are transmitted from multiple carriers.

The new features have some major impacts on UE RF requirements, including operating bands, transmitter characters (adjacent channel leakage power ratio [ACLR], unwanted emission mask (UEM), spurious emissions, etc.), and receiver characters (reference sensitivity, adjacent channel selectivity [ACS], blocking, intermodulation, etc.).

The requirements apply for UE equipped with two antenna connectors that support transmission on up to two antenna ports over one component carrier (as shown in Figure 2.50). UEs that support more than one feature (CA, UL-MIMO, enhanced DL-MIMO, and CoMP) will need to meet all of the separate corresponding requirements.

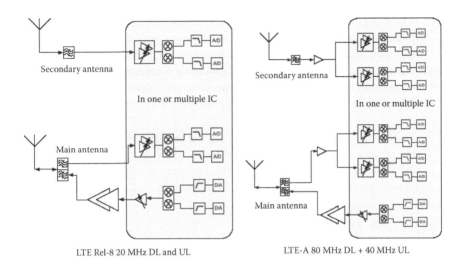

Figure 2.50   UE transceiver in CA.

Simultaneous transmission of a clustered PUSCH will generate additional intermodulation products (IMP), which cause increases in out-of-band (OOB) emissions and spurious emissions. Thus, maximum power reduction (MPR) is required to reduce these emissions. The UE may adopt a single wideband-capable (i.e., >20 MHz) RF front end and a single FFT.

The current LTE Rel-8 specification supports the following UE antenna configuration for the uplink:

- One antenna with one power amplifier (PA) (default configuration)
- Two transmit antennas but one PA (transmit antenna selection).

In LTE-A, the implementation of UL-MIMO will affect the architecture and physical aspects of both the eNB and UE. Support for UL-MIMO will have an effect on the design and implementation of multiple RF chains (PAs) and their associated antenna placements, and will create an increased demand on battery resources. Implementation of multiple power amplifiers is required in SU-MIMO-capable UEs, and possible optional advanced UE implementation includes the following:

1. *More than two antennas but one PA:* This antenna configuration includes UE autonomous antenna switching that can get link gain, especially if the eNB loses track of which antenna is used for UL transmission.
2. *Two antennas and two PAs:* This antenna configuration allows the exploitation of transmit spatial diversity, rank-2 spatial multiplexing, and rank-1/2 MIMO beamforming.

3. *Four antennas with four PAs:* This UE antenna configuration will enable up to four-stream transmission to achieve very high peak throughput. It may be more practical for customer premise equipment (CPE) and relay nodes.

4. *Antenna grouping:* This case involves a subset of K antennas ($K \geq 2$ PAs) among M antennas with $K < M$. This implementation is also envisioned as more feasible for CPE or relay nodes. This scheme is similar to the antenna-switching issue.

5. *Antenna virtualization:* In this case, the UE implements an antenna pattern instead of an omni or directional pattern, which has been widely used in smart antennas. The UE can use a fixed pattern or adjust the beam pattern according to the received DL signal as is possible in TDD operation.

2.8.3 Cubic Metric Property

The cubic metric (CM) and peak-to-average power ratio (PAPR) are important issues for both the downlink and uplink. The cubic metric increase translates to lower maximum transmit power and, therefore, lower coverage to cell-edge users that transmit at maximum or near-maximum power, especially in noise-limited environments. In 3GPP, the CM better reflects the required power amplifier back-off than does the PAPR.

Low CM was one of the important properties for the uplink multiple access design in LTE to increase coverage. SC-FDMA has been selected as the LTE Rel-8 uplink transmission scheme due to its outstanding PAPR performance, which finally maximizes radio coverage and minimizes power consumption when compared to other schemes with high PAPR statistics. The difference in the CM values between clustered SC-FDMA and OFDMA is around 2–3 dB. If noncontinuous resource allocation in a transmission power–limited environment is really required, low CM will be one of the most important criteria for the multicarrier transmission scheme.

With carrier aggregation, the single-carrier property in the UL is no longer preserved when transmitting on multiple carriers. In LTE-A, multiple simultaneous PUCCH transmission, simultaneous PUSCH and PUCCH transmission, and PUSCH transmission with noncontiguous resource allocation are supported. As a result, the CM (as the cubic power of the signal of interest compared to a reference signal) increases, which requires a larger back-off in the power amplifier, thereby reducing the maximum transmit power at the UE. An illustrative comparison is shown in Table 2.5 for transmit power loss compared to Rel-8, which assumes that the LTE-A UE uses the same PA for LTE-A as for LTE Rel-8. This substantial increase in the CM reduces the coverage when carrier aggregation is used, although it can be seen that the CM is in general still less than that for OFDMA (note that results for OFDMA are independent of modulation). In addition, the increase is most significant when the number of simultaneous component carriers goes from 1 to 2, and only increases gradually as more carriers are used. It should be noted

Table 2.5 Transmit Power Loss Compared to Rel-8

	Clustered DFT-S-OFDMA	*Nx SC-FDMA*	*OFDM*
QPSK	0–1.6 dB	0–2.0 dB	2.4 dB
16QAM	0–1 dB	0–1.3 dB	1.6 dB
64QAM	0–0.9 dB	0–1.2 dB	1.4 dB

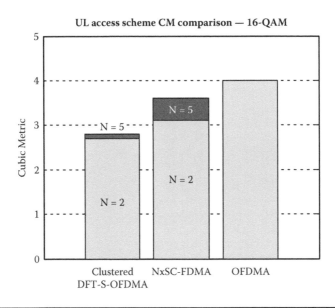

Figure 2.51   CM comparison of uplink access schemes for 16-QAM.

that since multicarrier transmission will usually be used for UEs in good channel conditions, there should be no loss of coverage for those users. On the other hand, users at the cell-edge will most likely be scheduled only on a single carrier. There is no difference in the CM in this case and coverage may actually be improved since the eNB has the ability to dynamically assign the users to the best UL carrier.

From a power-saving perspective, clustered DFT-S-OFDMA provides the smallest CM of the three methods, even when the UE maximum power must be backed off by the CM increase. The benefit is a function of the maximum number of clusters and the frequency selectivity of the channel. An illustrative comparison of the CM for these schemes is shown in Figure 2.51 for 16-QAM modulation (the results for QPSK and 64-QAM exhibit a similar trend).

Chapter 3

Collaborative Multipoint

3.1 General Description

Before Long Term Evolution Advanced (LTE-A), a single point for mobile communication was mainly used; each cell transmitted to its own user equipment (UE), and in the process created interference for the UE in adjacent cells. A single point often consists of multiple baseband, radio frequency (RF), and antenna systems, with each system being dedicated to serve one sector. Thus, intrasite functions well, but intersector coordination is typically limited. LTE-A should support higher downlink peak throughput and total sector throughput than LTE Rel-8. This requirement calls for an investigation of potential enhancements to the current single-point operation to include possible coordination among multiple points. *Multipoint* refers to transmission or reception from antennas that are not in close proximity (typically beyond the spacing of a few wavelengths; all antennas may be subject to different long-term fading). There are a few implementations that may all be referred to as multipoint operation, such as intercell coordination to serve UE at the cell boundary, multiple RF heads with centralized signal processing, evolved Node-B (eNB) and relay node coordination, eNB and home-eNB coordination, etc. The difference from traditional single-point operation is the larger-than-normal separation between antennas (or groups of antennas or distributed antennas). In multipoint operation, there are typically several participating points collectively serving a set of UE, each of which is attached to a single point and would otherwise suffer from cell-edge interference if these points were not coordinated; typically cell-edge UE may be better suited for multipoint. Therefore, UE can switch from single-point to multipoint operation or vice versa. The eNB can serve some of the UE attached to it in a traditional single-point fashion while serving other

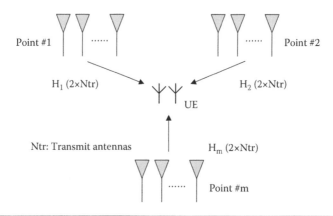

Figure 3.1 Downlink multipoint SU-MIMO, M points with Ntr antennas each serving a single UE with 2 receive antennas.

UE via multipoint coordination with other points. Figure 3.1 shows an example of multipoint collaborative multipoint (CoMP) scheme.

The basic idea of CoMP transmission/reception is to have multiple network nodes (remote radio unit [RRU], eNB) cooperate with distributed or centralized structures to act as the LTE-A target. CoMP involves multiple cells cooperating to transmit to a single UE to improve the coverage for high data rates and cell-edge and system throughput. Multiple coordinated points could collaboratively construct a larger cooperative multiple-input, multiple-output (MIMO) transmitter in a supercell, where the downlink transmissions are jointly configured to avoid intercell interference. CoMP transmission and reception has been considered for LTE-A as a method to improve coverage for high data rates, cell-edge throughput, and system throughput. The distributed or centralized structures can be formed using multiple eNBs: one eNB contains one or multiple cells, one cell contains one or multiple remote radio units, or one RRU contains one or multiple antennas.

3.1.1 Understanding CoMP

As defined in the Third-Generation Partnership Project (3GPP), downlink (DL) CoMP transmission implies dynamic coordination among multiple geographically separated transmission points. With respect to coordination position, coordination takes place everywhere in the cell or only at the cell edge. With respect to coordination nodes, CoMP can be divided into two types: intra-eNB CoMP and inter-eNB CoMP. With respect to coordination level, several schemes for CoMP have been proposed, including coordinated scheduling (CS)/coordinated beamforming (CBF), dynamic cell selection (DCS), and joint processing (JP).

For coordination everywhere in the cell, the coordinated multicell transmission happens for all UE in the cell regardless of the UE's position. This approach may provide a better performance at the expense of UE feedback and backhaul overhead compared with coordination only at the cell edge. For coordination only at the cell edge, single-cell transmission is used for UE in the cell center, while the multicell transmission only applies to UE at the cell edge. This approach may provide a good compromise between performance and reduction in the UE feedback and backhaul overhead.

With respect to coordination nodes, depending on whether all the cells in a CoMP cooperating set are controlled by the same eNB (site), there are two kinds of CoMP: intra-eNB CoMP and inter-eNB CoMP.

In an inter-eNB CoMP cooperating set, cooperation occurs only within sectors of different sites. This addresses the interference problem at the cell edge. Cooperation requires a high-speed, low-latency, site-to-site backbone connection. Static and dynamic clustering of CoMP cooperating sets is possible.

In an intra-eNB CoMP cooperating set (including intercell CoMP and intracell CoMP), cooperation occurs only within sectors of the same base station (BS) in one site. This scenario includes the CoMP among cells from the same eNB or among the distributed remote radio heads (RRH) from the same cell or same eNB. Cooperation does not require a high-speed, low-latency, site-to-site backbone connection, and the CoMP cooperating set size is the same as the number of sectors. For cells controlled by intra-eNB, the data can be easily available at each transmission point. Transfer latency for both data and control signaling between the transmission points can be very low. Coordinated scheduling for intercell CoMP and intracell CoMP is shown in Figure 3.2.

With respect to the type of coordinated users, we can also have two kinds of CoMP: single-user MIMO (SU-MIMO) and multiuser MIMO (MU-MIMO). CoMP can serve a single UE in the CoMP-SU-MIMO mode, or it can serve multiple instances of UE simultaneously in the CoMP-MU-MIMO mode.

With respect to coordination level, we have coordination as CS/CBF, DCS, and JP, which is shown in Figure 3.3. *Joint processing* means that data is available at each point in the CoMP cooperating set. Physical DL shared channel (PDSCH) data is available at each cell in the CoMP set and transmission occurs from one or multiple transmission points. Joint processing is also called *cooperative MIMO* (Co-MIMO) and provides coordinated transmission from multiple cells for active interference cancellation. CS/CBF means that data is only available at the serving cell, but user scheduling and beamforming decisions are made by using interference coordination among surrounding cells in the CoMP cooperating set.

Joint transmission (JP/JT) involves user-plane transmission from multiple points at one time, and PDSCH involves transmission from multiple points. Data sent to a single UE is simultaneously transmitted from multiple transmission points, which improves the received signal quality.

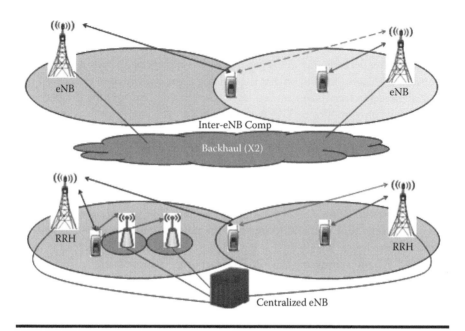

Figure 3.2   Coordinated scheduling for intercell CoMP and intracell CoMP.

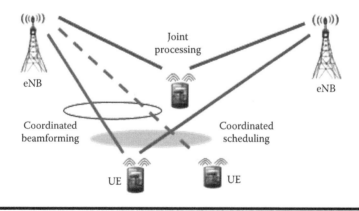

Figure 3.3   Several schemes for CoMP.

Dynamic cell selection (JP/DSC) involves user-plane transmission from one point at one time within a CoMP cooperating set. Dynamic cell selection causes the UE to be served by the most favorable transmission cell. The serving cell can be selected based on channel variation, resource availability, and so on. This is different from the handover in Rel-8, which is shown in Figure 3.4.

With coordinated scheduling/beamforming (CS/CBF) data is only available at the serving cell (data transmission is from that point), but user scheduling and

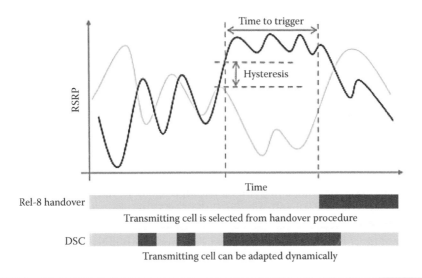

Figure 3.4 The difference between DSC and handover.

beamforming decisions are made with coordination among cells corresponding to the CoMP cooperating set. So CS/CBF can mitigate the spread of energy to victim UE while maximizing the gain for the desired UE. However, when JP/JT is compared to CS/CBF, CS/CBF requires less backhaul capacity.

3.1.2 The CoMP Set

3GPP standards define some new concepts, such as serving cell, CoMP cooperating set, and CoMP reporting set, etc. The serving cell is used for physical DL control channel (PDCCH) transmission, which is the same meaning as in LTE Rel-8. A CoMP cooperating set, which can be network predefined and user-centric by reference signal received power (RSRP) strength, directly or indirectly participates in PDSCH transmission to UE CoMP. A CoMP reporting set is a set of cells about which channel state information is reported by the UE.

3.1.2.1 Serving Cell

The serving cell is the cell for which RSRP is commonly the maximum for transmitting PDCCH assignments. In this cell, PDCCH is transmitted from the eNB and detected by the UE. The UE detects PDSCH from one cell. Although the actual traffic may be transmitted from multiple cells, the PDSCH transmitted in the other cells will follow the serving cell mechanism in demodulation reference signal (DMRS) configuration, resource allocation, and mapping (i.e., it manages the traffic as if it is from a serving cell). This definition applies to both the downlink and uplink.

3.1.2.2 CoMP Sets

The following describes the different types of CoMP sets.

■ **CoMP cooperating set:** This is a set of geographically separated points directly or indirectly participating in PDSCH transmission to UE. Note that this set may be, but need not be, transparent to the UE. CoMP transmission point(s) are subsets of the CoMP cooperating set. For JT, the CoMP transmission points are the points in the CoMP cooperating set. For DCS, there is a single transmission point at every subframe. This transmission point can change dynamically within the CoMP cooperating set. For CS/CBF, the CoMP transmission point corresponds to the serving cell. Basically, the network need not explicitly signal the CoMP transmission point(s) to the UE, and the UE reception of CoMP transmissions (CS/CBF, or JP with multicast-broadcast single-frequency network [MBSFN] subframes) is the same as that for non-CoMP (SU- or MU-MIMO). The CoMP transmission points are transparent to the UE in MBSFN subframes. But in normal frames, the CoMP cooperating set is an implementation issue and not visible to the UE. The main reason for the nontransparent CoMP transmission points in normal subframes for JP is because the cell-specific reference signal (CRS)/channel state information reference signal (CSI-RS) resource element in one transmission point may collide with the PDSCH resource element in another transmission point. This results in inaccurate channel estimation on those resource elements, and the control region in different points may be different from the serving cell. These issues could be solved by cell planning and control region coordination, but the impact on legacy UE and control capacity needs to be verified further. The signal-to-interference plus noise ratio (SINR) of the CoMP UE can be calculated based on the signal levels of all cells in the CoMP cooperating set (S_1, S_2, S_k), interference from all other cells ($I_k + 1 \ldots I_n$), and noise (N). If we assume that the strongest signal is S_1, and the second strongest signal is S_2, then the SINR is calculated as follows:

$$SINR_{k\,cell} = \frac{S_1 + S_2 + \cdots + S_k}{I_{k+1} + I_{k+2} \cdots + I_n + N}$$

■ **UE-specific CoMP cooperating points set:** The CoMP gain may be limited when the CoMP cooperating set is not large enough to include the potential received strong signal for JP. Within the selected CoMP cooperating set, CoMP transmission points are selected for joint transmission. The same resource block (RB) in each CoMP transmission point is allocated to CoMP UE for coherent or noncoherent combining. In general, the determination of the CoMP reporting set is UE-specific and is based on the wideband channel quality indicator

(CQI) or RSRP measurement from each cell. For example, the UE-specific CoMP cooperating set selection is dependent on the wideband CQI for the uth UE that is ranked in a descending order as $\overline{CQI}_{u,0} \geq \overline{CQI}_{u,1} \geq \cdots$. The serving cell is chosen as the cell with the highest wideband CQI, i.e., $\overline{CQI}_{u,0}$. The cell is selected in the uth UE's CoMP cooperating set when $\overline{CQI}_{u,i} \leq \overline{CQI}_{u,0} - Thr$,

where *Thr* is the predefined relative threshold. It is clear that there is a trade-off between the offset threshold and the number of cells within the CoMP reporting set. A higher threshold may lead to a larger number of cells being included in the CoMP reporting set, as well as a larger number of CoMP UEs, while a smaller threshold will result in a smaller CoMP reporting set and a relatively smaller number of CoMP UEs. A special case is to set the threshold = 0, where a UE is only connected to the strongest serving cell. In addition, the number of cells in the UE-specific CoMP cooperating set is no larger than the predefined maximum number, which is called the maximum size of the CoMP cooperating set.

■ **CoMP reporting set:** This is a set of cells about which channel state and statistical information related to their links to the UE are reported. The CoMP reporting set may be the same as the CoMP cooperating set. The actual UE reports may down-select cells for which actual feedback information is transmitted.

■ **CoMP measurement set:** This is a set of cells about which channel state and statistical information related to their links to the UE are measured. The CoMP measurement set may be the same as the CoMP cooperating set. The actual UE measurements may down-select cells for which actual feedback information is transmitted (reported cells).

■ **RRM measurement set:** This set supports radio resource management (RRM) measurements (already in LTE Rel-8) and therefore is not CoMP specific.

Figure 3.5 shows the relationships among different CoMP sets; that is, the serving cell ∈ CoMP transmission point(s) ∈ CoMP cooperating set ∈ CoMP reporting set ∈ RRM measurement set, (the ∈ symbol means "is a subset of or equal to").

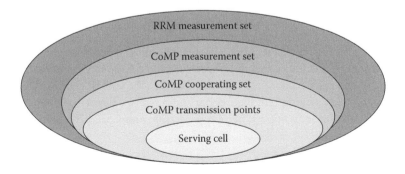

Figure 3.5   CoMP sets.

Table 3.1   CoMP Cooperating Set for JT and DSC

JT	DSC
1. PDSCH transmission from multiple points (part of or entire CoMP cooperating set) at a time	1. PDSCH transmission from one point at a time (within the CoMP cooperating set)
2. CoMP transmission points are the points in the CoMP cooperating set	2. A single point is the transmission point at every subframe; this transmission point can change dynamically within the CoMP cooperating set
3. Time domain–frequency domain QoS-aware packet scheduling (semistatic configuration of UE between reporting modes; dynamic configuration of PDSCH transmission points)	3. Time domain–frequency domain QoS-aware packet scheduling (semistatic configuration of UE between reporting modes; dynamic configuration of PDSCH transmission points)
4. RRH implementation provides best performance by good phase synchronization between transmission points	4. Multisite implementation with low-latency X2 interface
5. Multi-eNB implementation needs a stable external frequency reference	5. Could be combined with CBF

In CoMP transmission, the following procedure would be the most straightforward approach. Based on the results of the measurement set, the network decides to configure CoMP mode for UE. Moreover, some cells in the measurement set are determined to be in the reporting set (i.e., the CoMP reporting set is determined). Next, based on the CSI from the CoMP reporting set, the network determines which cell should transmit the PDSCH to the UE and which cell should avoid transmitting (i.e., the CoMP cooperating set is determined). For example, a CoMP cooperating set for JT and DSC is described in Table 3.1.

For CS/CBF mode, data is only available at the serving cell (data transmission is from that point), and user scheduling/beamforming decisions are made with coordination among cells in the CoMP cooperating set. The CoMP transmission point corresponds to the serving cell. Figures 3.6 and 3.7 give an explanation of how the CoMP cooperating set is applied in joint processing and CS/CBF mode.

3.1.3 CoMP Advantages

The main objective of coordinated transmission from multiple cells toward a single terminal is to turn intercell interference into a constructive signal. In a live

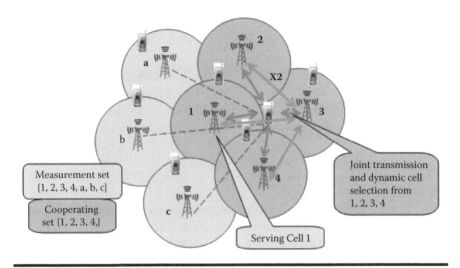

Figure 3.6   CoMP joint processing.

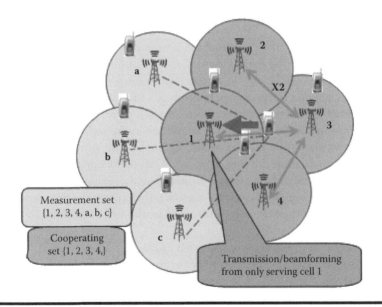

Figure 3.7   CoMP coordinated scheduling/beamforming.

network, if there are N UE instances, then there are $N-1$ interfering signals to cancel per UE. With CoMP, multiple cells are consolidated in one centralized unit. CoMP functions optimize performance across cells. Changing the interfering signals into constructive signals improves performance in the presence of intercell interference. For example, joint detection not only removes harmful interference but changes it into a useful signal, as illustrated in the following equation.

$$C = \log_2\left(1 + \frac{S}{N+I}\right) \Rightarrow C = \log_2\left(1 + \frac{S+I}{N}\right)$$

LTE Rel-8   CoMP

Therefore, we can conclude that intra-eNB joint transmission is an effective technique to improve the throughput at the sector border as well as system capacity as follows:

$$C_{\text{CoMP}} = BW^* \log_2\left(1 + (aS_1 + bS_2 + cS_3) / (I + \alpha I_1 - \beta I_2 - \gamma I_3 - \delta I_4 + N)\right)$$

where factors a, b, and c are the losses in combined power and less than 1; α is more intracell interference due to transmission to UE in other cells; and $\beta/\gamma/\delta$ is imperfect intercell interference cancellation.

3.2 CoMP Technology

CoMP is considered by 3GPP as a tool to improve coverage, cell-edge throughput, and system efficiency. The main idea of CoMP is as follows. When UE is in the cell-edge region, it may be able to receive signals from multiple cell sites and the UE's transmission may be received at multiple cell sites. If we coordinate the signaling transmitted from the multiple cell sites, the DL performance can be increased significantly. This coordination can be simple, as in the techniques that focus on interference avoidance, or more complex, as in the case where the same data is transmitted from multiple cell sites. For the UL, since the signal can be received by multiple cell sites, if scheduling is done from different cell sites, the system can take advantage of this multiple reception to significantly improve the link performance. In what follows, the CoMP architecture will be discussed, followed by different schemes proposed for CoMP.

CoMP communications can occur with intrasite or intersite CoMP; the characteristics of each CoMP architecture are summarized in Table 3.2. One advantage of intrasite CoMP is that significant information exchange is possible because this is communication within a site and does not involve the backhaul (connection between base stations). Intersite CoMP involves the coordination of multiple sites for CoMP transmission. Consequently, the exchange of information will involve a backhaul. This type of CoMP may put additional burden and requirements on the backhaul design. In each type of CoMP, the control of intercell interference via coordination of transmissions from multiple points can be realized by avoiding transmissions that mainly create interference, exploiting "good" channels toward the UE, and using/collecting energy from or at several points.

Table 3.2 Summary of the Characteristics of Each Type of CoMP Architecture

	Intra-eNB Intrasite	*Intra-eNB Intersite*	*Inter-eNB Intersite*
Information shared between eNB/sites	Vendor specific	CSI/CQI, scheduling info	Traffic, CSI/CQI, scheduling info
CoMP methods	CS, CBF, JP	CS, CBF, JP	CS, CBF, JP
Backhaul properties	Short distances between baseband and RRU; provides very small latencies and ample bandwith	Fiber-connected RRU provides small latencies and ample bandwith	Requires small latencies; bandwith dominated by traffic

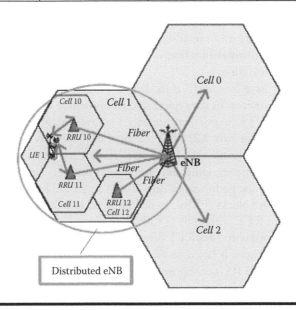

Figure 3.8 An illustration of intra-eNB CoMP with a distributed eNB.

An interesting CoMP architecture is the one associated with the distributed eNB depicted in Figure 3.8. In this particular illustration, the RRUs of an eNB are located at different locations in space. With this architecture, although the CoMP coordination is within a single eNB, CoMP transmission can behave like intersite CoMP instead.

3.2.1 CoMP Technical Requirements and Realization

The goal of CoMP technology is to effectively mitigate intercell interference (ICI) in LTE-A system. It has the potential benefit to improve the coverage of high data rates and to increase both the average cell throughput and cell-edge user throughput. With knowledge of coordinated cell channel state information (CSI) at the eNB, the transmitter can use various joint processing methods to cancel ICI completely. However, advanced algorithms, including transmitter joint preprocessing algorithms, receiver joint detection algorithms, and joint scheduling algorithms must be deployed to achieve CoMP benefits within the requirements of LTE-A system design and implementation.

CSI reporting and sharing is required for every cell in the cooperating set (per CoMP user). Scheduling complexity is increased since joint scheduler complexity grows with the number of cells and number of users in the CoMP area. The enhanced data transfer scheme over X2 has to be implemented in inter-eNB CoMP to share feedback info from all the serving cells to the joint scheduler, send transmit parameters from the joint scheduler to all the transmission points, and convey user data from the serving cell to all the transmission points.

The configuration and coordination scheme should be established within the CoMP measurement and transmission sets. UE measurements should be configured with low signaling and measurement overhead. Coordination among eNBs over X2 is the fundamental function of CoMP, while intra-eNB coordination is left to eNB implementation.

Another major issue for CoMP is the hybrid automatic repeat request (HARQ) process latency constraints. In the typical frequency-division duplexing (FDD) 8-ms round-trip time (RTT) scenario, there are only 3 ms available for eNB signal processing and communication over X2. JT/JP in the CoMP process needs two transactions over X2. The first transaction is to share UE CSI reports for joint scheduling, and the second is to share encoding and precoding parameters for joint transmission. Higher latency will cause performance degradation for all UE instances scheduled on same physical resource blocks (PRBs). In downlink an asynchronous HARQ process is designed for 7-ms RTT (soft limit) in CoMP, and in uplink a synchronous HARQ process is also designed with 7-ms RTT (hard limit). Moreover, configuring precoding and beamforming weights is based on old 7- to 17-ms UE channel estimation measurements in which CSI-RS is configured with a 10-ms periodicity, and the total time of UE processing, feedback transmitting, and eNB processing delay is 7 ms (in Rel-8).

Since in CoMP operation UE does not know about other UE, eNB has to estimate the effect on simultaneous transmission to multiple UE instances on SINR at the UE receiver after detection. UE CQI reporting is based on CRSs and CSI-RSs that are not precoded (no multiuser interference), so CQI compensation should be considered.

From coordinated to cooperative eNB or relay, the key components of the collaborative radiating element can be the sector, eNB, or even UE, performing coordination including intersite RRM with joint scheduling; intersite interference management with joint load and interference planning; avoidance, suppression, and orthogonalized intercell interference; and intersite precoding with joint codebook construction and precoding matrix indicator (PMI) selection. The coordination can be classified into several types: interference coordination only, distributed RRM, multicell MIMO, and full cooperation between the referent components.

1. **Interference coordination:** Intercell interference coordination and avoidance with eNB control on the X2 interface is similar to that in LTE Rel-8.
2. **Distributed RRM coordination:** Intercell interference is coordinated by interference-aware resource allocation, which is like flexible frequency reuse, uplink power control, and user scheduling (see Figure 3.9).
3. **Distributed RRM multicell MIMO coordination:** This is based on distributed RRM coordination, per eNB codebook design, but joint vector allocation is required. The two paired UEs will be served by several eNBs (uplink per eNB decoding), which is shown in Figure 3.10.
4. **Full cooperation:** Full eNB cooperation will share the full joint vector allocation, all CSI, and all data information between the referent components. Uplink and downlink joint processing has been shown to achieve capacity improvement under the backhaul constraint, which is shown in Figure 3.11.

A brief description of the CoMP operation image is provided in Figure 3.12. The selected serving cell may be changed with respect to the geometry value. The switch between single-cell and multicell operations is done with UE-specific RS, coordinated beamforming, joint transmission, and dynamic cell selection can be "transparent" to the UE, although each of them may require different feedback.

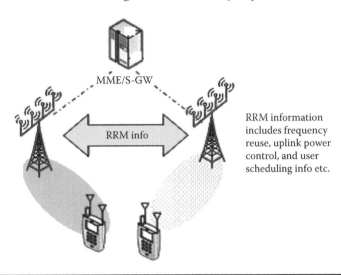

Figure 3.9 Distributed RRM.

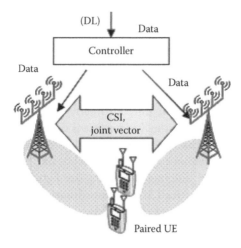

Figure 3.10 Distributed RRM multicell MIMO.

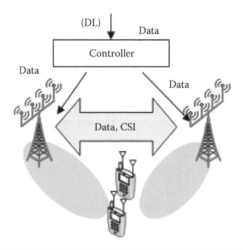

Figure 3.11 Full cooperation.

3.2.2 CoMP Cooperating Set Decision

The CoMP cooperating set decision principle considers the influence from comparative power of different candidate cells, and the relationship between comparative time delay and cyclic prefix (CP). The selection procedure of the CoMP cooperation set includes three steps.

First, the UE's serving cell should be determined; RSRP commonly is the strongest and is noted by $RSRP_{max}$. Then, several cells are selected as candidates in the UE's RSRP list, whose RSRP should be stronger. Calculate the comparative difference

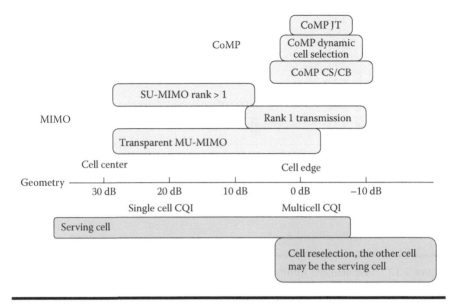

Figure 3.12 Geometry and single-cell and multicell operations.

values between the RSRPs and $RSRP_{max}$ as follows: $\Delta RSRP_i = RSRP_{max} - RSRP_i$. If $\Delta RSRP_i < RSRP_{threshold}$, the corresponding cell i can be selected as the candidate cell, where $RSRP_{threshold}$ is a predetermined threshold value. Finally, the candidate cells are refined to ensure the comparative time delay among cells is less than the CP. The comparative time difference values are calculated as follows: $\Delta TA_i = abs(TA_i - TA_{base})$, where TA_i is the respective timing advance (TA) from the UE to the candidate cells. If $\Delta TA_i + \tau_{max} < CP$, the corresponding cell i can be selected as the final candidate cell, where τ_{max} is the maximum path delay. Figure 3.13 gives the average number of candidate cells for each user based on the different $\Delta RSRP$.

The other method for determining the CoMP cooperating set is through the path loss difference between the user's serving cell and the closest nonserving cell instead of $\Delta RSRP$. With a preconfigurable path loss difference threshold value, both the user type (center or edge users) and user number can be easily controlled based on the signal attenuation to achieve balance between complexity and CoMP gain. To further limit the intercell signaling overhead, we can also predefine the maximum size of cluster M, so that only the nearest neighboring cell sites will be in the user's cluster.

3.2.3 Multiplexing between LTE and LTE-A UE

There are two multiplexing schemes for LTE UE and LTE-A CoMP UE: frequency-division multiplexing (FDM) and time-division multiplexing (TDM).

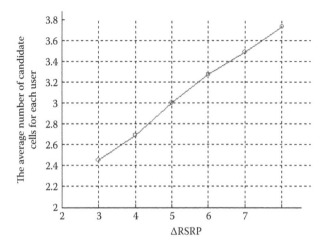

Figure 3.13   The average number of candidate cells for each user.

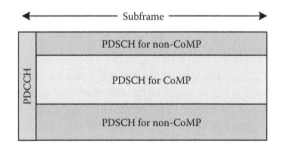

Figure 3.14   FDM of LTE and LTE-A CoMP UE.

The FDM scheme allows LTE UE and LTE-A UE to coexist in a subframe where different sets of RBs are allocated to different types of UE, as illustrated in Figure 3.14.

Under this scheme, the LTE-A-specific features should be transparent to LTE UE and system operations should be built on current LTE architectures.

In TDM, LTE UE and LTE-A UE are separated into different subframes—LTE-only and LTE-A-only—which are subframes where LTE-A-only subframes can be used to conduct CoMP transmission. Figure 3.15 is an illustration of the TDM scheme.

Under this scheme, the LTE-A-only subframes appear as MBSFN subframes to Rel-8 LTE UE. In those LTE-A-only subframes, backward compatibility issues will vanish and more design freedom will be available for performance enhancement of LTE-A systems.

Several methods can be considered to perform different operations for LTE-A CoMP UE. One simple way is to create a coordinated multipoint frequency zone

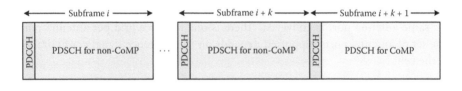

Figure 3.15 TDM of LTE and LTE-A CoMP UE.

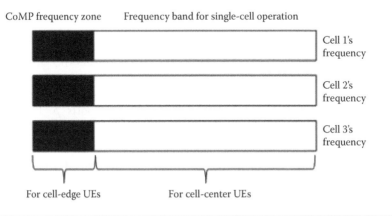

Figure 3.16 Possible frequency allocation of CoMP-SU-MIMO.

where all the cell-edge UEs belong to a CoMP cluster and are jointly scheduled using CoMP SU-MIMO mode while the cell-center UEs are scheduled by individual cell. (A cluster is defined as a number of points for each user to realize multicell cooperative reception.) The division of the system bandwidth is illustrated in Figure 3.16.

3.3 Reference Signal Design

In LTE Rel-8, channel estimation and demodulation all use the same set of RSs. To enhance performance, many new features, such as CoMP and MIMO, are introduced in LTE-A. Reference signals are further defined into two categories: one for channel measurement (CSI-RS) and the other for demodulation (DMRS). In LTE-A, DMRS should be precoded in the same way as data, which makes these reference signals dedicated reference signals.

In total, five types of DL reference signals are defined in LTE-A: cell-specific reference signals (CRS), MBSFN reference signals, UE-specific reference signals demodulation reference signal (DMRS), CSI reference signals (CSI-RS), and positioning reference signals (PRS).

There is one reference signal transmitted per downlink antenna port. An antenna port is defined such that the channel over which a symbol on the antenna port is conveyed can be inferred from the channel over which another symbol on the same antenna port is conveyed. There is one resource grid per antenna port. The set of antenna ports supported depends on the reference signal configuration in the cell:

■ Cell-specific reference signals support a configuration of one, two, or four antenna ports and are transmitted on antenna ports $p = 0$, $p \in \{0, 1\}$, and $p \in \{0, 1, 2, 3\}$, respectively. Reference signals are dispersed over each resource block.

■ MBSFN reference signals are transmitted on antenna port $p = 4$.

■ UE-specific reference signals are transmitted on antenna port(s) $p = 5$, $p = 7$, $p = 8$, or one or several of $p \in \{7, 8, 9, 10, 11, 12, 13, 14\}$.

■ Positioning reference signals are transmitted on antenna port $p = 6$.

■ CSI reference signals support a configuration of one, two, four, or eight antenna ports and are transmitted on antenna ports $p = 15$, $p = 15, 16$, $p = 15, \ldots, 18$, and $p = 15, \ldots, 22$, respectively.

3.3.1 Reference Signal Configuration in an LTE-A Network

LTE-A eNBs should always support legacy LTE UE. The Rel-8 common reference signal is also used for LTE-A UE to detect Physical Control Format Indicator Channel (PCFICH), Physical Hybrid ARQ Indicator Channel (PHICH), PDCCH, physical broadcast channel (PBCH), and PDSCH (TxD only). The downlink reference signal evolution in LTE is depicted in Table 3.3.

DMRS and the CSI-RS-based approach are adopted in LTE-A. The main motivation of introducing these two reference signals is to reduce RS overhead. Dedicated reference signals could be allocated on a per-UE basis, be adaptively assigned on a per-subframe basis depending on the layers transmitted, and thus avoid always transmitting the same number of CRSs so as to reduce overhead. DMRS is used for demodulation of PDSCH only (except TxD) and is transmitted only in an RB allocated for a UE in every subframe. There are 12 resource elements for DMRS up to rank-2, and 24 resource elements up to rank-8. CSI-RS is used for measurement and obtaining channel estimates that are transmitted by puncturing PDSCH resource elements in a duty cycle; whenever the current subframe is carrying a CSI-RS reference signal, the extracted CSI-RS-based channel estimate is used to obtain the CQI over the allocated PDSCH RBs. Meanwhile, CSI-RS overhead can be made very small, for example, less than 1% for 8Tx antenna support. The function of reference channels in LTE-A is shown in Figure 3.17.

LTE-A will support independent antenna configuration. Although the LTE-A antenna port is larger than 4Tx, the Rel-8 antenna port can be defined as less than 4Tx, and any combination is possible between the number of LTE-A CSI-RS ports and the number of CRS ports.

Table 3.3 Downlink Reference Signal Evolution of LTE

	Rel-8	*Rel-9*	*Rel-10*
CRS	CRS (antenna ports [APs] 0–3) for UE measurements, reporting (CQI, PMI), and demodulation (PDSCH, PDCCH, etc.)	CRS (APs 0–3) supported	CRS (APs 0–3) supported for legacy UE, but no sense in configuring many CRS ports simultaneously with DMRS operation
CSI-RS	N/A	N/A	Intracell CSI-RS for PMI, CQI reporting for 1/2/4/8-Tx; resource element (RE) muting to enhance intercell orthogonality
UE-specific RS (DMRS)	UE-specific RS (AP 5)	DMRS (APs 7–8) for DL dual-layer beamforming	DMRS (APs 7–14) for DL SU-MIMO (rank 1-8) & MU-MIMO (rank 1-2)
Other RS	AP 4 for MBSFN reference signals for MBSFN demodulation	PRS for positioning	—

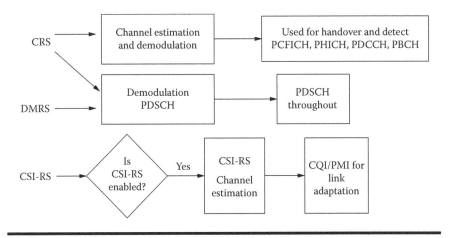

Figure 3.17   The function of the reference channel in LTE-A.

3.3.2 DL Reference Signal Design: CRS

In LTE Rel-8, CRSs are transmitted in every subframe and over the entire frequency band. Up to four groups of different CRSs can be transmitted within a cell, with each CRS group corresponding to one of up to four cell-specific antenna ports, referred to as antenna ports 0 to 3. CRSs can be used for downlink channel estimation for coherent demodulation of the physical channel transmitted from antenna ports 0 to 3, and UE measurements are based on the reference signal strength or quality for handover decisions in connected mode, and so on. The CRS can also be used to derive channel state information for the corresponding antenna ports. The channel state information can be used to assist link adaptation, precoder-matrix/vector selection, and so on.

CRS is cell specific and appears in every normal subframe. CRS is a full bandwidth transmission and spreads resource elements for RS uniformly over all resources in the system bandwidth in a staggered way. This leads to the CRS design of antenna ports 0–3 for unicast and port 4 for MBSFN. Note that the resource block size is defined partly based on CRS spacing, such that each resource block has the same CRS pattern.

CRS for antenna ports 0–1 and CRS for antenna ports 2–3 have different densities due to overall overhead restriction, and CRS transmission for two transmit antennas (antenna ports 0–1) is usually considered to be the default mode. Depending on the number of antenna ports, overheads incurred from CRS in LTE Rel-8 are as follows: 4.76% for 1 port, 9.52% for 2 ports, and 14.29% for 4 ports.

In Rel-8, CRS can be shifted in frequency to reduce the amount of CRS-to-CRS interference. The reuse factor (frequency shift) is 3 and is tied to cell ID; only 3 different shifts are available, which can constitute a rather low reuse factor.

3.3.2.1 LTE-A CRS Design Goals

Rel-8 of LTE provides CRS for up to 4 antenna ports. In order to support higher-order MIMO operation, LTE-A UE should be able to estimate the channel possibly from 8 antenna ports. Therefore, LTE-A has defined UE-specific DMRS for up to 8 streams, an extension from the 2-stream DMRS on Rel-9. A low-overhead CSI-RS that supports up to 8Tx antennas (0.96% for 8Tx) is to be used that is not UE specific to enable channel information feedback.

To support 8 transmit antennas at the eNB for spatial multiplexing, 8 cell-specific reference signal (CRS) patterns are needed for channel estimation at the UE. As in Rel-8, they are used for channel quality measurement and rank-adapted spatial multiplexing, including the closed loop and open loop precoding recommendation in LTE, as well as for data demodulation. Some design considerations in DL CRS for 8Tx transmission are as follows: The LTE Rel-8 CRS for antenna ports 0 to 3 should not be affected, especially for antenna ports 0 and 1, which are used for measurements. Therefore, code-division multiplexing (CDM) should not be done between CRSs for antenna ports 0 to 3 and 4 to 7. Performing such CDM may also affect the Rel-8 UE as the channel and noise estimation algorithms may not anticipate the code-division multiplexing with other reference signals. For data demodulation, sufficient CRS density in both the frequency and time domains is needed for tracking channel variation. If used only for assisting closed loop transmission, the CRS density may not need to be as high. So, one challenge in CRS design for 8 antennas is to support a mixture of both LTE-A UE and Rel-8 UE, where the latter can only support channel estimation with up to 4 transmit antennas. Several considerations for mapping and signaling when 8 CRSs are supported in a subframe for PDSCH transmissions are described in the following text.

The common RS overhead is always the same irrespective of the ranking for PDSCH transmissions. The overhead in LTE seems to be tolerable since it does not exceed 15% even for 4Tx antennas. However, if we keep the current FDM/TDM common RS structure for 8Tx RS design, the pilot overhead increment is inevitable to maintain the channel estimation performance. If we use the same density as for antenna ports 2 and 3 in LTE for the new antenna ports (4 through 7), it would result in an overhead of 24%! The most effective way to reduce the RS overhead is to separate RS for demodulation, from RS for CQI and compute the measurement.

Besides, in LTE-A, we need to also take into account the downlink dedicated pilot for rank-adaptive spatial multiplexing when the RS structure is discussed for 8Tx transmission since the dedicated pilot only needs to be employed with the rank number of antenna ports to be estimated. In addition, the 8Tx RS structure should support legacy UE, which implies that current 2Tx or 4Tx common RS structure should be kept if the LTE zone is not defined.

3.3.3 DL Reference Signal Design: DMRS

In LTE Rel-8, demodulation RS (UE-specific RS) can be used for downlink channel estimation for coherent demodulation of PDSCH; rank 1 DM-RS-based transmission is already supported (i.e., Rel-8 port 5). In LTE-A, up to eight different UE-specific and precoded reference signals corresponding to up to eight layers can be transmitted from the UE point of view. In a given subframe, the UE-specific reference signals are only transmitted within the resource blocks that are used for PDSCH transmission to the specific UE within this subframe. The structure for transmission from multiple antenna ports is an extension of the structure for a single antenna port. The UE-specific DMRS densities are 12 REs/RB (resource elements/resource block) for the total rank of up to 2, and 24 REs/RB for the total ranks of 3–8.

The reference signal is transmitted in each virtual antenna port (i.e., layer or stream), and only in scheduled RBs and the corresponding layers. The antenna port numbers were defined in different LTE releases, that is, 3GPP Rel-8, port 5, Rel-9, port 7 and port 8, and LTE-A, ports 7–port 14. The design principle of LTE-A is an extension of the concept of Rel-8 UE-specific RS (used for beamforming) to multiple layers; RSs on different layers are mutually orthogonal and channel estimation accuracy should be reasonable within allocated time/frequency resources.

RS and data are subject to the same precoding operation, thus there is no need to transmit precoding information, and channel estimation is per-PRB based.

3.3.3.1 DMRS Design

Forward compatible DMRS patterns for rank 1 and rank 2 designs from LTE Rel-9 dual-layer beamforming (including patterns, spreading, and scrambling) adopt CDM between two layers. For rank 2, DMRS for the first layer and the second layer are multiplexed by means of CDM by using orthogonal cover code (OCC) over two consecutive resource elements in the time domain, which is shown in Figure 3.18. OCC was introduced as a multiplexing enhancement scheme in addition to cyclic shift (CS). With the enhancement, different orthogonal codes can be covered on DMRSs between two slots of a subframe for different layers. The reason for choosing CDM involves network performance, flexibility control signaling overhead, and compatibility with other RS and control channel regions. The main advantage is that CDM is more flexible and convenient to support MU-MIMO and CoMP. If FDM is used in MU-MIMO/CoMP, which is shown in Figure 3.19, UE needs to know the location of the DMRS used for other co-scheduled UEs, which may increase signaling overhead in the downlink. If a UE pairing fails, then the unused DMRS resource elements will waste resources. This problem does not appear if CDM is used on the DMRS. Another advantage of CDM is power sharing that can be kept identical between layers on each RE for both data REs and DMRS REs. Depending on the power allocation to different layers or users in MU-MIMO, this may save additional control information for power offsets between DMRS and data with higher-order modulation. If there are different transmit weights in the

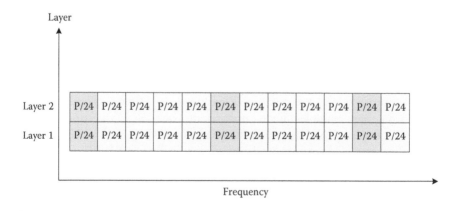

Figure 3.18　DMRS EPRE of the CDM pattern.

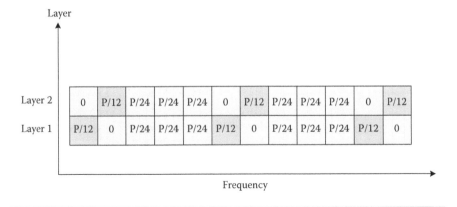

Figure 3.19   DMRS EPRE of the FDM pattern.

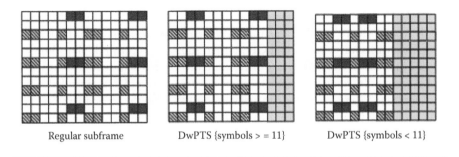

Figure 3.20   Baseline DMRS pattern for ranks 1–2. (DwPTS—downlink pilot time slot)

DMRS REs and the PDSCH REs in FDM, the intercell interference differs at the DMRS REs and the PDSCH REs, which will impact the minimum mean squared error (MMSE) detection performance. For example, assume the total transmission power for one orthogonal frequency-division multiplexing (OFDM) symbol inside one RB is P. For two-layer transmission, the PDSCHs of different layers are multiplexed by spatial code; therefore the PDSCH energy per resource element (EPRE) for each layer is $P/24$.

The main characteristics of rank 1 and rank 2 DMRS are as follows:

- 12 REs are enough for dual-layer beamforming; rank 1 and rank 2 should use the same pattern; and the UE will need to have two kinds of channel estimation schemes for the UE transmission ranks 1 and 2, which is shown in Figure 3.20.
- Two orthogonal sequences are used to multiplex up to two layers of transmission. The orthogonal sequence is {1, 1}, {1 –1}. This will support up to two layers transmitted to a single UE or one layer for two UEs with orthogonal reference signals for each layer.

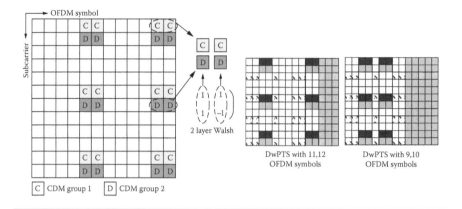

Figure 3.21 Baseline DMRS pattern for ranks 3–4; length-2 OCC mapping.

DMRS pattern for rank 3 and rank 4: Hybrid CDM + FDM (two CDM groups are defined in the frequency domain) DMRS patterns are adopted for rank 3 and rank 4 transmission with normal CP (normal subframe, downlink pilot time slot [DwPTS]). The OCC length is 2 with same orthogonal sequence as ranks 1–2, and there are 24 resource elements in an RB (see Figure 3.21); extended CP is not supported. The DMRS for the first layer and that for the second layer are also multiplexed by means of CDM by using OCC over two consecutive resource elements in the time domain. The DMRS for the third layer and that for the fourth layer are multiplexed by means of CDM by using OCC over two consecutive resource elements (D in Figure 3.21) in the time domain. The DMRS for the first and second layers and for the third and fourth layers are multiplexed by means of FDM.

DMRS pattern for ranks 5–8: Similar to ranks 3 and 4, hybrid CDM + FDM DMRS patterns are adopted for ranks 5–8 transmissions with normal CP (normal subframe, DwPTS). Extended CP is not supported in this case; same location with the same density (24 resource elements per PRB) as rank 3 and rank 4. The main difference between rank 3 and rank 4 is that the length of OCC in the time domain is 4 for both CDM groups. Figure 3.22 shows the DMRS pattern for ranks 5–8.

The main characteristic of ranks 5–8 DMRS is that it reuses the rank 3–4 pattern, the DMRS patterns for lower number of layers should be a nested property as subsets of the patterns for a higher number of layers. Because the number of layers from different cells is different, when one cell chooses to transmit a lower number of layers and another cell chooses to transmit a higher number of layers, if this nested property is not fulfilled, then the interference among the DMRSs and PDSCH from different cells is really hard to handle. So, the pattern of rank 5–8 extends the length of OCC from 2 to 4 in the time dimension. The DMRS resource element positions of a given layer are independent of the total number of

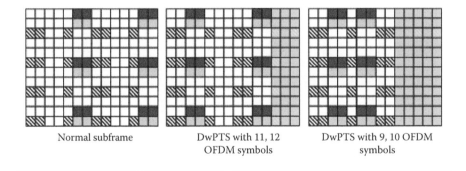

| Normal subframe | DwPTS with 11, 12 OFDM symbols | DwPTS with 9, 10 OFDM symbols |

Figure 3.22   DMRS pattern for ranks 5–8, length-4 OCC mapping.

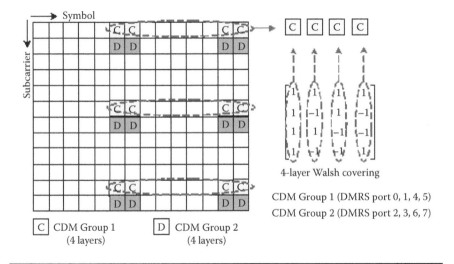

Figure 3.23   C and D denote a CDM group for multiplexing up to four layers.

transmitted layers and the only difference in channel estimation is the despreading operation with OCC 4 or OCC 2. The DMRS of four layers are multiplexed using orthogonal sequences over four resource elements in time. The DMRS resource elements are divided into two groups, shown in Figure 3.23, and the length of the OCC in the time domain is 4 for both CDM groups.

To conclude, Figure 3.24 shows the DMRS mapping pattern and multiplexing scheme for up to eight-layer transmission using a simple extension of the Rel-8 pattern. The DMRS mapping position for a lower-rank transmission is designed to be a subset of the mapping position for a higher-rank transmission to minimize the change in the resource element position for data symbols. The CDM is first used for rank 2 transmission. Then, FDM is applied to a higher-rank transmission when the insertion density becomes double (equal to 24 REs/RB).

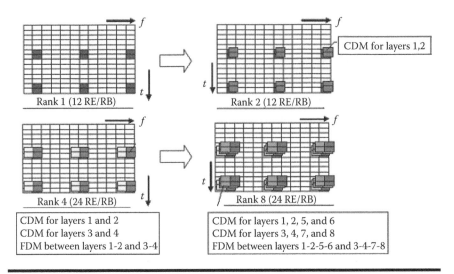

Figure 3.24 DMRS mapping pattern and multiplexing scheme for up to 8-layer transmission.

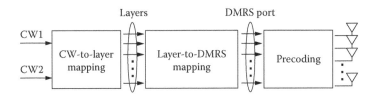

Figure 3.25 Layer-to-DMRS mapping in LTE-A.

The DMRS density of 24 REs/RB with CDM is a final choice for rank 4–8 transmission. The main feature of LTE-A DMRS is as follows:

■ DMRS ports 0–7 are defined with 2 CDM groups such that the DMRS ports {0, 1, 4, 6} and {2, 3, 5, 7} belong to the first and second CDM groups, respectively
■ Orthogonal code multiplexing within a CDM group: Length-2 OCC up to rank 4, length-4 OCC for ranks 5–8

3.3.3.2 Layer-to-DMRS Port Mapping

Due to the introduction of DMRS for data demodulation, a new mapping between transmission layers and DMRS ports should be introduced (see Figure 3.25).

In LTE-A, in case of rank 1–2 transmission, up to 2 codewords (CW) will share one CDM group and the layer-to-DMRS port mapping defined for dual-layer

Table 3.4   Layer-to-DMRS Port Mapping

Rank	Codeword	Layer	DMRS Port
1	CW-0	Layer 0	Port 0
2	CW-0	Layer 0	Port 0
	CW-1	Layer 1	Port 1
3	CW-0	Layer 0	Port 0
	CW-1	Layer 1, Layer 2	Port 2, Port 3
4	CW-0	Layer 0, Layer 1	Port 0, Port 1
	CW-1	Layer 2, Layer 3	Port 2, Port 3
5	CW-0	Layer 0, Layer 1	Port 0, Port 1
	CW-1	Layer 2, Layer 3, Layer 4	Port 2, Port 3, Port 5
6	CW-0	Layer 0, Layer 1, Layer 2	Port 0, Port 1, Port 4
	CW-1	Layer 3, Layer 4, Layer 5	Port 2, Port 3, Port 5
7	CW-0	Layer 0, Layer 1, Layer 2	Port 0, Port 1, Port 4
	CW-1	Layer 3, Layer 4, Layer 5, Layer 6	Port 2, Port 3, Port 5, Port 7
8	CW-0	Layer 0, Layer 1, Layer 2, Layer 3	Port 0, Port 1, Port 4, Port 6
	CW-1	Layer 4, Layer 5, Layer 6, Layer 7	Port 2, Port 3, Port 5, Port 7

beamforming in Rel-9 can be reused. In addition, for transmissions beyond rank 2, DMRS ports are independently mapped onto layers according to rank, so the DMRS port is defined with three parameters, including a CDM group, OCC index, and power level according to the rank. In this case, a CDM group is tied to a codeword so that a codeword always contains DMRS ports from a CDM group, as shown in the Table 3.4. Rank is independent of DMRS allocation; rank N includes Layer 0– Layer $N-1$, Layer N corresponds to DMRS port n, $n = 0, \ldots, 7$.

3.3.3.3 OCC Mapping for DL DMRS

In LTE-A, up to 8-layer transmission is supported and the new OCC mapping scheme is applied. A length-4 OCC mapping should be designed to be backward compatible with the mapping scheme for Rel-9; rank 3–8 OCC mapping is a superset of that for ranks 1–2 for co-scheduling Rel-9 UE and LTE-A UE for MU-MIMO.

The OCC mapping scheme should maintain two-dimensional orthogonality, time domain orthogonality, and orthogonality from two adjacent same-CDM groups in the frequency domain to achieve better performance in a time/frequency selective channel. In addition, the OCC mapping scheme should be exploited regarding the OCC length-4 to achieve peak power randomization. Perfect peak power randomization must be achieved for ranks 1–2, and partial peak power randomization must be achieved for ranks 3–4.

3.3.3.3.1 OCC Mapping Pattern

The binary phase shift keying (BPSK) alphabet is adopted for the OCC mapping pattern of normal CP, length-4 Walsh sequences, defined as follows:

$$W_4 = \begin{pmatrix} 1 & 1 & 1 & 1 \\ 1 & -1 & 1 & -1 \\ 1 & 1 & -1 & -1 \\ 1 & -1 & -1 & 1 \end{pmatrix} = (a \ b \ c \ d)$$

Ranks 5–8, OCC = 4 codes

Symbols a, b, c, and d represent the element of the Walsh cover code. The design criteria for length-4 OCC mapping is to keep time and frequency domain orthogonality, that is, a–d are mapped to four resource elements in the time domain and a–d are mapped to the closest four resource elements in the frequency domain, which is shown in Figure 3.26.

The OCC mapping pattern for extended CP is shown in Figure 3.27. A mechanism similar to that applied in Rel-9 is used with the same pattern in even PRBs and odd PRBs.

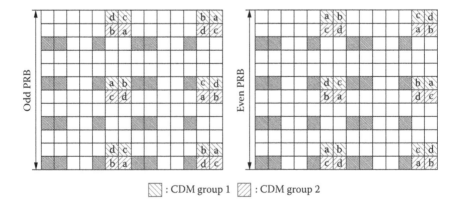

□ : CDM group 1   ▨ : CDM group 2

Figure 3.26   OCC mapping pattern for normal CP.

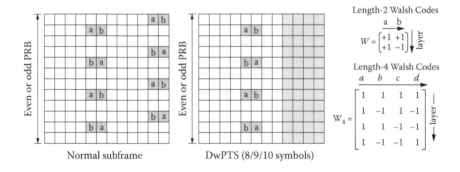

Figure 3.27 OCC mapping pattern for extended CP.

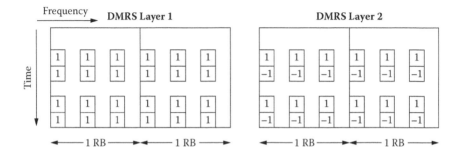

Figure 3.28 Simplest method of the length-2 OCC mapping scheme.

The DMRS for each layer utilizes different CDM code to orthogonalize the DMRS. For example, the simplest method of assigning CDM code for each DMRS layer is to assign $\{+1, +1\}$ code to the first layer and $\{+1, -1\}$ to the second layer for all the CDM resource element sets within the allocated RBs. The length-2 OCC mapping scheme as shown on the right in Figure 3.28 was accepted to achieve a better peak power randomization effect, which is necessary for full power utilization.

The DMRS sequence for each layer is multiplied with a precoding element and multiplexed together. This means that for certain precoding matrix row vectors, such as $\{+1, +1\}$ or $\{+1, -1\}$, the DMRS sequence values are combined and transmitted onto a physical antenna port. As a result of combining CDM codes to a physical antenna port, certain precoded resource elements may have twice the power and certain precoded resource elements may have zero power. Figure 3.29 shows an example of a two-frequency subcarrier DMRS sequence prior to precoding and after precoding.

From Figure 3.29 (left), we see that all the DMRS resource elements within a physical antenna port in certain OFDM symbol may have twice the power in two-layer transmission or zero power. In Figure 3.29, w_1 and w_2 represent the precoding weights for the first and second layers at the eNB transmit antenna,

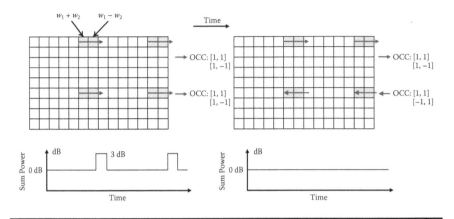

Figure 3.29 **Peak-to-average power ratio (PAPR) analysis of mapping scheme of orthogonal cover codes.**

respectively. Therefore, in the worst case, $w_1 = w_2$, the transmission power of the DMRS for the 6th and 13th OFDM symbols becomes twofold compared to that for data signals, while the transmission power for the 7th and 14th OFDM symbols becomes zero. If we decide to multiplex 4 layers in a CDM manner, then DMRS resource elements in certain OFDM symbols have four times the power or zero power. This can be a critical issue for an eNB with high-cost power amplifiers (PAs). From this perspective, it is beneficial to randomize the CDM codes so that precoded DMRS values change across the frequency by alternately reversing the mapping direction for the orthogonal cover codes in the frequency domain (see Figure 3.29, right). However, by employing the reversing scheme, the resource elements with different transmission powers alternately appear in the frequency domain; therefore, the peak transmission power is averaged.

3.3.3.4 Sequence Generation

LTE-A supports co-scheduling of up to 4 UE instances for MU-MIMO mode. Some additional features are necessary to provide additional UE-specific reference signals, such as additional UE-specific reference signals distinguished by scrambling sequences.

For antenna port 5, the UE-specific reference signal sequence is $r_{n_s}(m)$ defined by

$$r_{n_s}(m) = \frac{1}{\sqrt{2}}(1 - 2 \cdot c(2m)) + j\frac{1}{\sqrt{2}}(1 - 2 \cdot c(2m+1)), \quad m = 0, 1, ..., 12N_{RB}^{PDSCH} - 1$$

where N_{RB}^{PDSCH} denotes the bandwidth in resource blocks of the corresponding PDSCH transmission. The pseudo-random sequence $c(i)$ generator is initialized with $c_{init} = (\lfloor n_s/2 \rfloor + 1) \cdot (2N_{ID}^{cell} + 1) \cdot 2^{16} + n_{RNTI}$ at the start of each subframe. We can see that the DMRS sequence is a kind of cell-specific sequence, since the UE ID is excluded from the initialization of the pseudo-random sequence generator. In MU-MIMO mode, if the UE could know the DMRS sequence of co-scheduled UE, the UE could use linear minimum mean square estimation (LMMSE) equalization to cancel the inter-UE interference for better demodulation performance.

If we consider a use case for MU-MIMO for LTE-A, a UE-specific scrambling sequence is applied to the DMRS, it is very difficult to perform channel estimation for co-scheduled UEs since the UE does not know the UE IDs for other UE. The UE receiver may perform interference rejection combining (IRC) or advanced interference cancellation assisted by the channel estimation using the co-scheduled OCC assigned to co-scheduled UE. To solve this problem, LTE-A adopts the application of a scrambling sequence where the initialization factor is given by a higher-layer signaling as shown in the following text.

In LTE-A, for any of the antenna ports $p \in \{7, 8, \ldots, \upsilon + 6\}$, the reference signal sequence $r(m)$ is defined by

$$r(m) = \frac{1}{\sqrt{2}}(1 - 2 \cdot c(2m)) + j\frac{1}{\sqrt{2}}(1 - 2 \cdot c(2m+1)),$$

$$m = \begin{cases} 0,1,\ldots,12N_{RB}^{max,DL} - 1 & \text{normal cyclic prefix} \\ 0,1,\ldots,16N_{RB}^{max,DL} - 1 & \text{extended cyclic prefix} \end{cases}$$

The pseudo-random sequence generator shall be initialized with $c_{init} = (\lfloor n_s/2 \rfloor + 1) \cdot (2N_{ID}^{cell} + 1) \cdot 2^{16} + n_{SCID}$ at the start of each subframe, where n_s is slot index, n_{SCID} is scrambling sequence index. For antenna ports 7 and 8, n_{SCID} is given by the scrambling identity field in the most recent downlink control information (DCI) format 2B or 2C associated with the PDSCH transmission. Based on the dynamically allocated DMRS ports indicated by DCI format 2B or 2C, a flexible and unified SU/MU-MIMO transmission mode will be adopted. If there is no DCI format 2B or 2C associated with the PDSCH transmission on antenna port 7 or 8, the UE shall assume that n_{SCID} is 0. For antenna ports 9 to 14, the UE shall assume that n_{SCID} is 0. For single-user transmission, the same scrambling sequence is used for two layers. For MU-MIMO transmission, the scrambling sequence used for different users can be the same or different depending on the indicated n_{SCID}. For rank>3 transmission, the two code division multiplexing (CDM) groups use the same scrambling sequence.

3.3.3.4.1 Overhead Analysis

RS overhead associated with Rel-8 CRS and LTE-A reference signals are summarized in Table 3.5.

Table 3.5 Measurement RS Overhead Assuming Normal CP

Number of Antenna Ports	Rel-8 CRS		LTE-A DMRS
	Unicast Subframes	MBSFN Subframes	
1	4.76%	1.19%	7.14%
2	9.52%	2.38%	7.14%
4	14.29%	3.57%	Rank 1/2: 7.14% Rank 3/4: 14.28%
8	n/a	n/a	Rank 1/2: 7.14% Rank 3/4/5/6/7/8: 14.28%

3.3.3.5 PRB Bundling

From the previous discussion, we know that the density for UE RS patterns up to rank 8 is 24 resource elements per RB, and the density of UE-RS per antenna port is 3 resource elements per RB. With such a constraint on the UE RS overhead, the channel estimation losses may be significant.

PRB bundling was proposed to improve channel estimation performance for higher-rank UE RS on such a UE RS overhead. PRB bundling means that UE may assume that precoding granularity is multiple RBs, although UE is still allowed to perform single-RB channel estimation. PRB bundling is beneficial to small-delay-spread or less-frequency-selective channels. When PRB bundling is introduced, the single channel estimation operation is based on more samples and may involve more correlation coefficients. In PRB bundling (shown in Figure 3.30), a few contiguous PRBs will be scheduled to one UE and the same precoding vector is used for these contiguous PRBs, so PRB bundling can also be called *precoding granularity*. Then the UE could perform joint channel estimation across these contiguous PRBs to achieve higher channel estimation accuracy. Since higher-rank transmissions are not applicable for very frequency-selective channels, we believe that bundling of few contiguous resource blocks can be used for higher-rank transmission to obtain a reasonable trade-off between channel estimation loss and overhead.

It should be noted that bundling can improve channel estimation performance, but it adds constraints on the scheduler and implies a large precoding granularity as the bundling size increases. Usually, the larger the precoding granularity, the worse the precoding gain that can be achieved. RB bundling will require the eNB to perform some downlink precoding vector across bundled RBs, which reduces the precoding flexibility at the eNB. The reduced precoding flexibility may lead to a performance degradation of the system. Figure 3.31 gives the 16 quadrature

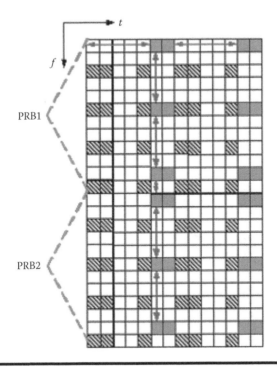

Figure 3.30 PRB bundling for higher rank.

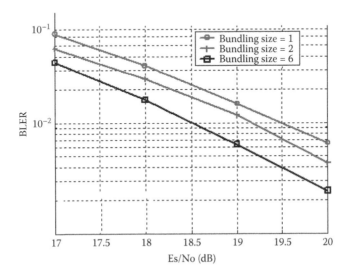

Figure 3.31 Performance of PRB bundling for rank 8 transmission with OCC length-4. (BLER—block error rate)

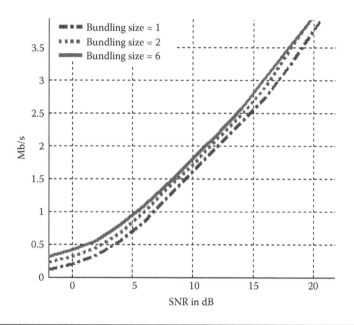

Figure 3.32 Link throughput comparison for different sizes of precoding granularity.

amplitude modulation (16-QAM) PDSCH data performance analysis of 1, 2, and 6 PRB bundling for rank 8 transmission with OCC length-4 under the extended pedestrian A (EPA) channel model.

The link throughput comparison for the TU6 channel (30 km/h) by 10-MHz bandwidth with good channel estimation is illustrated in Figure 3.32. The link throughput simulation is conducted under link adaptation and rank adaptation. We can see that for a low PDSCH signal-to-noise ratio (SNR) regime (–2 to 2 dB), 2-PRB-based channel estimation and demodulation provides a gain of around 1.5 dB. At a relatively high SNR regime (14+ dB), 2-PRB-based channel estimation and demodulation provides a gain of around 1 dB.

The number of bundling RBs is equal to the resource block group (RBG) size in LTE-A systems. Since downlink resource allocation is based on RBG, the size of RB bundling could potentially be based on RBG as a function of the total system bandwidth. A UE configured for transmission mode 9 in LTE-A for a given serving cell may assume that precoding granularity is multiple resource blocks in the frequency domain when precoding matrix indicator/rank indicator (PMI/RI) feedback is configured. Fixed system bandwidth-dependent precoding resource block groups (PRGs) of size P' partition the system bandwidth, and each PRG consists of consecutive PRBs. The UE may assume that the same precoder applies on all scheduled PRBs within a PRG. The PRG size that a UE may assume for a given system bandwidth is given in Table 3.6.

Table 3.6 PRG Bundling Size

System Bandwidth (N_{RB}^{DL})	PRG Size (P') (PRBs)
≤10	1
11–26	2
27–63	3
64–110	2

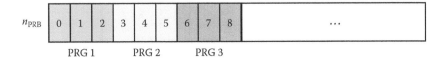

Figure 3.33 Example of PRGs.

The bundling size is essentially the PRG size. A PRG size of 1 means that a UE can only assume per-PRB-based precoding. In practice, we have a trade-off between the performance gain of PRB bundling and implementation complexity. An example of PRG of size 3 ($Q = 3$) across the system bandwidth is shown in Figure 3.33.

It is important to mention that RB bundling will require the eNB to perform the same downlink precoding vector across bundled RBs, which reduces the precoding flexibility at the eNB. Downlink precoding granularity is closely tied to the feedback granularity, so the reduced precoding flexibility may lead to a performance degradation of the system and should be carefully studied to support LTE-A systems. Where the network is performing RB-based precoding, RB bundling is not possible. When the network is performing precoding over a group of RBs, RB bundling provides nonnegligible gains.

3.3.3.6 Power Allocation of DMRS

Power utilization on DMRS resource elements is not fully exploited if the eNB keeps using the power allocation method of rank 1–2, where equal power for data and DMRS is assumed. For transmission beyond rank 2, two additional CDM groups will be allocated to DMRS ports. The number of layers multiplexed in the RS is different between the RS resource element and the data resource element. The power level could be different assuming each resource element has the same power spectral density (PSD) regardless of the type of resource element. The DMRS resource element will face different power utilization than the data resource element. Figure 3.34 shows two examples (rank 3 and rank 4) to describe the power inefficiency issue, where the same power for the DMRS resource element and data resource element is assumed for each layer.

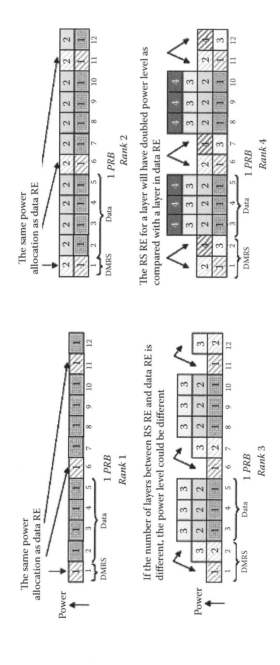

Figure 3.34 Analysis of power efficiency issue for LTE-A ranks 1–4.

There are two alternatives proposed for DMRS power allocation: rank-dependent power allocation and rank-independent power allocation. For rank-dependent power allocation, the total power of each CDM group is equal. The energy per resource element (EPRE) of each DMRS port is different for odd ranks, and the power ratio between DMRS EPRE and PDSCH EPRE varies with rank. The power ratio should signal to the UE for MU-MIMO. For rank-independent power allocation, EPRE for each port is averaged by the total power, and the power ratio between the DMRS EPRE and the PDSCH EPRE is fixed. For ranks 1–2, the power ratio between the DMRS EPRE and the PDSCH EPRE is 0dB. For ranks 3–8, the DMRS is FDM+2 CDM groups multiplexed, the data is only CDM, and the power ratio between the DMRS EPRE and the PDSCH EPRE is 3dB fixed boosting.

Considering full utilization of the transmission power and minimization of the extra complexity at UE side, the rank-independent power allocation scheme was adopted by LTE-A.

3.3.4 DL Reference Signal Design: CSI-RS

3.3.4.1 CSI-RS Design

In Rel-8, CRS is used for both channel measurement and demodulation for transmission modes 1–7. In LTE-A, due to the introduction of 8 transmission antennas and potential extension to multicell CoMP, the overhead of CRS (and possible extension) is no longer tolerable. Furthermore, in order to support LTE-A components such as CoMP, additional design principles of CSI-RS should be considered. UE needs to measure the channels or other metrics for feedback based on the CSI-RSs from all cells in a CoMP measurement set, which differs from Rel-8 where the UE needs to measure the channel based on the CRSs from only the serving cell. Therefore, LTE-A will support two types of downlink reference signals: DMRS and CSI-RS:

- DMRS for up to 8 layers of PDSCH demodulation; UE specific and precoded.
- CSI-RS for up to 8 antenna port CSI (CQI/PMI/RI) estimation; cell specific and not precoded.

In LTE-A, CSI-RS reference signals that target CSI estimation are cell specific, sparse in frequency and time, and punctured into the data region of the normal/MBSFN subframe, and there is no mixed use of Rel-8 CRS and LTE-A CSI-RS for a configured LTE-A CSI measurement. In other words, when an LTE-A UE is configured in any of the transmission modes 1 to 8, it uses only CRS for channel estimation for all Rel-8 CSI feedback modes. When a LTE-A UE is configured in transmission mode 9, it uses only CSI-RS (1, 2, 4, or 8 CSI-RS ports) for channel estimation for all CSI feedback modes. The general design characteristics of CSI-RS are described in the following text.

The CSI-RS is a cell-specific reference signal of full bandwidth transmission and transmits in each physical antenna port or virtualized antenna port used for measurement or for deriving feedback on channel quality and spatial properties purposes only.

The CSI-RS pattern is cell specific, is sparse in frequency and time, and has a larger reuse factor than CRS. The pattern of the CSI-RS depends on the number of antenna ports, system time, and physical cell ID. Channel estimation accuracy can be relatively lower than DMRS. The CSI-RS for different antenna ports of the same cells needs to be orthogonally multiplexed.

Generally UE must reach an acceptable measurement accuracy based on CSI-RSs, but CSI-RSs may experience more serious interference since the received power from multiple cells is, in general, lower than the CRS power from the serving cell. Accordingly, it should be better for the CSI-RSs of the multiple cells to be orthogonal to each other. There are ample opportunities to provide a much higher reuse factor since the CSI-RSs are sparser than the CRSs (only 3 different shifts) in time and frequency. The CSI-RSs of antenna ports of same cell are orthogonally multiplexed in TDM/FDM fashion. Within the CoMP clusters, intercell orthogonality of the CSI-RS is supported by the following methods: CDM among cells, time shift within a subframe, frequency shift within a subframe, subframe offset, cell-specific hopping pattern, or combinations of these items, as shown in Figure 3.35.

DMRSs do not have any impact on Rel-8/9 UEs because they are confined to LTE-A scheduled resources (PRBs), but it is not the case for CSI-RSs, which need to be transmitted periodically over the full bandwidth in dedicated subframes. The CSI-RS is transmitted by puncturing the data RE on both the LTE Rel-8/9 and LTE-A PDSCH. The PDSCH RE that collides with the CSI-RS RE in other cells should be muted for the measurement of intercell interference so that the UE can see the other cell CSI-RS clearly without data interference. The CSI-RS is regarded as the data RE for LTE UE. Some performance impact on legacy UE is inevitable by loss of information due to puncturing and interference from the CSI-RS.

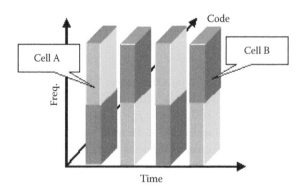

Figure 3.35   CSI-RS combination of time, frequency, code multiplexed.

Multicell CSI-RS design is considered to future-proof CoMP. In the CoMP scenario, UE should be able to measure the channels with acceptable accuracy from all cells in the CoMP measurement set where the UE may need to estimate the channels of nonserving cells. The serving cell's PDSCH as well as other cells' PDSCHs could then be muted in an attempt to raise the SINR on the CSI-RS to improve the estimation of nonserving cell channels. In addition to TDM or FDM solutions, RE data puncturing (muting) is adopted for neighbor cell measurement. To achieve CSI-RS time–frequency orthogonality among those multiple cells within a cluster of CoMP cells, muting of REs was proposed, where the cells in the CoMP cluster transmit their CSI-RS. PDSCH muting for the CSI-RS target is used to achieve intercell CSI-RS quasi-orthogonality and improves the intercell channel measurement accuracy and the CoMP UE throughput performance in comparison with CSI-RS without PDSCH muting and CRS. Muting is beneficial for CoMP schemes that require intercell measurement in the future release. An example of the CSI-RS and PDSCH muting patterns for each cell within the cluster are assigned according to Figure 3.36. The rectangular elements in Figure 3.36, marked by different tones, correspond to the REs that are occupied by the CSI-RS of each cell. The crossed rectangular elements indicate the muted REs of each cell.

Muting configuration is cell specific and is signaled via higher layers. PDSCH muting is performed over a bandwidth that follows the same rule as the CSI-RS; a UE may assume DL CSI-RS EPRE is constant across the DL system bandwidth and constant across all subframes until different CSI-RS information is received. The intrasubframe location of the muted RE is indicated by a 16-bit bitmap. Each bit corresponds to a 4-port CSI-RS configuration. All REs used in a 4-port CSI-RS configuration set to 1 are muted (zero power assumed at UE), except for the CSI-RS REs that belong to this CSI-RS configuration. When muting of PDSCH REs is configured, an LTE-A UE assumes PDSCH rate matching around the muted REs. One value of subframe offset and duty cycle is signaled for all the muted resource elements, using the same encoding as for the subframe offset and duty cycle of the

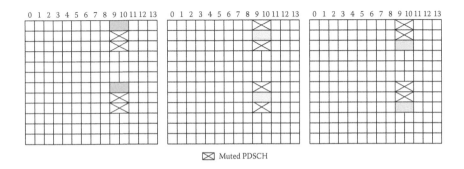

Figure 3.36 CSI-RS and PDSCH muting patterns for three cooperating cells.

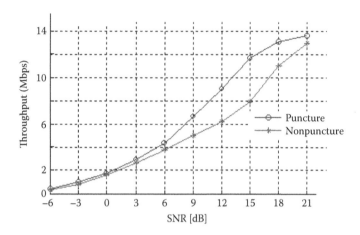

Figure 3.37   PDSCH performance gain with data puncture at REs collides with CSI-RSs in the neighbor cells.

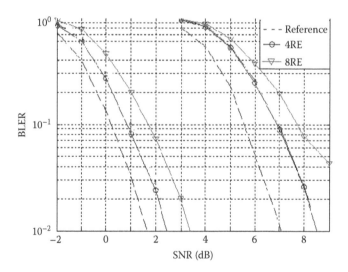

Figure 3.38   Impact data puncturing on the link performance of the LTE legacy UE.

CSI-RS. Figure 3.37 shows the concrete impact of the CSI estimation accuracy on the CoMP JP throughput performance due to data puncturing in the CoMP scenario.

For LTE-A UE, the muting location should be informed to avoid performance degradation. However, while CSI-RS punctures the data region of legacy UE, increasing the overhead of the CSI-RS can severely degrade the performance of legacy UE if too many data REs are punctured. Figure 3.38 shows the link simulations to evaluate the impact of data puncturing on the performance of the LTE legacy UE. It is always better to minimize the total RE punctures as much as possible, while achieving enough accuracy of spatial information feedback.

From Figure 3.37 we can see that data puncturing at the REs collides with the CSI-RSs in the neighbor cells, which is an effective way of reducing intercell interference and guarantees CSI measurement quality for the coordinated cells in a CoMP measurement set. Figure 3.38 shows the PDSCH throughput loss of LTE legacy UE with different numbers of punctured REs per PRB. Higher puncturing densities per antenna port can incur large losses in UE performance, especially for modulation and coding schemes (MCSs) with high coding rates. It seems that puncturing beyond four REs per RB may be tolerable and there is a need to be cautious with respect to more punctured resource elements per PRB.

To conclude, the CSI-RS is a new type of reference signal targeting to CSI estimation only that supports up to eight cell-specific antenna ports. The CSI-RS is sparse in time and frequency, for example, every 6th subcarrier in one OFDM symbol per frame and around 0.12% overhead per antenna port. The CSI-RS patterns for a lower number of layers should be a subset of the patterns for a higher number of layers (i.e., nested structure of 2/4/8 ports). The characteristics of the CSI-RS pattern for normal CP are described as follows:

- CDM-T with OCC length-2 CSI-RS patterns are supported for 8 antenna ports.
- The density of 1 port is 2 RE/port/PRB; the density of 2/4/8 ports is 1 RE/port/PRB; the 1 CSI-RS port pattern has the same position as 2 ports. Such density translates into a frequency spacing of 6 subcarriers between CSI-RS REs, which is similar to Rel-8 CRS spacing.
- To avoid antenna port 5 and CSI-RS collision of the same cell, the other three time division duplexing (TDD) mandatory patterns are added; these patterns can be used in the initial deployment of LTE-A, when legacy UE dominates in the network.
- The duty cycle of the CSI-RS patterns should be configurable in a semistatic way to a limited set of values, for example, {2, 5, 10} ms through higher-layer signaling, and should avoid the conflict of the CSI-RS with the resource elements carrying system information (see Figure 3.39).

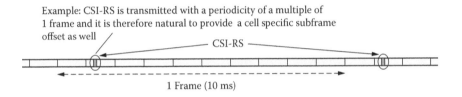

Example: CSI-RS is transmitted with a periodicity of a multiple of 1 frame and it is therefore natural to provide a cell specific subframe offset as well

CSI-RS

1 Frame (10 ms)

Figure 3.39 **Example of the duty cycle of the CSI-RS patterns (the CSI-RS density is 1 symbol every 10 ms per antenna port in the time domain).**

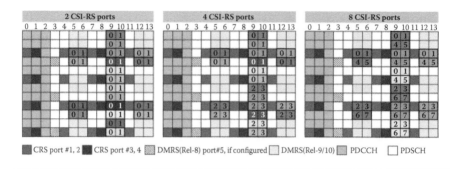

Figure 3.40   CSI-RS pattern; normal CP FDD and TDD mandatory.

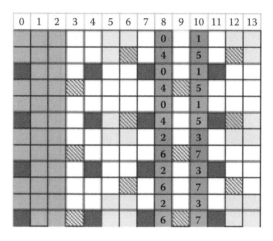

Figure 3.41   CSI-RS pattern of normal CP TDD mandatory.

The four CSI-RS patterns for normal CP supported for FDD and time-division duplexing (TDD) are shown in Figures 3.40 and 3.41. CDM is the baseline as the CSI-RS port multiplexing for each pair of CSI-RS ports, and CDM time domains are used for CSI-RS port pair multiplexing. Each different tone in the figure represents a different CSI-RS pattern, nested corresponding patterns for 4, 2 antenna ports.

Three CSI-RS patterns supported for TDD and optionally for FDD are shown in Figure 3.40.

3.3.4.2 CSI-RS Overhead Analysis

In LTE-A, the lowest considered CSI-RS resource element density is 1 resource element per PRB per antenna port. This provides acceptable link-level performance for SU-MIMO transmission, while ensuring minimal Rel-8 UE performance degradation. For CSI-RS density in the frequency domain, to mitigate the legacy performance impact, only one subcarrier is used per PRB for 2, 4, and 8 antenna ports. The eNB transmits all the CSI-RSs for every antenna port within the same subframe. For CSI-RS density in the time domain, assuming 10-ms periodicity, CSI-RS overhead can be calculated as 0.12% per antenna port [1/840], 1 symbol every 10 ms per antenna port [1/140] and 1 subcarrier every 6 subcarriers per antenna port [1/6]). We can also achieve 0.96% for 8 antenna ports.

3.3.4.3 CSI-RS Power Allocation

In Rel-8 CQI calculations, UE is informed of the ratio of PDSCH EPRE to CRS EPRE (ρ_A). In LTE-A, CSI calculation is based on the CSI-RS, and the UE is informed of the ratio of PDSCH EPRE to CSI-RS EPRE (P_C) via higher layer signaling.

P_C is the assumed ratio of *PDSCH EPRE* to *CSI-RS EPRE* when the UE derives CSI feedback and takes values in the range of [–8, 15] dB with a 1-dB step size.

3.3.4.4 CSI-RS Sequence Generation

The CSI-RS sequence reuses the CRS sequence, with the length as half of the CRS length. The reference signal sequence is $\eta_{l,n_s}(m)$ defined by:

$$\eta_{l,n_s}(m) = \frac{1}{\sqrt{2}}(1 - 2 \cdot c(2m)) + j\frac{1}{\sqrt{2}}(1 - 2 \cdot c(2m+1)),$$

$$m = 0,1,...,N_{RB}^{max,DL} - 1$$

where n_s is the slot number within a radio frame and l is the OFDM symbol number within the slot. The pseudo-random sequence $c(i)$ generator will be initialized with $c_{init} = 2^{10} \cdot \left(7 \cdot (n_s + 1) + l + 1\right) \cdot \left(2 \cdot N_{ID}^{cell} + 1\right) + 2 \cdot N_{ID}^{cell} + N_{CP}$ at the start of each OFDM symbol, where

$$N_{CP} = \begin{cases} 1 & \text{for normal CP} \\ 0 & \text{for extended CP} \end{cases}$$

The CSI-RS configuration is transmitted to the UE by dedicated signaling (including zero Tx). The content of CSI-RS signaling includes the following:

■ CSI-RS port number: 2 bits
■ CSI-RS configuration: 5 bits

- Duty cycle and subframe offset: 8 bits (may be updated after the 4-ms duty cycle is introduced)
- UE-specific power offset: 5 bits.

3.3.4.5 Antenna Mapping of CSI-RS

LTE Rel-8 UE is not capable of estimating eight downlink channels. From the Rel-8 UE perspective, an 8-antenna eNB must be able to function as a legacy eNB with up to 4 antennas. On the other hand, to utilize the transmit power from all available transmit power amplifiers (PA), an antenna mapping is required for LTE Rel-8 antenna ports to the transmit PA (here called "antennas") when there are fewer configured LTE Rel-8 antenna ports than the number of antennas. As an example, an antenna virtualization scheme can be exploited to make 8 antennas appear as 4 at the UE, which is shown in Figure 3.42 (left). Figure 3.42 (right) exemplifies the virtual antenna mapping scheme with the

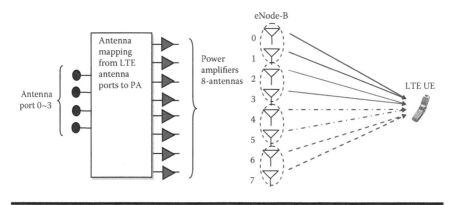

Figure 3.42   Mapping from LTE antenna ports to eight LTE transmit antennas.

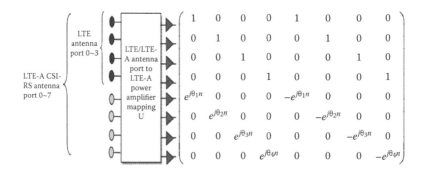

Figure 3.43   Antenna mapping from LTE antenna ports to PA.

antenna pairs of 0–1, 2–3, 4–5, and 6–7. Alternatively, different antenna pairing for virtual antenna mapping is also possible, such as the antenna pairs of 0–4, 1–5, 2–6, and 3–7.

In this case, the broadcast channel (BCH) for the LTE system still configures 4 antennas to LTE UE, whereas the BCH for an LTE-A system configures 8 antennas to LTE-A UE.

For LTE-A, CSI antenna ports are introduced to enable measurements for LTE-A UE. Up to 8 new CSI antenna ports need to be defined to support measurements from an 8-antenna eNB. It is possible for the eNB implementation to support multiple antenna setup in a network, with the number of antenna port for Rel-8 CRS less than the number of antenna ports for LTE-A CSI-RS. Although the antenna mapping (an illustrative scheme is shown in Figure 3.43) is an implementation issue, it has some impact on the definition of the new CSI-RS in LTE-A.

3.3.4.5.1 Combinations of Rel-8 CRSs and LTE-A CSI-RS Antenna Ports

In an LTE-A network, the maximum number of Rel-8 CRS antenna ports is 4, and the maximum number of LTE-A CRS antenna ports is 8. LTE-A UE can be configured by eNB to use either Rel-8 CRS or LTE-A CSI-RS for CSI measurement. A combination of Rel-8 CRS and LTE-A CSI-RS antenna ports is an important issue in designing the network. An antenna configuration with two antenna ports per cell (shown in Figure 3.44) is a typical application in the current cellular network. Therefore, LTE-A CRS antenna ports must support up to 8 and Rel-8 CRS antenna ports must support 2, which may be the baseline for Rel-8 LTE employing 2 Tx MIMO and employing 4–8 higher-order SU/MU-MIMO for advanced UE.

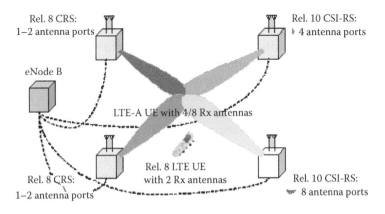

Figure 3.44 Example of Rel-8 LTE with 2 Tx MIMO and LTE-A higher-order MIMO using distributed antennas.

3.3.5 UL Reference Signal Design: General

3.3.5.1 UL RS in Rel-8

3.3.5.1.1 DMRS

A UE-specific DMRS is used for uplink channel estimation and coherent demodulation, and is transmitted once every 0.5 ms slot. Single antenna transmission, cyclic shift-based* CDM multiplexing is possible for UL MU-MIMO, and the cyclic shift index is conveyed to a UE in the UL grant. Eight orthogonal cyclic shifts can be used as a maximum. The current LTE UL DMRS is transmitted from 1 antenna for single-input multiple-output (SIMO), and occupies the same bandwidth as the resource allocation size. UL DMRS adopts 30 base sequence groups with up to 2 base sequences per group, and supports 12 equally spaced cyclic time shifts of a base sequence (spacing = 5.55μs). Three bits of DMRS cyclic shift offset is indicated in UL grant PDCCH DCI format 0. Cyclic shift hopping is always enabled for physical uplink shared channel (PUSCH) and physical uplink control channel (PUCCH) and is used to separate the DMRSs for different UE participating in the MU-MIMO operation with equal bandwidth allocation in the uplink (different cyclic shifts for orthogonality). To maintain DMRS orthogonality among different UEs, the transmission bandwidth of those UEs paired for MU-MIMO has to be identical, which can reduce scheduling flexibility. PUSCH DMRS base sequence and hopping pattern coordination is supported in LTE-A and sequence group hopping on a slot boundary is already supported in Rel-8.

3.3.5.1.2 Sounding Reference Signal (SRS)

The SRS is one set of uplink reference signals in LTE. The SRS is used for uplink channel state estimation at the network side to assist uplink scheduling, link adaptation, uplink power control, and antenna selection, and to assist in maintenance of synchronization and the downlink transmission (e.g., the downlink beamforming in the scenario with UL/DL reciprocity), especially in time division duplexing (TDD). Uplink SRSs are transmitted every Nth subframe, counting the uplink pilot time slot (UpPTS). The SRS requires that each transmit antenna transmits SRSs with a period shorter than the channel coherence time and at a bandwidth significantly larger than the channel coherence bandwidth. Multiple SRS users can be code-multiplexed and transmitted on the same bandwidth even without

---

* Cyclic shift (CS) separation is the primary multiplexing scheme of UL DM-RS. In Rel-8, there is only one CS index indicated by PDCCH for each UE, a 3-bit CS separation indicator for the DM-RS is adopted in the UL scheduling grant (DCI format 0). It can support the maximum of 8 orthogonal RS resources with equal CS separation and may be sufficient to support SU-MIMO for up to four TX antenna ports per UE. It would not be sufficient to use only CS for the intra-cell interference mitigation for an operation scenario with 4×4 UL SU-MIMO, MU-MIMO and CoMP reception in LTE-A.

PUSCH (or PUCCH) scheduled. This provides the channel estimation for any part of the spectrum and makes it possible to schedule UL transmission on resource blocks with instantaneously good channel quality. The SRS is a prerequisite for UL frequency-selective scheduling to improve the system throughput. The SRS is a UE-specific measurement reference signal. The SRS resource is defined in cell-specific manner, and UE is configured to transmit SRSs of a minimum 4 PRB within the cell-specific SRS resource. Up to 4 levels of tree-based SRS bandwidth (shown in Figure 3.45) are defined, and SRS hopping can be configured per eNB.

There are different ways of multiplexing SRS signals, such as sounding the channel in FDM or CDM fashion. SRS users also can use different combs (either even or odd subcarriers). The period of SRS transmission can be configured so that in a slow varying wireless channel, it is not necessary to transmit SRS every subframe. In this case it is possible to multiplex users in a TDM scheme. CDM users employ a base sequence with different cyclic shifts, with a fixed shift interval. The cyclic shift is selected such that it can accommodate large delay spreads. Rel-8 SRS design features are as follows:

- Interleaved frequency-division multiple access (IFDMA) comb SRS structure reusing DMRS base sequences; same as PUCCH base sequence group.
- Configurable UE-specific SRS bandwidth (BW) is restricted to be a multiple of 4 RBs (contiguous); support frequency hopping to be enabled allowing cyclically sounding the entire cell-specific SRS bandwidth with different frequency starting positions and different transmission instances.
- Eight equally spaced cyclic time shifts per comb are supported (spacing = 4.17 µs); no cyclic shift hopping for SRS is supported.

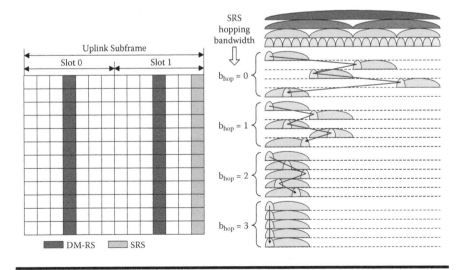

Figure 3.45   SRS position and hopping BW for tree-based SRS structure.

3.3.5.2 UL RS in LTE-A

In LTE-A, UL reference signals should be enhanced to support new features efficiently. The number of DMRS resources used should be equal to the number of layers to enable proper channel estimation and satisfy the backward compatibility requirement in the target to enable performance improvement with less complexity. Therefore, the motivation for using a DMRS enhancement in LTE-A are as follows:

- UL SU-MIMO with a maximum of 4 layers, and simultaneous SRS transmission from multiple transmit antennas: Does it need orthogonality between different layers? Does it need a signaling design with backward compatibility?
- MU-MIMO issues: For intracell and intercell MU-MIMO, it is required that good orthogonality be guaranteed between paired UE. Does it need orthogonality with unequal bandwidth allocation for paired UE?
- UL-CoMP issues: Does it need orthogonality between CoMP UEs and normal UEs with backward compatibility?

Based on these items, some enhancement of the UL DMRS in terms of introducing OCC and/or IFDMA seems to be necessary. Some candidate schemes for enhancement of the UL DMRS are shown in Figure 3.46.

For the candidate schemes for the UL DMRS of LTE-A, we should consider how to combine these different types of orthogonal resources. Table 3.7 shows a comparison of different orthogonal RS multiplexing schemes, such as cyclic shift, OCC, and IFDMA, from the viewpoints of backward compatibility, performance improvement, and MU-MIMO- and CoMP-specific issues.

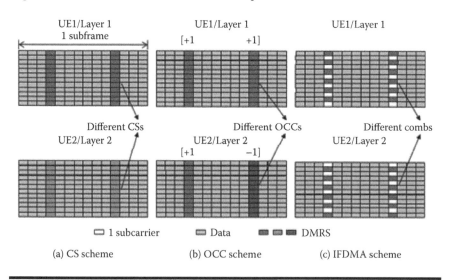

Figure 3.46 Candidate schemes for enhancement of the UL DMRS.

Table 3.7 Comparisons among IFDMA, CS, and OCC

Comparison		IFDMA	Cyclic Shift	OCC
Backward Compatibility	Paired with LTE UEs	No	Yes	Yes (OCC[+1,−1])
	Indication via signaling	Comb info	No	OCC info
	Resources allocation limitation	Yes (PRB number is a multiple of comb number). Limits the number of supportable BW allocations as the number of allocated RBs should be divisible by the number of combs.	No	No
Performance Improvement	Diversity gain from intrasubframe frequency hopping	Yes	Yes	No
	Orthogonality	Same as LTE	Same as LTE, but loss due to delay spread	Better than LTE (loss in moving scenarios due to Doppler frequency shift)
	Orthogonal resources	Same as LTE	Same as LTE	Doubled if combined with CS

(Continued)

Table 3.7 (*Continued*) Comparisons among IFDMA, CS, and OCC

Comparison		IFDMA	Cyclic Shift	OCC
	Orthogonality with sequence/ group hopping	Yes	Yes	No
Support of MIMO, CoMP	SU-MIMO	Yes	Yes	Yes
	MU-MIMO with unequal bandwidth	Yes	No, only supports equal bandwidth	Yes
	UL CoMP	Yes	Yes, same root sequence (already exists in LTE)	Yes, no sequence/ group or CS hopping

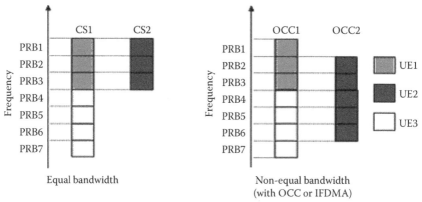

Figure 3.47 Equal vs. unequal bandwidth allocation.

Based on the comparisons in Table 3.7, we can see that different cyclic shifts of the DMRS sequence can be considered as a good baseline scheme to support SU-MIMO, MU-MIMO, and CoMP, but this allows only MU-MIMO multiplexing between users with equal bandwidth allocation (see Figure 3.47). The major advantage of OCC is the capability to increase the number of orthogonal RS resources, support different transmit bandwidths for different UE, and obtain higher scheduling flexibility while the drawback is that orthogonality is destroyed in a high-mobility case. OCC also cannot guarantee orthogonality for

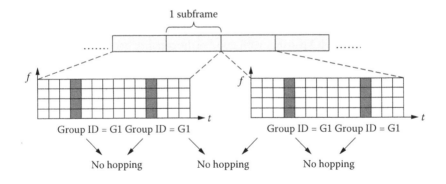

Figure 3.48 Intra- and intersubframe disabling SGH.

MU-MIMO with unequal bandwidth allocation when slot-level sequence/group hopping* is enabled. IFDMA is precluded since it does not support pairing LTE UE and LTE-A UE, the number of PRB has to be a multiple of the comb number, and interference suppression will be reduced by 10log10 (number of combs), which leads to worse channel estimation, particularly for cell-edge UE.

It should be noted that because constant amplitude zero autocorrelation (CAZAC) sequence hopping is cell specific and applies to both PUSCH and PUCCH, disabling it will degrade the performance for all transmissions. Absence of sequence hopping can cause significant performance degradation as the cross-correlations among CAZAC sequences of different lengths can be large.

System-level simulation results show that MU-MIMO with unequal bandwidth allocation (with OCC or IFDMA) can bring evident gain over equal bandwidth allocation. As OCC is beneficial for unequal resource allocation for MU-MIMO, OCC is introduced in LTE-A in addition to primary cyclic shift multiplexing for SU-MIMO because it helps in a high-SNR region with 3 or 4 layers and supports unequal bandwidth allocation.

In Rel-8, the RS sequence is defined by a cyclic shift of a base sequence and the DMRS cyclic shift offset is indicated in the PDCCH UL grant. The DMRS cyclic shift for the other SU-MIMO layers is implicitly determined with maximal separation that minimizes cross-talk between estimated channels of the different layers, for example, 6 CS separation between layers for 2-layer transmission. We also can modify PDCCH CS offset mapping to support possible assignment of each of the 12 CSs with SU-MIMO and maximal separation. The base sequences are divided into 30 sequence groups (SGs). The eNB can enable or disable sequence hopping and group hopping (SGH) (see Figure 3.48) slot by slot with a cell-specific signaling that aims at improving the resistance of UL DMRS toward, for example, intercell interference. If SGH is enabled, ICI randomization gain can be obtained, but OCC cannot guarantee DMRS orthogonality between UEs; if SGH is disabled, ICI randomization

---

* CAZAC sequence hopping is enabled for Rel-8, SPS, PUCCH, and other UL transmissions.

gain cannot be obtained. So UE-specific SGH should be introduced to enable most UE to benefit from ICI randomization gain and enable LTE-A UE to have better flexibility in MU-MIMO pairing, with good orthogonality between UE instances.

Therefore, a different sequence group for each CC or a different cyclic shift sequence for each CC is suitable and a detailed design for LTE-A UL DMRS is needed.

3.3.6 UL Reference Signal Design: DMRS

3.3.6.1 UL DMRS Design

For multiple transmit antennas in LTE-A, DMRSs can be extended to nonprecoded DMRSs and precoded DMRSs. The precoded RS can provide lower RS overhead with beamforming gain, which may provide better channel estimation performance compared to that of nonprecoded RSs. The nonprecoded UL DMRS solution would be unnecessarily wasteful of RS resources and was adopted in Rel-8. It is used to estimate channels from each transmit antenna and may need multiple cyclic shift allocations for a UE to estimate spatial channels of each antenna port, thus requiring Nt (i.e., number of transmit antennas) cyclic shift allocations for a single UE. In this case, uplink MU-MIMO and intercell interference coordination seems to have a further restriction as compared with LTE due to multiple cyclic shift allocations for a single UE. Therefore, as a baseline, UL precoded DMRS is adopted in LTE-A, which should be precoded with the precoder assigned for PUSCH data transmission (i.e., signaled on the UL grant). Consequently, the receiver can directly estimate the precoded channel and therefore save computational resources as well. The number of UL DMRS resources used should be equal to the number of layers to enable proper channel estimation. UL precoded DMRS is precoded to estimate channels from each layer and may need the layer number of the cyclic shift, which means that if the transmission rank is one, then only one cyclic shift value is needed irrespective of Nt when a precoding is employed. Therefore, the number of cyclic shift values can be kept for other UE as much as possible.

The precoding applied for the demodulation reference signal is the same as the one applied for the PUSCH. Cyclic shift separation is one of the primary multiplexing schemes of the demodulation reference signals with a good orthogonal property. Up to four uplink DMRSs can be transmitted from a UE. The instantaneous bandwidth of the uplink DMRS equals the instantaneous bandwidth of the corresponding PUSCH transmission. The structure for the case of transmission from multiple antenna ports is an extension of the structure for the case of a single antenna port.

Cyclic shift separation for DMRS multiplexing, up to 4 layers, can be separated by means of different cyclic shift. The advantages of utilizing cyclic shift include backward compatibility and good orthogonality. Since the maximum number of cyclic shifts is 8 in Rel-8, utilizing different CSs to distinguish different antenna ports and different layers is possible, with the maximum supported delay spread* of

---

* In the large delay spread environment, there would be a limit on the available cyclic shifts.

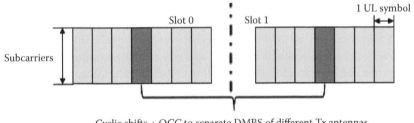

Figure 3.49 Code division multiplexed UL DMRS.

66.67 µs/4 = 16.67 µs. Different from Rel-8 DMRS, the OCC sequence is supported to further keep the orthogonality among UE instances and reduce interlayer interferences in LTE-A without increasing UL grant signaling overhead. As we know, OCC can improve orthogonality among DMRSs from UE transmitter antennas with SU-MIMO and allow spatial multiplexing (MU-MIMO) for PUSCH transmissions with different bandwidths (see Figure 3.49). This is a complementary way of mitigating interlayer interference with SU-MIMO.

Note: Length-M reference signal sequence: $r(n)$, $n = 0, \ldots, M - 1$

Cyclic shift of the reference signal sequence can be generated by

$$r^{(\alpha)}(n) := e^{j\frac{2\pi\alpha}{M}n} \cdot r(n), n = 0, \ldots, M - 1$$

where $r(n)$ is the Length $- M$ reference signal sequence.

The orthogonality between DMRSs for PUSCH from different cells could be supported by scheduling with selected sequence shift pattern f_{ss}^{PUSCH}. Also, in order to improve the intercell interference randomization for MU-MIMO with different bandwidth pairing, OCC can be used for MU-MIMO.

3.3.6.2 CS/OCC Mapping

In LTE-A, UL DMRS patterns may be designed by combining CS and OCC. CS/OCC mapping should consider backward compatibility, SU-MIMO performance, MU-MIMO flexibility, and complexity.

- **Backward compatibility:** Cyclic shift values for the first layer will follow the Rel-8 design.
- **SU-MIMO performance:** Cyclic shift spacing between layers is at least 3 to create high orthogonality.
- **MU-MIMO flexibility:** All MU-MIMO scenarios can be supported.
- **Complexity:** A simple nested design is proposed to simplify the configurations.

As in Rel-8, the 3-bit cyclic shift field is contained in the PDCCH DCI; zero-autocorrelation codes are used as DMRS sequences, so different cyclic shifts of DMRS sequences can be used as orthogonal reference signals. In LTE-A, the CS values between the layers will be as far apart as possible to minimize interlayer interference and have no additional signaling to indicate the CS value of each layer. Meanwhile, the PDCCH DCI will indicate both CS and OCC for all transmission layers in LTE-A (see Table 3.8). The introduction of OCC {1, 1} and {1, –1} will potentially increase the multiplexing capacity from 12 to 24, improve orthogonality among DMRSs from UE transmitter antennas with SU-MIMO, and allow spatial multiplexing for PUSCH transmissions with different MU-MIMO bandwidths.

This UL DMRS design scheme entails the separation of 4 layers with 4 cyclic shifts and with 2 time orthogonal codes (see Table 3.9). It is obvious that multiplexing with both cyclic shifts and orthogonal codes is more robust and is needed at higher SNRs, especially when cyclic shift separation is ¼ of the OFDM symbol and less. When the

Table 3.8 Mapping of the Cyclic Shift Field in Uplink-Related DCI Format to $n_{DMRS,\lambda}^{(2)}$ and $[w^{(\lambda)}(0) \quad w^{(\lambda)}(1)]$

Cyclic Shift Field in Uplink-Related DCI Format	$n_{DMRS,\lambda}^{(2)}$				$[w^{(\lambda)}(0) \quad w^{(\lambda)}(1)]$			
	$\lambda=0$	$\lambda=1$	$\lambda=2$	$\lambda=3$	$\lambda=0$	$\lambda=1$	$\lambda=2$	$\lambda=3$
000	0	6	3	9	[1 1]	[1 1]	[1 –1]	[1 –1]
001	6	0	9	3	[1 –1]	[1 –1]	[1 1]	[1 1]
010	3	9	6	0	[1 –1]	[1 –1]	[1 1]	[1 1]
011	4	10	7	1	[1 1]	[1 1]	[1 1]	[1 1]
100	2	8	5	11	[1 1]	[1 1]	[1 1]	[1 1]
101	8	2	11	5	[1 –1]	[1 –1]	[1 –1]	[1 –1]
110	10	4	1	7	[1 –1]	[1 –1]	[1 –1]	[1 –1]
111	9	3	0	6	[1 1]	[1 1]	[1 –1]	[1 –1]

Note: $w^{(\lambda)}(m)$ is the OCC orthogonal sequence.

Table 3.9 Illustration of DMRS Resource Allocation for Multiplexing up to 4 Layers

Cyclic Shift	C1	C2	C3	C4
OCC {+1,+1}	Layer 1		Layer 3	
OCC {+1,–1}		Layer 2		Layer 4

time orthogonal covering code is used, the cyclic shift separation of ½ the OFDM symbol is maintained. For example, Layer 1 and Layer 3 have a large cyclic shift separation, and similarly for Layer 2 and Layer 4. They both occupy the same number of cyclic shifts and maintain twice the cyclic shift separation. Thus, it will be more robust than frequency selectivity in the channel and "leakage" between cyclic shifts.

The example of DMRS for multiple layers belonging to the same UE is shown in Figure 3.50. The OCC explicitly indicated is used for the 1st and 2nd layers and another OCC is used for the 3rd and 4th layers. The twelve CS resources use the same approach as in the acknowledgment/negative acknowledgment (ACK/NACK) resource assignment in the PUCCH.

3.3.6.3 Mapping of UL RS Sequence

For clustered discrete Fourier transform spread OFDM (DFT-S-OFDM), DMRS should be transmitted in the same band of data transmission, and discontinuous RB allocation within a component carrier (CC) is also necessary for reference signal transmission. One issue is the DMRS sequence design for frequency-noncontiguous resource allocation. The base CAZAC sequence is used in Rel-8 LTE according to the whole allocation size and split into clusters; the DMRS sequence design is also for non-contiguous resource allocations. Figure 3.51 shows an example of the mapping of an

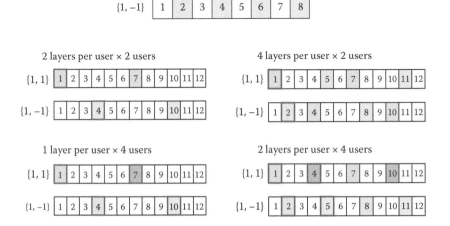

Figure 3.50 **OCC resource allocation using cyclic shift. (Left) 4-layer multiplexing per cell. (Right) 8-layer multiplexing per cell.**

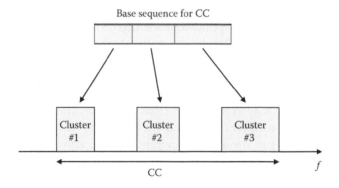

Figure 3.51 Mapping of RS sequence for frequency noncontiguous resource allocation.

RS sequence whose length corresponds to the number of allocated subcarriers within which a CC is generated. The base sequence is divided into the size of each cluster. The divided base sequence is mapped to the DMRS resource of each cluster, respectively.

3.3.7 UL Reference Signal Design: SRS

3.3.7.1 SRS Design

As mentioned, channel state information needed for closed loop precoding can be obtained only through SRS measurements, and the performance of precoded UL SU-MIMO and CoMP will largely depend on the performance of SRS-based channel sounding. LTE supports transmissions in only one component carrier and from only one UE transmitter antenna in each subframe. These attributes lead to a simple SRS transmission structure where the respective time, frequency, and code resources for each UE instance can be allocated through higher-layer signaling. However, in LTE-A uplink, there is support for UL transmission in multiple component carriers and up to 4 UE transmitter antennas, which necessitates a different management of SRS resources, SRS overhead, and activation of SRS transmissions. The amount of SRS overhead depends on the SRS transmission interval and SRS bandwidth. With a 10-ms SRS transmission interval and full-band SRS, the relative overhead is approximately 0.7%.

SRS capacity can be calculate using this expression:

$$SRS\ Capacity = \frac{System\ SRS\ BW}{UE\ Specific\ SRS\ BW} \times N_{CS} \times N_C$$

$$\times \frac{UE\ Specific\ SRS\ Periodicity}{Cell\ Specific\ Subframe\ Configuration\ Period}$$

where N_{CS} is the number of cyclic shifts and N_C is the number of frequency combs. In Rel-8, at most 8 cyclic shifts can be supported per $_{SRS}Bandwidth$ and frequency comb, depending on the maximum delay spread seen in the cell.

So, in order to support uplink SU-MIMO with up to four antennas, UE needs to transmit uplink SRSs from each configured antenna, and the performance of SRS-based channel sounding may become a limiting bottleneck for widespread UL MIMO and CoMP usage. The motivation for SRS enhancement is described afterward.

3.3.7.1.1 Sounding Capacity Enhancement

3GPP TR 36.913 specifies that an LTE-A system should be able to support at least 300 active users without discontinuous reception (DRX) in a 5-MHz bandwidth. LTE Rel-8 UE needs only 1 SRS resource,* which does not support UEs with multiple transmit antennas, but LTE-A and future UE need multiple SRS resources; the reasons include UL MIMO, CoMP, and multiple component carrier usage.

- **UL MIMO:** Maximum 4 resources for 4 antennas, thereby decreasing the number of supportable UE instances by four times compared with LTE Rel-8. To minimize the sounding delay, LTE-A UE supports concurrent transmission of SRSs via multiple TX antennas.
- **UL CoMP:** Multiple cells reserve resources for each CoMP UE. Sounding should be reliably received at all cells in a cooperating set that requires coordination of sounding resources between cooperating cells to avoid intercell interference.
- **UL carrier aggregation:** Each UE needs SRS resources for each component carrier and it is desirable to specify different sounding configurations for multiple component carriers.
- **DL SU-/MU-MIMO/CoMP:** On-demand of SRS Tx for downlink scheduling to compute long-term channel statistics for downlink transmission.

Multiplexing capacity of SRS in the same symbol can be a significant bottleneck for system performance optimization because the total SRS resource can be insufficient for LTE-A UE with multiple transmit antennas. The SRS resource shortage would result in longer latency and lower system performance. In LTE, at most 16 SRSs with maximum SRS bandwidth can be multiplexed in each uplink subframe, thanks to two transmission combs and eight usable orthogonal sequences by cyclic shift. The approaches that can extend the SRS capacity include extension

---

* For TDD, the maximum number of cell-specific SRS transmissions can vary from 2–10 per radio frame, depending on the UpPTS length and the UL-DL configuration. For FDD, there is only one symbol which can be scheduled for SRS in a normal subframe and semi-static configuration of SRS resources is used.

of cyclic shift resource and root sequence,* shorter sounding latency, and changing of sounding comb structure.

3.3.7.1.2 Sounding Performance Enhancement

The transmission power of LTE Rel-8 UE is limited to 23 dBm, and LTE-A and future UE total transmission power may be still limited to 23 dBm. If each antenna is limited to less power than 23 dBm, cell-edge UE will have worse sounding performance, so SRS needs precoding gain for better performance to realize intercell interference cancellation. On the other hand, Rel-8 SRS design cannot support sufficient intercell channel estimation due to nonorthogonality between SRS sequences for different cells.

3.3.7.1.3 Backward Compatibility

LTE-A SRSs should not conflict with or cause significant interference to existing periodic SRSs for legacy UE.

We can develop SRS design principles of LTE-A using these enhancements. The baseline for the SRS in LTE-A operation is nonprecoded and antenna specific. For multiplexing of the sounding reference signals, it is better to reuse Rel-8 SRS principles (CS separation, IFDM separation) with some possible modifications.

SRS should be configured per component carrier in case of carrier aggregation. When a UE is configured with two or more UL component carriers, the eNB needs to know the channel/interference conditions on the additional component carriers before scheduling on other UL component carriers. Such information can help the Medium Access Control (MAC) scheduler to decide if it is worthwhile to increase a user's throughput by aggregating additional component carriers.

SRS for all the configured transmit antenna ports should be transmitted within one single-carrier OFDM (SC-OFDM) symbol of the same subframe. Both cyclic shift and comb could be used for multiplexing different antennas and the SRS transmission bandwidth and frequency resource block positions are the same for all antenna ports. Actually, further discussion is needed to see if we can narrow down the configuration parameters and how to use time cycling to multiplex the SRS from antenna ports in practice.

In LTE-A, a new dynamic aperiodic SRS is supported for sounding capacity enhancement in addition to the periodic SRS that was used in LTE Rel-8. In aperiodic sounding, triggering is at least by PDCCH UL grants, and one-slot aperiodic

---

* In Rel-8, the SRS sequence is defined by the phase shift of the CAZAC root sequence and one root sequence per cell is used. The SRS multiplexing capability is limited by the number of phase shifts supported within the symbol interval with the tolerance of maximal delay spread. The SRS multiplexing capacity is set to 8 in Rel-8 for a CAZAC root sequence. Increasing the root sequence number in one cell in LTE-A is the simplest way to achieve more SRS capacity, but requires reasonable cell planning.

SRS transmission is supported. The number of transmission ports for aperiodic and periodic SRS resources are independently configurable if multiple antenna SRS is supported for periodic SRS.

3.3.7.2 Aperiodic Sounding

Aperiodic SRS is a solution for maximizing the SRS capacity and minimizing the SRS overhead because SRS is not transmitted when it is not needed. Aperiodic SRS can provide the eNB with the flexibility to dynamically schedule UE for one-slot SRS transmission on a need basis. It is proposed to allow dynamic physical layer activation/deactivation of SRS transmission from multiple antennas/component carriers to allow timely response to changing channel and traffic conditions. In order to minimize the overhead, the SRS enhancements need to work on top of existing SRS resources and introduce aperiodic SRS. Aperiodic SRS and periodic SRS can be jointly used and the eNB can capture the channel information with minimal delay.

In LTE-A, aperiodic SRS transmission will reuse the Rel-8 (time/frequency/code) SRS resources and transmit in cell-specific SRS subframes. Cell-specific SRS configuration parameters are applicable to both periodic and aperiodic sounding, but UE-specific SRS configuration parameters, such as SRS bandwidth, starting position, transmission comb, and cyclic shift, could be different between periodic and aperiodic sounding. These parameters for periodic SRS are radio resource control (RRC) configured and the SRS transmissions are semipersistent. For aperiodic SRS, it is first configured via RRC signaling, similar to periodic SRS. There can be separate SRS resources reserved for periodic and aperiodic SRS within the existing SRS resources. Aperiodic SRS is transmitted only after it is triggered. A UE will commence aperiodic SRS transmission in subframe $n + k$ ($k \geq 4$) upon detection of a positive SRS request in subframe n, which is shown in Figure 3.52.

3.3.7.2.1 Aperiodic Sounding Duration

Currently, one-slot SRS is supported in LTE-A, and multislot aperiodic SRS (A-SRS) transmission is not supported in LTE-A. Similar to aperiodic CQI in LTE

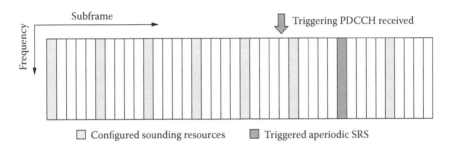

Figure 3.52   Aperiodic SRS.

Figure 3.53 Collision avoidance using cyclic shift.

Rel-8, users will be triggered to transmit A-SRS, and DCI format 4 and 0 can both be used for A-SRS triggering.* A-SRS is always transmitted on the same component carrier as the scheduled PUSCH. For DL assignment triggering (DCI 1a, 2b, 2c), A-SRS is transmitted on the UL carrier, which is system information block (SIB)-2 linked to the DL carrier in which the PDSCH is transmitted. For each type of triggering, independent RRC-configured parameters are used.

- DCI Format 4: Two new bits for an SRS request indicate 3 sets of RRC-configured A-SRS transmission parameters; a state of "1" indicates no A-SRS activation. The possible parameters are transmission comb, cyclic shift, bandwidth, frequency domain position, hopping bandwidth, duration (if multislot SRS is supported), and number of antenna ports. The parameters of cyclic shift, comb, or frequency domain position could be used to avoid collision. One example of using cyclic shift to avoid collision is most straightforward and is shown in the Figure 3.53.
- DCI Format 0: One new bit (RRC-configured) indicates A-SRS activation. A-SRS activation is not supported in UE common search space.
- A UE may be configured with SRS parameters for trigger type 0 and trigger type 1 on each serving cell. The following SRS parameters are serving cell specific and semistatically configurable by higher layers for trigger type 0 and for trigger type 1: transmission comb, starting physical resource block assignment, duration, SRS bandwidth, frequency hopping bandwidth, cyclic shift, number of antenna ports, and so on.

The set of antenna ports used for sounding reference signal transmission is configured independently for periodic and each configuration of aperiodic sounding. A UE configured for SRS transmission on multiple antenna ports of a serving cell shall transmit SRSs for all the configured transmit antenna ports within one single-carrier frequency-division multiple access (SC-FDMA) symbol of the same subframe of the serving cell. The SRS transmission bandwidth and starting physical resource block assignment are the same for all the configured antenna ports of a given serving cell. Figure 3.54 is an example where the UE transmits an aperiodic SRS for each transmit antenna simultaneously. The transmission timing is determined by the reception timing of the UL grant.

---

* Triggering of aperiodic sounding: A UE shall transmit a sounding reference symbol on per serving cell SRS resources based on two trigger types: Trigger type 0: higher-layer signaling, and Trigger type 1: DCI formats 0/4/1A for FDD and TDD and DCI formats 2B/2C for TDD.

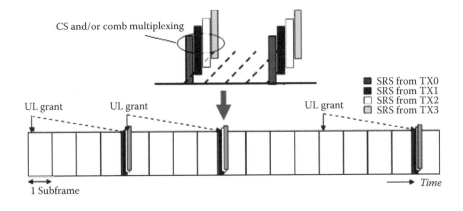

Figure 3.54 Simultaneous transmission from all transmit antennas.

The set of {2, 5, 10} ms is supported for UE-specific A-SRS periodicity. UE is not expected to receive multiple triggering for the same A-SRS transmission opportunity.

3.3.7.3 Multicell SRS Detection

In LTE Rel-8, intra-eNB SRS coordination is supported by coordinating the usage of the SRS resources, such as cell-specific SRS subframes, SRS sounding region, comb, and so on. UE can be assigned an orthogonal set of resources with the SRS resources of the coordinating UE not assigned to other UE in the cell. As the Rel-8 SRS design is not sufficient to avoid intercell interference to support the CoMP that is shown in Figure 3.55, it is essential to coordinate the SRS configuration among cells in order to support CoMP deployment in LTE-A. Multicell SRS coordination would configure the SRS sequences orthogonally within a cluster of cells. The coordination of SRS configuration would minimize the SRS cross-interference and thus provide more reliable channel state estimation. Figure 3.56 gives an example of multicell SRS detection; the SRS detection in the nonserving cell is interfered with by the UL signals in that cell. Usually, the detection performance of SRS from CoMP UE is poor at cooperative points due to cross-correlation between SRS sequences from different cells.

The orthogonality between CoMP UE and LTE (non-CoMP) UE can be achieved by coordinating SRS resources including the TDM, FDM, and CDM methods. Some considerations for reliable multicell SRS detection are similar to UL DMRS for CoMP.

- The solution of TDM/FDM between CoMP UE and non-CoMP UE with SRS scheduling information coordination will avoid the SRS of a CoMP UE being interfered with by data or SRS in the cooperative points. This method has no standard impact, but creates reduced resource utilization efficiency.
- In the TDM/FDM region, allocating cluster-specific sequence shift patterns for CoMP UE within the CoMP cell cluster has higher resource utilization

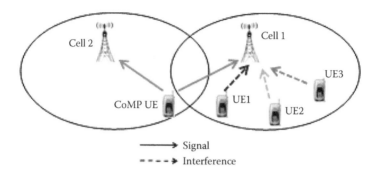

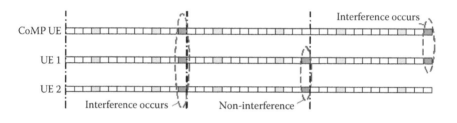

Figure 3.55 SRS intercell interference.

Figure 3.56 SRS transmission in a multicell SRS detection scenario.

efficiency to enable the same SRS bandwidth with different cyclic shifts to be allocated to UE from different cells.

■ Another alternative to multicell SRS detection reliability is to improve SRS capacity by increasing available sounding resources, increasing cyclic shift/transmission combs, sounding via DMRS, and sounding via PUSCH PRB.

In general, with the introduction of uplink MIMO and other features, SRS capacity could be a bottleneck for system-level performance. More efficient sounding management, such as dynamic aperiodic sounding, can create even more detection reliability. Improved SRS capacity and multi-cell coordination could be used together for both single cell and multi-cell SRS detection.

3.4 CoMP Feedback

In Chapter 1 we discussed the feedback and backhaul requirements of joint processing (JP) and coordinated scheduling/coordinated beamforming (CS/CBF) schemes, including the fact that coordinated beam switching (CBS) requires feedback overhead that is similar to Rel-8 CQI and requires very limited and infrequent exchange over X2 interface. JP requires the same feedback as CBF,

Table 3.10 Requirement Comparison of Different CoMP Feedback

Scheme	Description	Feedback
CBS	None	CQI of serving cell
CBF	Network coordination CBF	CQI; CSI of multiple cells
	Intrasite CBF	CQI; CSI of multiple cells
JP	Intrasite coherent JP	CQI; CSI of multiple cells

including precoder, CQI/CSI, and so on. Table 3.10 gives a simple analysis of the requirements of different CoMP feedback.

Regardless of how the actual downlink coordinated transmission is carried out, the network will need some information related to the downlink channel conditions measured by the UE. In LTE Rel-8, there are 5 feedback types or formats to support downlink SU-MIMO, and PMI/CQI/RI implicit feedback is supported using PUCCH-based periodic reporting and PUSCH-based aperiodic reporting. However, a direct extension of the traditional single-cell PMI/CQI/RI reporting alone may not be sufficient to support CoMP. One of the most important aspects of LTE-A CoMP operation is the feedback design, which provides the eNB with adequate spatial information and should also incur only low overhead in terms of the proportion of system resources and UE power consumption. Generally speaking, feedback design should target the most competitive CoMP scheme that is capable of providing significant cell average and cell-edge throughput gains with reasonable feedback overhead and certain backhaul requirements. Enhanced feedback could be optimized as much as possible to reuse these channels, even though the existing feedback channels may be modified to increase feedback capacity in the LTE-A specification. Since multiple cells participate in the transmission, the information needed for closed loop operation at the network side increases linearly with the number of the cooperating cells. For an FDD system, the information is mainly obtained by UE feedback, which will be a heavy burden for the uplink channel. The CoMP measurement set is defined as the set of cells about which channel state information related to their link to the UE is fed back. The CoMP measurement set may be the same as the CoMP cooperating set, and the actual UE reports may down-select reported cells for which actual feedback information is transmitted. Due to the use of LTE-A UE-specific DMRS, the CoMP UE is capable of receiving and demodulating PDSCH that is associated to only a single serving cell. Therefore, it is possible for some CoMP and non-CoMP single-cell transmission schemes to share the same feedback mechanism and enable the eNB to decide the actual transmission scheme. With this case, the non-CoMP single-cell feedback is just a special case of CoMP feedback when the reporting set shrinks to a single cell.

In an LTE-A system, the three main categories of CoMP feedback mechanisms are as follows:

■ Explicit channel state and statistical information feedback: The channel as observed by the receiver, without assuming any transmission or receiver processing; that is, UE directly measures the Nt x Nr downlink channel and/or interference matrices and reports the instantaneous or statistical information.

■ Implicit channel state and statistical information feedback: Feedback mechanisms that use hypotheses of different transmission and/or reception processing, for example, CQI, PMI, and RI. In Rel-8, the PUCCH and PUSCH are used for transmitting periodic and aperiodic PMI/CQI/RI information, which are forms of implicit channel information feedback, respectively.

■ For Time-Division Long-Term Evolution (TD-LTE), UE transmission of SRSs can be used for CSI estimation at the eNB to exploit channel reciprocity. The SRS for channel sounding is mainly used for uplink adaptation and MIMO precoding control in FDD, but it can also be used for adapting the DL CoMP transmission in TDD. Although explicit feedback can provide accurate information concerning the channel transfer function, it significantly increases the overhead in the UL. Thus, to minimize the overhead, SRS feedback can be applied.

3.4.1 Explicit Feedback

Explicit feedback mainly includes direct channel coefficient feedback, spatial channel covariance matrix feedback, and principal eigenvector feedback. Accurate channel information is fed back to the eNB, which is beneficial for the eNB to choose a more accurate precoding matrix for the scheduled UE to eliminate interuser interferences. But the feedback burden is huge and the modulation and coding scheme (MCS) level is not easily determined accurately because of the lack of receiver processing information. On the contrary, in implicit feedback the PMI/CQI/RI information is fed back to the eNB, so it has relatively small feedback overhead. However, if there is a co-scheduled UE and the UE cannot know the precoding matrix, CQI mismatch problems will exist.

For example, first, the UE may measure the downlink channel (H_i, where i being the index of the cell in the CoMP reporting set) based on the CSI-RS of the corresponding cell. The UE then reports the set of matrices $\{H_1, H_2, \ldots H_N\}$. This scheme could apply equally to both joint processing and coordinated beamforming. Alternatively, for coordinated beamforming, the UE may report a downlink channel associated with the serving cell $\{H_1\}$ and aggregated interference $H_2 + \ldots + H_N$, which may be used to derive the recommended PMIs for the interfering cells.

Actually, explicit feedback of the channel matrix (H) can represent the full CSI and is applicable for all operation modes including coherent/noncoherent JP and CS/CBF. The kth user received signal is represented by $y_k = H_k W_x + n_k$. The n_k is

k^{th} user's additive Gaussian noise with each element having a variance equal to σ^2. The **W** indicates the precoding matrix and x indicates the transmit signal. In CoMP, the size of H can be large and it is not practical to instantly feed back H.

Explicit feedback of channel covariance $(R = \mathbf{H}^H\mathbf{H})$ is usually used for noncoherent JP and CS/CBF. The transmit covariance matrix of the channel **H** is averaged over both time and frequency as $R = (\text{sum } \{\mathbf{H}n^H\mathbf{H}n\})/M$, $n = 0, 1, 2, \ldots, M-1$, where M is the span of frequency sub-bands and subframes over which averaging is performed. Averaging in the frequency domain can be on a wideband or sub-band basis, over one subframe (or a longer period), and R is an instantaneous correlation estimated based on an instantaneous channel estimated from CSI-RS in a subframe. If accumulated over a longer period of time, it eventually converges to statistical correlation. The correlation matrix can be deemed as a compressed or averaged "channel" from a set of channel response matrices. A wideband covariance matrix reduces the impact of channel estimation errors, but it comes at the cost of increased mismatch with the frequency-selective channel and reduced frequency scheduling gain. The covariance information is easier to estimate for the receiver than the full channel information and a better estimate can be obtained due to longer estimation intervals. CSI consists of short-term and long-term information, which require different feedback periodicities. The spatial channel covariance matrix or its largest eigenvalue and corresponding eigenvectors can represent long-term spatial feedback. Spatial covariance contains more subspace information but also will incur higher feedback overhead as expected.

Explicit feedback of the channel eigenvector matrix (from singular value decomposition of H) is also used for noncoherent JP and CS/CBF. The singular value decomposition (SVD) of the channel matrix is represented by $\mathbf{H} = \mathbf{U}\Sigma\mathbf{V}^H$ and is shown in Figure 3.57. **V** denotes the right singular vectors of channel ($V = [V_1\ V_2\ V_3\ V_4]$, where V_1, V_2, V_3, and V_4 are the four element eigenvectors) and the eNB obtains V by PMI and additional analog feedback. We can use the SVD of the channel matrix H to separate the channel into N orthogonal "pipes" depending on channel rank; each eigen channel can use a different MCS depending on the SNIR. Finally, we can get output Y according to this equation:

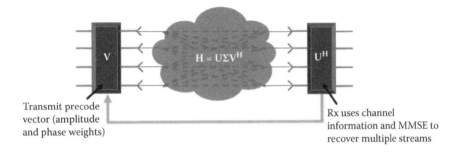

Transmit precode vector (amplitude and phase weights)

Rx uses channel information and MMSE to recover multiple streams

Figure 3.57   SVD of the channel matrix.

$$y = U^H HVx + U^H n = U^H U \Sigma V^H Vx + U^H n, \quad D^2 = \begin{bmatrix} \lambda_1 & 0 & 0 & 0 \\ 0 & \lambda_2 & 0 & 0 \\ 0 & 0 & \lambda_3 & 0 \\ 0 & 0 & 0 & \lambda_4 \end{bmatrix}$$

$$= Dx + U^H n$$

where n is the noise vector, D is a diagonal matrix so that the elements of x are separated at the receiver, and $\lambda_1, \lambda_2, \lambda_3,$ and λ_4 are the eigenvalues.

3.4.2 Implicit Feedback

Implicit channel feedback reflects the recommended RI/PMI/CQI, which is aligned with the feedback paradigm as in Rel-8. The UE reports the recommended MIMO transmission format based on certain knowledge of the CoMP schemes. The CSI-RS from the CoMP reporting set is used to derive the implicit channel information, which may be derived based on the UE's assumption of the downlink CoMP scheme. For JP transmission mode, the UE reports the recommended RI/PMI for the reporting set, assuming coherent or noncoherent combining. For coordinated beamforming transmission mode, the UE reports the recommended RI/PMI for the serving cell, as well as PMIs for the nonserving cells in the reporting set. The PMIs for the nonserving cells are optimized along with the serving-cell PMI to reduce the co-channel interference and improve the cell coverage and throughput.

Implicit CSI feedback is based on a predefined set of precoding codebooks, which are a set of matrices calculated offline and known at the eNB and UE. Codebook-based feedback is a good approach to reduce feedback overhead, similar to PMI in Rel-8, including multicell CQI/PMI/RI and single-cell CQI/PMI/RI. Different PMI selection criteria for implicit feedback could, for example, be maximum likelihood, minimum singular value, mean squared error, and capacity. To some extent, PMI is a coarse approximation of covariance and is a very effective compression technique for SU-MIMO. This is a short-term CSI for instantaneous information that could be used to reconstruct more efficient channel information for multicell scheduling. It is worth mentioning that Rel-8 PMI can be deemed as an approximation of the dominant eigenvector(s) and covariance matrix, but LTE-A UE cannot simply follow the Rel-8 SU-MIMO feedback framework and reports implicit SU-MIMO PMI/CQI for much higher throughput. Multicell CQI/PMI/RI* and single-cell CQI/PMI/RI are defined as follows:

---

* **Rank indicator (RI):** A preferred transmission rank (number of data streams).

 Precoding matrix indictor (PMI): The index of the UE recommended precoding matrix in the rank-r codebook. For LTE Rel-8, a single PMI is reported for each frequency sub-band, corresponding to the RI report.

 Channel quality indicator (CQI): Quality of the channel (e.g., in the form of supportable data rate, SNR). The reported CQI is associated with the reported PMI.

- **Multicell CQI/PMI/RI:** One CQI for each of the selected subsets based on that given multicell hypothesis.
- **Single-cell CQI/PMI/RI:** One or several CQIs are reported based on one or multiple hypotheses of the transmission coordination.

Implicit feedback has the advantage of good backward compatibility and minimizes the impact on the LTE-A specifications. The drawback of codebook-based implicit feedback may result in the quantization error of channel information.

3.4.3 *Difference between Explicit and Implicit Feedback*

The main difference between explicit feedback and implicit feedback is whether the UE receiving process and capability are considered in feedback calculation, that is, whether to use hypotheses of different transmission or reception processing. It is obvious that the overhead of explicit feedback is overall larger than implicit feedback, and typically grows linearly with the number of transmit antennas and the number of receive antennas. On the other hand, implicit CSI feedback can provide reasonable performance gain with acceptable feedback overhead under a single-cell scenario in various transmission modes.

In practice, there are several distinct ways of feeding back the CSI: integrated, individual, and mixed feedback (i.e., combinations of full or a subset of these three are possible). On the other hand, unlike in the single-cell SU-/MU-MIMO operation, multiple PMI feedbacks (describing different channels linking to the reporting UE) are essential to the success of DL CoMP. For both DL CoMP CS/CBF and DL CoMP JP, the network needs to know the channel information related to the links from the cells in the CoMP measurement set to the reporting UE.

In order to support feedback of channel quality, choice of frequency bands, or channel spatial information (e.g., codebook) for up to 8 layers, additional reference signals will be broadcasted from the eNB. These channel state information reference signals will not be used for demodulation and will be designed to be sparse in time and frequency with an overhead of 1% or less. The CSI-RS design will support up to 8 transmit antennas and will potentially enable a UE at the cell edge to measure CSI-RSs transmitted from adjacent cells for CoMP support. The power, density, and placement of CSI-RS will have to be designed carefully as it may interfere with the data transmissions of a Rel-8 or Rel-9 user in the same subframe.

Because the explicit state information varies at a faster rate than implicit information (e.g., RI/PMI), which is a relatively longer-term reflection of the channel (e.g., overall geometry, channel correlation, ability to support multiple streams, and beam direction), the sensitivity of explicit state information to air interface/X2 delay and UE speed should be borne in mind. If channel feedback cannot fit in a single reporting subframe, multiple reporting subframes are needed and may further add to the sensitivity to channel variation. In general, explicit channel feedback is able to provide the highest possible CoMP gain theoretically, since the eNB can design the CoMP coordination at the utmost flexibility. This should be treated as the CoMP performance upper bound, given the availability of reliable/fast channel feedback and X2 interface.

In order to enhance Rel-8 PMI-based implicit feedback for MIMO/CoMP, we also considered explicit feedback as an enhancement to LTE Rel-8 implicit feedback.

3.4.4 Consideration of Feedback

For the CoMP schemes that require feedback, individual per-cell feedback is considered as the baseline. Complementary intercell feedback might also be needed. This implies that the feedback contents for multiple cells are in a per-cell basis (multicell codebook is excluded), although the UE reports to the serving cell only, while the CSI information to other cells is shared among multiple cells by the serving cell. The CBF/CS scheme may need only individual per-cell CSI, while the intercell feedback is mainly used for the coherent JP scheme.

For example, the serving base station for UE1 is eNB1 and the serving base station for UE2 is eNB2. Let N_T be the number of transmit antennas at the eNBs, and N_R be the number of receive antennas at the mobile users. Also let H_{11}, H_{12}, H_{21}, and H_{22} be the respective channel gains. Then the received signal at UE1 and UE2 can be represented by:

$$Y_1 = H_{11}X_1 + H_{21}X_2 + N_1$$

$$Y_2 = H_{12}X_1 + H_{22}X_2 + N_2$$

where Y_i is the $N_R \times 1$ vector of received signal at mobile user i, X_i is the $N_T \times 1$ vector of transmitted signal at base station i, and N_i is the $N_R \times 1$ additive white Gaussian noise (AWGN) vector, which is shown in Figure 3.58.

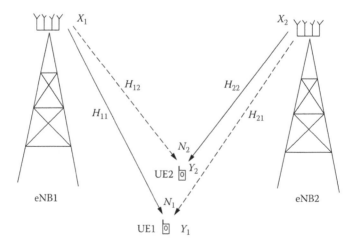

Figure 3.58 **Two eNBs communicate with two UEs simultaneously.**

In this model, we have four channel gain matrices, which are, H_{11}, H_{12}, H_{21}, and H_{22}. Furthermore, we have two sets of data: one set is intended for UE1 while the other set is intended for UE2. In LTE Rel-8 there is no channel knowledge sharing between the two eNBs; each eNB knows only the information related to the channel between itself and its serving UE. That is, eNB1 knows only information related to H_{11} while eNB2 knows only information related to H_{22}. In this way, we can apply intercell interference coordination to mitigate interference. The basic idea of this technology is to schedule cell-edge users in different frequency bands so that the information is mitigated. Now we can have two levels of information sharing of CoMP feedback—partial channel knowledge sharing and total channel knowledge sharing—between the two eNBs. In the case of total channel knowledge sharing, the information related to H_{11}, H_{12}, H_{21}, and H_{22} is shared between the two eNBs. In the case of partial channel knowledge sharing, a subset of the information related to H_{11}, H_{12}, H_{21}, and H_{22} is shared between the two eNBs. For example, eNB1 may know only information related to H_{11} and H_{12} while eNB2 knows only the information related to H_{21} and H_{22}. Also, if we consider the JP scheme, the content of channel knowledge sharing should include user data as well.

In a total channel knowledge sharing scenario (e.g., a multicell MIMO or JP scheme), the precoding matrices are jointly designed such that the intended signals are orthogonal to the interference signals. In a partial channel knowledge sharing scenario (e.g., coordinated beamforming), UEs send feedback information about the least interfering PMI of the interfering eNB and suggest that the interfering eNB use it. In this sense, the interference can be mitigated by limited coordination. Another similar method is the PMI coordination technique; a set of PMIs, instead of the least-interference PMI for coordinated beamforming, is recommended for use in the interfering eNBs. With the increased overhead, PMI coordination improves the scheduling flexibility and results in higher overall network throughput. The potential system architecture of coordinated beamforming, where three interfering sectors are coordinated, is depicted in Figure 3.59. There is a master scheduler (which may be located within an eNB) that locally schedules multiple cells together relying on the high-bandwidth low-latency backhaul.

Note: The other beam coordination scheme is called *null-forming by zero forcing*, which means that a neighboring cell tunes its transmit beam in the null direction of the reported PMI at the reported sub-band.

For multicell coordination, the master scheduler may be defined in a virtual eNB, which may be located at one of cooperating eNBs, and has control over the behavior of multiple eNBs involved in the joint transmission. As the UE moves around cells, a set of cells is dynamically chosen for coordination from among cells near the UE, which can be done based on the UE measurement reports. A set of eNBs is controlled by a master eNB that shares necessary information, such as data and scheduling information, to jointly schedule and configure the transmission for

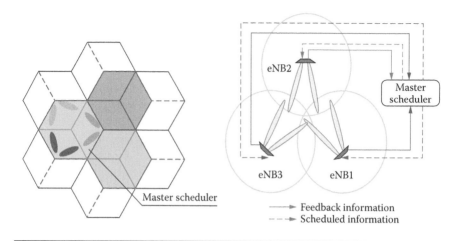

Figure 3.59 **System architecture of coordinated beamforming.**

the UE. Thus, a master eNB covers a geographical area where the joint transmission takes place under the control of its associated eNBs.

If the master cell and its coordinated cells belong to the same eNB, there are no problems such as delay and overhead associated with intra-eNB communication. In contrast, if signaling occurs between cells belonging to different eNBs, it is important to minimize signaling delay and overhead. The master cells send out coordination requests to the dominant interferers, which contain PMI recommendation or restriction information for each PRB. Upon receiving the coordination request, the coordinated cells try to follow the requests as much as possible.

In Figure 3.59, the UE calculates and reports CQI/PMI/RI in the uplink based on the measurement of reference signals from the active cells. The master scheduler collects channel quality information from the active eNBs, makes a decision on scheduling of the UE, and transfers the scheduling information to the active eNBs. The active cells then transmit data simultaneously to the UE in accordance with the scheduling information.

3.4.5 Feedback of CSI

Compared to Rel-8, there are a number of factors to modify the CSI feedback in the design of LTE-A. For example, the LTE-A CoMP and MIMO scheme is different from Rel-8 MIMO with respect to the eNB's processing. Because the dimensionality of the channel state is greatly increased compared to Rel-8, it becomes important to utilize the underlying MIMO channel structure and long-term characteristics in the codebook and feedback design. In Rel-8, the CQI reported by the UE is calculated on the hypothesis of the transmission mode and the precoding vector is based on the single UE reporting. In LTE-A, the UE pairing is the key operation and the precoding vector of each UE is based on the multiple UEs reporting. Based on the reported

CSI, the eNB can reduce the interference to other users by precoding. In addition, different DL CoMP transmission schemes require different CSI feedback contents. For CS/CBF, only per-cell CSI feedback is needed. For JT with global precoding (global transmit spatial correlation matrix), both the per-cell CSI feedback and the intercell CSI feedback (including relative phase and amplitude) will be needed to enable multicell global CoMP precoding. For JT with local CoMP precoding, the per-cell CSI feedback is required. Based on the reported CSI, coordinated eNBs can enhance the UE's signal strength by joint processing transmitted data.

3.4.5.1 CQI Feedback

The LTE Rel-8 CQI report (given per codeword), is based on measurement of the CRS, which can be configured in three bandwidth modes: wideband, UE selected sub-band, and higher layer–configured sub-band. Each CQI* report relates to a specific reference resource in time and frequency to assist the eNB in selecting an appropriate modulation and coding scheme (MCS) to use for the downlink transmission. For LTE-A, CSI-RS-based CQI is also being considered by the eNB. The measured CQI is designated to assist the eNB in selecting the appropriate MCS to use for the downlink transmission and also to determine the aggregation level and the transmission power of PDCCH. In LTE-A CoMP scenarios, there are two distinct ways of feeding back the CQI in the measurement set: individual feedback and integrated feedback.

Individual feedback means that the CQI of each cell is individually fed back. Individual feedback requires more feedback overhead. However, it allows the eNB to schedule cells with good channel quality for the UE's data transmission.

Integrated feedback means that an integrated CQI of all the cells in the set is fed back. Integrated CQI can better reflect the channel quality of the joint processing. However, this feedback approach will impose some restrictions on scheduling, since it is difficult to recover the individual CQI of each cell from the integrated CQI. The feedback overhead is reduced due to integration.

In a live network, for single-layer joint processing, in order to reduce the complexity of specification and overhead of control signaling, reports of CQI/PMI should be constrained to a semistatic cell set. In this semistatic approach, the network informs the UE of the cell set for CQI/PMI reporting. The cell set is UE-specific.

3.4.5.2 PMI Feedback

The PMI feedback can be seen as a simplified method of channel quantization algorithm. For different transmission techniques, the precoding matrix of each

---

* The reported CQI is not a direct indication of the SINR; instead, it indicates the highest modulation scheme (quadrature phase-shift keying [QPSK], 16-QAM, 64-QAM) and channel coding rate value that allows the UE to decode received DL data with a transport block error rate not higher than 10%. Thus, the CQI information of the UE takes into account not only the prevailing radio channel quality but also the characteristics of the UE's receiver.

Table 3.11 Summary of PMI Feedback Schemes

	No PMI	*Single PMI*	*Multiple PMIs*
Feedback overhead	No	Low	High
Application scenario	TDD; Angles of arrival (AoA)-based transmission scheme; open loop transmission	Global precoding; Single-frequency network (SFN) precoding with same precoder	Both joint processing and coordinated scheduling
CQI feedback	Individual; integrated; mixed	Integrated	Integrated; individual; mixed

cell could be calculated either on a per-cell basis or as a whole, when the UE experiences severe interference from neighbor cells and needs to report interfering beam information. The beam information can be single entity or multiple entities that represent the best PMI with the least interference, or the worst PMI with largest interference, or the quantized effective channel vector. In TDD configuration, the eNB could acquire instantaneous downlink channel state information via uplink measurement, and the downlink precoding matrix can be obtained based on the information without feedback. For single PMI, global precoding transmit antennas of cooperating cells should be treated as a whole. The precoding matrix is selected from the "large" codebook, and thus a single PMI is fed back. This is efficient in terms of feedback overhead and performance. For multiple PMIs, the precoding matrices are required on a per-cell basis, so the UE needs to feed back a PMI for each cell individually. The codebook of single-cell transmission can be reused. The feedback overhead increases linearly with the number of cooperating cells. The summary of PMI feedback schemes can be found in Table 3.11.

The other important issue of CSI feedback is feedback timing. In Rel-8/9 LTE, UE can measure the downlink channel status on every subframe using CRS, and thus the channel feedback delays for UE have the same feedback periodicity, even though the UE may have different feedback timing offsets. In LTE-A, however, for a given CSI-RS duty cycle, the channel feedback delays for UE with different feedback timing offsets can be different since the time interval between the channel measurement and its feedback varies according to the feedback timing offset, which will induce different performance.

3.4.6 PMI and RI Limitation and Coordination

When LTE and LTE-A UE are located at the cell edge and scheduled within the same cell-edge band, both LTE and LTE-A UE at the cell edge are protected by the intercell interference coordination (ICIC)* scheme based on intercell relative narrowband transmit power (RNTP) signaling, and LTE-A CoMP UE is further protected by additional PMI and RI limitation and coordination.

To support PMI limitation and coordination, the LTE-A CoMP UE should report PMI restriction and recommendation information to its serving cell too. Through multicell coordination, the neighboring cell is requested either to use the recommended precoder or not to use the restricted precoder. PMI recommendation is more effective than PMI limitation in suppressing interference. In order to avoid excessive feedback overhead, the PMI information can usually be limited to one or two strongly interfering cells.

To support PMI coordination, the LTE-A UE is required to feed back the following information to its serving cell: CoMP UEs report PMI restriction/ recommendation information to their serving cell. A typical example of such feedback includes the preferred PMI indices for each frequency sub-band:

$$(\mathbf{w}_{b_i,ik}, \mathbf{w}_{b_j,ik}) = \arg\max_{\mathbf{w}_i, \mathbf{w}_j} \log\left(1 + \frac{\left\|\mathbf{H}_{b_i,ik}\mathbf{w}_{b_i,ik}\right\|_2^2 P_i}{N + \left\|\mathbf{H}_{b_j,ik}\mathbf{w}_{b_j,ik}\right\|_2^2 P_j}\right)$$

$$= \arg\max_{\mathbf{w}_i, \mathbf{w}_j} \left\|\mathbf{H}_{b_i,ik}\mathbf{w}_{b_i,ik}\right\|_2^2 P_i - \left\|\mathbf{H}_{b_j,ik}\mathbf{w}_{b_j,ik}\right\|_2^2 P_j$$

where $\mathbf{w}_{b_i,ik}$ is the preferred precoding index for serving cell b_i to be used for UE k, and $\mathbf{w}_{b_j,ik}$ is the recommended precoding index for neighboring cell b_j. Through multicell coordination, UE could report not only its own PMI, but also PMIs that would cause it minimum interference, and the neighboring cell is requested either to use the recommended precoder or not to use the restricted precoder. PMI recommendation is more effective than PMI restriction in suppressing interference. Figure 3.60 gives a simple explanation of PMI coordination.

The CQI feedback should also reflect PMI coordination. When estimating CQI, the UE assumes that PMI coordination is done as it has been reported, or the UE is informed previously by the serving cell about the actual PMIs used by the strong interference cells.

Rank coordination is similar with PMI limitation and coordination. Therefore, the UE feeds back the interfering rank that is the most harmful or beneficial to itself. For UE associated with the serving eNB, the terminal transmits its preferred RI to the eNB and also transmits to the serving eNB the rank of the interfering RI or interfering eNB RI that minimizes or maximizes its performance. If the interfering eNB RI minimizes the performance, the UE requests the interfering eNBs to restrict the use of that transmission rank or recommends that the interfering eNBs transmit the number of layers corresponding to the interference RI.

---

* Refer to Chapter 8.

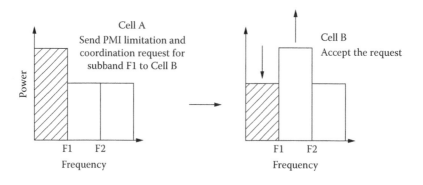

Figure 3.60   Inter-eNB signaling for PMI coordination.

The neighboring eNBs will exchange all the collected rank information from multiple UEs to coordinate their limitation of the interference RIs from all UE.

In a live network, one approach for avoiding interference is separation of LTE and LTE-A UE into different time frequency resource regions and application of interference coordination. In another approach, both LTE and LTE-A UE, when they are located at the cell edge, are scheduled within the same cell-edge band. LTE and LTE-A UE instances at the cell edge are protected by the ICIC scheme based on intercell RNTP signaling, and LTE-A CoMP UE is further protected by additional PMI and RI limitation and coordination.

3.5 DL CoMP

In the general sense, downlink CoMP implies dynamic coordination between downlink transmissions from multiple geographically separated transmission points. Examples of such coordination include dynamic coordination in scheduling between transmission points (e.g., allowing for dynamic interference coordination), joint transmission to UE from multiple transmission points with dynamic selection of the transmission weights, and fast switching of the transmission to the UE between the transmission points (in a sense, a special case of joint transmission). In principle, the best serving set of users will be selected so that the transmitter beams are constructed to reduce the interference to other neighboring users, while increasing the served users' signal strength.

In the category of joint processing, data to single UE are simultaneously transmitted from multiple transmission points to improve the received signal quality and cancel interference for other UE. Data intended for a particular UE instance is jointly processed at different cells. As a result of the joint processing, the received signals at the intended UE will be coherently or noncoherently combined. In CS/CBF, data to a single UE are transmitted from one of the transmission points while the scheduling decisions are coordinated to control the interference generated in a set of coordinated cells. Both downlink CoMP categories are shown in Figure 3.61.

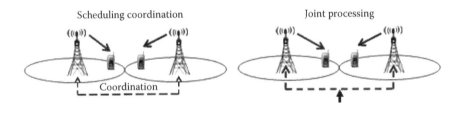

Figure 3.61 DL CoMP categories.

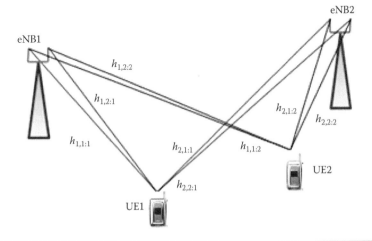

Figure 3.62 Generic model of a two-site CoMP.

Actually, DL CoMP processing can be formed in three steps. Step 1: Identify victim UE and cooperating eNBs based on some threshold and by serving eNB and feedback from UE. Step 2: The UE estimates power/channel from interfering eNBs and sends feedback from the UE to the serving eNB. Step 3: Scheduling by the network. Because transmit parameters (beam, MCS, UE) depend on the transmit parameters of a cooperating eNB, all cooperating eNBs for all UE in a single cell will need to be scheduled jointly. By coordination between downlink transmission multipoints, each point will choose a suitable set of transmit parameters to guarantee the desired power in itself and the minimum interference to neighbor cells.

3.5.1 DL CoMP Mathematical Model

Joint processing and coordinated scheduling/beamforming can be represented in a generic model (see Figure 3.62) where the cooperating set consists of two neighboring base stations: eNB1 and eNB2. Two UEs, UE1 and UE2, are in the CoMP serving area. For simplicity (without losing the generality), each UE shown in Figure 3.62 has only one receive antenna and each base station has two transmit antennas. We use

$h_{i,j:u}$ to denote the complex coefficient of the channel connecting the jth antenna of ith eNB and the uth UE, assuming single-path fast fading. Out of all combinations, there are a total of eight channel coefficients in this two-site CoMP.

In the case of joint transmission, data is available at both eNBs to achieve the network precoding gain for each UE. The transmit weight applied at the jth antenna of ith eNB for the uth UE is denoted as $w_{i,j:u}$. The optimum weight vector is chosen from the total candidate set $\mathbf{W}_4$ (4 indicates that each vector has four elements), so that the inner product of the weight vector and the channel vector for the same UE are maximized mathematically as

$$
\mathbf{W}_{u,\mathrm{JP}} =
\begin{bmatrix}
w_{1,1:u} \\
w_{1,2:u} \\
w_{2,1:u} \\
w_{2,2:u}
\end{bmatrix}_{\mathrm{JP}}
= \underset{\mathbf{w}_{u,\mathrm{JP}} \in \mathbf{W}_4}{\arg\max} \sum_{i,j=1}^{2} w_{i,j:u} h_{i,j:u}
$$

For multiuser joint transmission, we also want to pair UEs to minimize the co-channel interference as follows:

$$
[u1,u2]_{\mathrm{pair,JP}} = \underset{u1,u2 \in U}{\arg\min} \sum_{i,j=1}^{2} w_{i,j:u1} w_{i,j:u2}^{*}
$$

In the case of coordinated scheduling/beamforming, user data is available only at the serving cell and there is no precoding gain across eNBs. Transmit weight itself depends only on the user in the serving cell and is chosen from the total candidate vector set $\mathbf{W}_2$ (2 indicates that each vector has two elements) as follows:

$$
\mathbf{w}_{u,\mathrm{CS/BF}} =
\begin{bmatrix}
w_{u,1} \\
w_{u,2}
\end{bmatrix}_{\mathrm{CS/BF}}
= \underset{\mathbf{w}_{u,\mathrm{CS/BF}} \in \mathbf{W}_2}{\arg\max} \sum_{j=1}^{2} w_{u,j} h_{u,j:u}
$$

Similar to multiuser joint transmission, UEs in the CoMP area sharing the same resource need to be carefully paired in order to reduce the other cell interference, as follows:

$$
[u1,u2]_{\mathrm{pair,CS/BF}} = \underset{u1,u2 \in U}{\arg\min} \sum_{j=1}^{2} w_{u1,j} w_{u2,j}^{*}
$$

Transmit weight is determined at the UE, which feeds back the PMI. User pairing is determined by eNB, that can be based on the correlation properties of the codebook to avoid strong coupling between power allocation, PMI, RI, and so on.

The difference between joint scheduling and joint beamforming lies in the spatial correlation assumption between the same eNB antennas. Joint scheduling assumes widely spaced vertical or cross-polarized antennas, whereas beamforming implies a highly correlated antenna array to actually form the physical beams.

3.5.2 Joint Processing

In the category of joint processing/reception, data to a single UE instance is simultaneously transmitted from multiple transmission points to improve the received signal quality and cancel interference from other UE. This technique is particularly beneficial for multi-transmit antenna cells because it enables spatial interference nulling as well as transmit channel gain combining across multiple cells. In this category, data intended for a particular UE is jointly processed at different cells and eNBs should exchange corresponding control and/or data information to support joint processing. After sharing the channel information among the eNBs, the serving and coordinated eNBs can calculate their individual precoding weight vectors/matrices to jointly serve some specific UE. Efficient joint processing relies on low-latency broadband backhaul to support data transfer to cells that jointly serve a UE, mechanisms to deliver transmit-side channel state information and HARQ feedback to the scheduler, as well as mechanisms to convey scheduling decisions, spatial precoding parameters, resource allocation, and MCS and HARQ information to the appropriate transmitters/receivers of the UE signal. These considerations make joint processing suitable for intercell or intra-eNB cooperation as well as cooperative transmission within a set of RRHs interconnected by high-speed broadband links.

Joint processing yields simultaneous transmission of packets to one or more UEs by multiple cells, thereby providing benefits of coherent intercell energy combining as well as interference nulling. Since multiple points transmit data signals to a specific UE, which may not belong to their serving cells, the related information (e.g., scheduling information, channel measurement information, data) should be known before the actual transmission. In terms of the manner of the combination of signals from multiple cells at the UE, joint processing can be classified as coherent transmission and noncoherent transmission. For noncoherent transmission, the precoding matrix of each cell is calculated on a per-cell basis. For coherent transmission, the cells should require knowledge of the channel phase difference between access points, and the PMI could be fed back either individually (phase correction + precoding) or as an aggregate (global precoding). Coherent transmission provides an efficient way of changing interference into received signals, while transmit diversity approaches only boost power without mitigating interference. This advanced technique is particularly beneficial for cell-edge throughput, and it is anticipated to be the dominating application of CoMP.

The key idea of coherent transmission of CoMP joint processing is to have transmitted signals from different cells coherently combined over the air. Under the class of coherent transmission, the network obtains channel state information of all the cooperating cell sites. By adjusting the phase of the transmitted signal according to the available CSI, the signal arriving at the intended UE could be combined coherently. In addition to the single-cell precoding gain and power gain, array gain and diversity gain can be attained by coherent transmission. First, the cooperating cell sites decide on a cell-edge UE based on the downlink average received power level within the predefined threshold from the maximum average received power. The cooperating sites then select the best precoding vector for the cell-edge UE so that the instantaneous received power after coherent combining of the signals from multiple cells with joint processing is maximized. Finally, the cell-edge UE measures the received SINR after coherent combining and feeds back the measured value to the network. So the benefits of CoMP may come from the aspect that power of some interference resources becomes useful signals. For a given UE performing intra-eNB joint transmission, the SINR of the UE is computed as follows:

$$SINR = \frac{\sum_{c=0}^{n} P_c}{\sum_{i} P_i + N}$$

where c and i denote the average received power of the sectors in the cooperating set and the interfering cells, respectively, and N is the noise power.

As an example, Figure 3.63 shows a basic model of JT CoMP where the data vector **d** is transmitted to a CoMP UE from two CoMP transmission points. In the model shown, $\mathbf{x}_i$, $\mathbf{w}_i$, and $\mathbf{H}_i$ represent the modulated signal, the precoder, and the DL channel between the UE of interest and the ith CoMP transmission point, respectively.

The JT CoMP transmission strategy is to assume that the data payload sent from each CoMP transmission point in the transmission time interval (TTI) is generated using the parameter redundancy version (RV) generated for the transport block (TB). The actual over-the-air transmitted signals from each CoMP transmission point shown in Figure 3.63 may be the same or different, i.e., x1 = x2 or x1 ≠ x2. Both CoMP strategies provide the opportunity for constructive combining of the

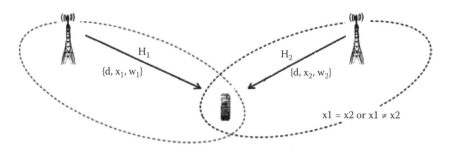

Figure 3.63   Joint transmission CoMP.

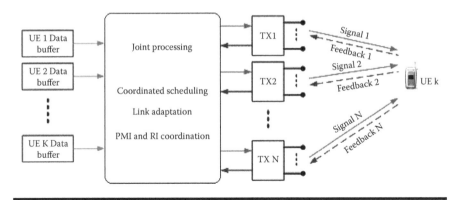

Figure 3.64   Illustration of downlink multicell coordination.

signals received by the CoMP UE and can further benefit from the possibility of improved coding gain through the reception of more parity bits.

In contrast, noncoherent transmission does not make use of the relationship of CSI among the cooperating cells, and therefore the signal arriving at the UE is unable to do coherent combining. In addition, the single-cell precoding gain and power gain, and extra diversity gain can be obtained by noncoherent transmission.

An illustration of downlink multicell coordination is given in Figure 3.64. The joint processing UE scheduling and link adaptation is performed in a super-cell controller, which generates the assigned transport block. The link adaptation includes MCS, rank adaptation, and precoding adaptation. Naturally, the cooperative link adaptation parameters and the generated TB are shared within the backhaul network and forwarded to the participating cells by the supercell controller. Each point receives the link adaptation parameters along with the TB and performs the channel encoding, rate matching, symbol mapping, layer mapping, and precoding based on the associated link adaptation parameters.

3.5.3 Coordinated Scheduling and Beamforming

The main concept of coordinated scheduling is to perform downlink beam sweeping with a predefined pattern in every cell, and multiple eNBs collaborate to mitigate intercell interference. The serving cell will choose an orthogonal beam or simply reduce transmit power on some resources in order to improve the SINR of the UE experiencing interference at the cell edge. JP differs from coordinated scheduling and beamforming by requiring both data and CSI sharing among cells in the serving eNB set. Compared to JP, the backhaul load of coordinated scheduling and beamforming is much lower since only channel info and scheduling decisions need to be shared among eNBs. This makes coordinated scheduling and beamforming a desirable solution in scenarios with limited backhaul capacity.

Beam patterns of different cells are synchronized across time/frequency resources and UEs periodically feed back channel quality that is seen on different resources and therefore corresponds to different combinations of serving and interfering beams. Each cell schedules UEs according to the channel and interference conditions, thereby achieving opportunistic beamforming and interference avoidance simultaneously.

For coordinated scheduling, each eNB begins by assigning the available resources in its own cell; it performs independent scheduling based on the available channel estimates and the current interference situation. Each eNB determines which UE should transmit on which PRBs and at which power level. Then, the generated resource allocation tables are exchanged between a certain set of cooperating cells via the X2 interface. Upon reception of resource allocation information from the other cooperating cells, the serving eNB is aware of the UEs that will transmit data in the cooperating cluster during the intended transmission interval of the UEs for which the link adaptation remains to be done. If the eNB has appropriate CSI from these interfering UEs, it can accurately predict the interference caused by them and use this to predict the interference level for the UEs scheduled in its own cell.

The coordinated scheduling scheme mainly relies upon availability of periodic/aperiodic channel quality feedback from a possibly large number of UEs and can be supported by UE PMI set reporting. PMI set reporting is a form of collaboration among multiple eNBs to mitigate ICI by restricting strongly interfering PMIs and using the best PMI in a restricted codebook subset in the neighboring cells. UEs feed back serving cell PMIs and strongly interfering PMI sets of neighboring cells to the serving cell. Then, neighboring cells restrict the strongest interfering PMI(s) and use the best PMI in a restricted codebook subset.

The following example involves two eNBs that systematically cycle through a fixed set of narrow beams while beamforming their traffic channels and their RSs (see Figure 3.65). The UE reports CQIs based on the beamformed RS. In this mode, each eNB would determine its own beam cycling pattern from either a predefined pattern or based on its load and user distribution, and then

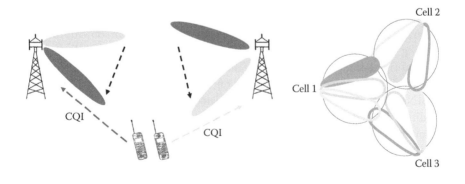

Figure 3.65   Multiple eNBs systematically cycle through a fixed set of beams.

communicate this information over the backhaul to a master controller. These cycling patterns can change, and it is expected that they will be a function of traffic distribution.

Coordinated beam scheduling can be treated as a special class of coordinated beamforming to some extent. In coordinated beam scheduling, the eNB in each cell selects a beam pattern independently of the out-of-cell interference due to the UE scheduled in other cells. Multicell negotiation is limited to the periodicity and order of the pattern, and the beam in each cell is purely a function of the cycling pattern. Usually, coordinated beam scheduling is suitable for low UE mobility and buffer full traffic better.

Coordinated beamforming is a single-cell beamformed transmission that takes into account not only the spatial downlink channel to the desired UE, but also cooperative interference reduction to a UE served by adjacent cells via proper beam selection. The purpose of coordinated beamforming is to manage interference by some means of beam cooperation among points. Coordinated beamforming implies coordination of scheduling decisions and transmit beam selection to reduce interference caused to UEs scheduled in the neighbor cells, while every UE receives data from a single serving cell. Coordinated beamforming benefits a relatively large population of UEs that have a small number of dominant interferers. A coordinated beam switching scheme to solve the flashlight effect was proposed, which requires the transmission points in a coordinated cluster to synchronously cycle through a fixed set of selected narrow beams, and the switching can happen in both time and frequency. Figure 3.66 provides an example of intra-eNB intercell coordination with two cooperating cells.

In Figure 3.66, S_1 and S_2 represent the received signal power from cells 1 and 2 belonging to the same eNB, respectively. $I + N$ denotes the total noise plus interference signal power from cells outside the coordination. In this case, the CQI for each cooperating cell is represented as

$$CQI_1 = \frac{S_1}{S_2 + I + N}, \; CQI_2 = \frac{S_2}{S_1 + I + N}$$

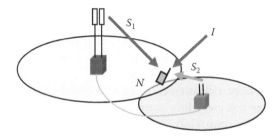

Figure 3.66 **Example of intra-eNB intercell coordination with two cooperating cells.**

In each CQI, the received signal power from the other cooperating cells is included as interference. Meanwhile, CQI after CS and CBF can be represented as

$$CQI_{CS} = \frac{w_1 S_1}{I + N}, \; CQI_{CBF} = \frac{w_1 S_1}{w_2 S_2 + I + N},$$

where w_1 denotes the beamforming gain for cell 1 estimated at the eNB. For CS, S_2 is removed by a muting operation at cell 2. For CBF, w_2 denotes the beamforming gain for cell 2 estimated at the eNB (e.g., $w_2 = 0$ if perfect zero-forcing is assumed). Such beamforming gain can be derived from the long-term covariance matrix, for instance.

In practice, CBS mode may be selected in a fully loaded cell, and CBF mode may be selected for a lightly loaded cell, particularly if the traffic is bursty. This will have the additional benefit of reducing the feedback overhead in the more heavily loaded cell by using the CBS mode.

Beamforming techniques based on cooperation between cells can be distinguished into two types according to different scenarios: a single site with multiple cells and multiple cell sites. A single site with multiple cells is normally a conventional sectored cell arrangement and needs only minimal changes to network signaling specifications. Multiple cell sites should be controlled by different eNBs and network specifications needed to support cooperation. Coordinated beamforming will combine the multiantenna mode of LTE Rel-8 with beamforming as the downlink transmit mode of CoMP. One typical method is transmitting a different row of signal after multiantenna processing within different cells. After being weighted with different UE-specific beamforming vectors in different cells, each row of the signal is transmitted in the direction of the UE. The general procedure is given in the following text using as two-eNB CoMP scenario:

- Two eNBs determine a candidate set of UEs attached to both eNBs on a semistatic basis. For coordinated precoding and UE pairing, the eNBs request that those candidate UEs feed back information to be used in CoMP mode.
- The UE then estimates the DL channel to the anchor eNB and the other interfering eNBs. These estimated channels could be used for either receiver demodulation processing, link adaptation purposes, or both.
- The UE feeds back information related to desired channel and interference channels to its anchor as defined in the CoMP mode of operation. The eNBs in CoMP operation share the feedback information as collected from candidate UEs in each cell.
- The scheduler of participating eNBs performs final UE selection/pairing and also computes the precoding matrices based on throughput optimization goals and fairness constraints.

A simple example of coordinated beamforming between two eNBs is shown in Figure 3.67. It is assumed that UE 1 has been scheduled by eNB 1, while at the same

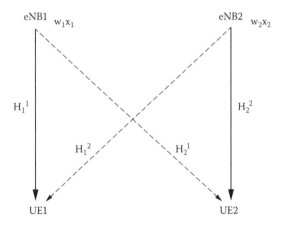

Figure 3.67 A simple illustration of CBF between two cells.

time UE 2 has been scheduled by eNB 2. Let H_i^j be the channel between eNB j and UE i. For simplicity and without loss of generality, all UEs have the same number (Nr) of receive antennas and both eNBs have the same number (Nt) of transmit antennas.

The input–output relationship may be written compactly as

$$y_1 = \sqrt{P_1} H_1^1 w_1 x_1 + \sqrt{P_2} H_1^2 w_2 x_2 + n_1$$

$$y_2 = \sqrt{P_2} H_2^2 w_2 x_2 + \sqrt{P_1} H_2^1 w_1 x_1 + n_2$$

where P_1, P_2 are transmit power from eNB 1 and eNB2, y_1, y_2 are the Nr*1 received vectors at UEs 1 and 2, w_1, w_2 are the Nt * 1 beamforming vectors for eNBs 1 and 2, x_1, x_2 are the information symbols for UEs 1 and 2, and n_1, n_2 are the Nr*1 random AWGN noise vectors seen by UEs 1 and 2, respectively.

For such a CBF scheme, the UE selection decision at eNB 1 would depend on the UE selection decisions at other eNBs, which are coordinating with eNB 1. In particular, eNB 1 should have to obtain knowledge of P_1, H_1^1, and H_2^1 and eNB 2 must obtain knowledge of P_2, H_2^2, and H_1^2. Each eNB should determine the precoder by itself. For example, each cell decides which UEs to schedule and the corresponding transmit precoding assuming no coordination between eNB 1 and eNB 2. Then, each cell will revisit its decision on the UEs to schedule, their transmit precoding, and transmit power, based on decisions made by other cells in iteration. Finally, a new CQI is computed based on the precoding, power, and UE decisions in neighboring cells, so the scheduling decision in a given cell is not only a function of the utility metric of the users scheduled by that cell, but also the utility metric of victim users that have been scheduled by other cells after iteration. The iterative method of CS/CBF of coordinated scheduling and beamforming is shown in Figure 3.68.

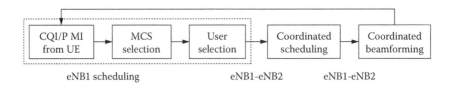

Figure 3.68 Iterative method of coordinated scheduling and beamforming.

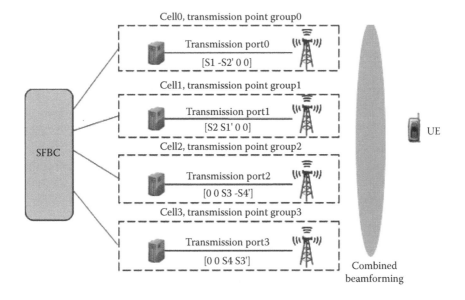

Figure 3.69 An example of the transmit diversity structure of CoMP when there are four transmission ports and the total number of service cells and coordinate cells is more than the number of transmission ports.

3.5.3.1 CoMP Transmission and Transmit Diversity

When a CoMP transmission is working in an open loop transmit diversity scheme, it can be combined with beamforming. In this case, different rows of transmit diversity signal are transmitted from different cells with weighted vectors based on each transmission port. Assume that CoMP transmission is based on m transmission ports, and the total number of service cell and coordinate cells is n. If $m < n$, we can group the n cells into m groups, and each group corresponding to a transmission port. If $m \geq n$, we can group the antennas of service cells and coordinate cells into m groups with beamforming at each transmission port. An example of four transmission ports according to these two cases is given in Figures 3.69 and 3.70.

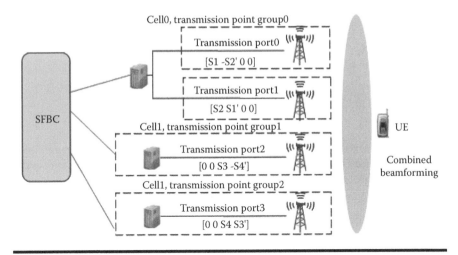

Figure 3.70 An example of the transmit diversity structure of CoMP when there are four transmission ports and the total number of service cells and coordinate cells is less than the number of transmission ports.

Finally, let's conclude the CoMP procedure for CS/CBF by an example as follows:

- The UE provides feedback to the serving eNB based on measurements of CSI-RS transmitted from cooperating eNBs that correspond to a CoMP cooperating set.
- The serving eNB decides transmission parameters (precoding weight, MCS) and schedules the UE. First, the serving eNB shares the feedback information received from the UE with its cooperating subset through backhaul. Then, the determination of transmission parameters procedure may be performed in an iterative way across cooperating eNBs depending on the scheduling information and transmission parameters of the cooperating set.
- The serving eNB conveys scheduling information and transmits the PDSCH to UE using UE-specific DMRS.
- The UE performs PDSCH demodulation using UE-specific DMRS.

3.6 UL CoMP

In DL CoMP, specifications are required for both eNB and UE, while less specification effort is foreseen to support UL CoMP. UL CoMP implies a possibility for the joint processing of signals of one user being received at multiple geographically separated points (see Figure 3.71). In general, a UE does not need to be aware of the network nodes that are receiving its transmission or the processing that is carried out on the corresponding received signals, either at these nodes or, alternatively, at

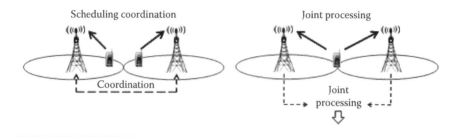

Figure 3.71 The PUSCH is received at multiple cells, and scheduling is coordinated among the cells in uplink CoMP.

a central node. What a UE needs to know is how any downlink signaling associated with uplink transmission (scheduling grants, HARQ acknowledgments, and/or power-control commands) is being provided.

UL coordinated multipoint reception scheduling decisions can be coordinated among cells to control interference. It should be noted that in different instances, the cooperating units can be separate eNBs, remote radio units, relays, and so on. Moreover, since UL CoMP mainly impacts the scheduler and receiver, it is primarily an implementation issue. Consequently, the evolution of LTE will likely define only the signaling needed to facilitate multipoint reception.

UL CoMP processing can also be done in three steps, similar to downlink: Step 1: From SRSs received on cooperating eNBs, the network will identify UEs that benefit from CoMP and related eNBs. Step 2: The up-to-date SRS measurements should be shared between cooperating eNBs. Joint scheduling will be done over the cooperating eNBs, and the scheduling information needs to be shared among cooperating eNBs. Step 3: The received signal information (from I/Q samples to hard bit decisions) should be shared among cooperating eNBs.

UL JP-CoMP implies joint processing of the signals received at multiple points to improve (especially) cell-edge user throughput by UL CoMP macro diversity reception. The joint processing in Figure 3.72 is transparent to the UE; it is an implementation matter, that has very limited impact on the specifications, it depends on the cooperation strategy and the requirements for the backhaul are different. In intrasite coordination, no backhaul is required. For coordination of different sites we need a change in the X2 interface definition. In the UL CoMP, the received data at each CoMP reception point can potentially be transferred to the serving cell for joint processing. Users transmit data to all receive points and the receive points will forward the received data to the serving cell before or after decoding it. The resolution level of information exchanged via the X2 interface can generally be I/Q sample (per antenna) level, soft bit level, code block level, or transport block level.

In UL CoMP macro diversity reception, the network will make the decision on whether eNBs should make the macro diversity reception based on UE

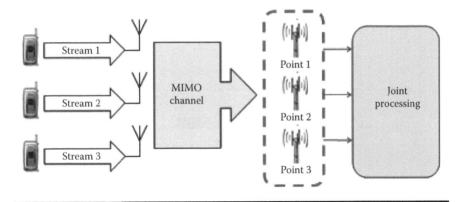

Figure 3.72   UL JP-CoMP.

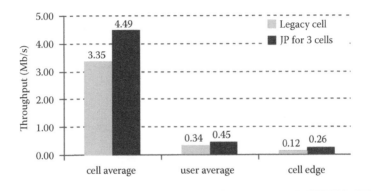

Figure 3.73   Throughput comparison of legacy cell and UL JP for three cells.

measurements. Assuming that maximum ratio combining (MRC) is used to combine the signals received from both the serve link and selected macro links with ideal interference cancellation, the SINR can be achieved according to this equation:

$$SINR = SINR_{\text{ServerLink}} + \sum_{n=0}^{N} SINR_{\text{Comp Link\_n}}$$

As shown in Figures 3.73 and 3.74, for the ideal IC-based intersite MRC case, there are bit throughput gains for the macro diversity in comparison with the legacy cell.

An uplink CoMP JP system can be viewed as a composite MIMO system, and is shown in Figure 3.75. Each OFDM subcarrier has an associated CoMP MIMO channel matrix H. This solution should take into account the difference in the timing between cells and calculate the corresponding phase shift and window function

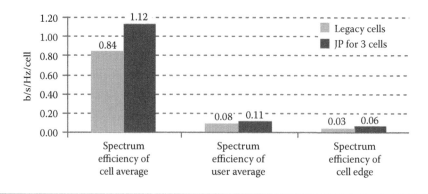

Figure 3.74 Spectrum efficiency comparison of legacy cell and UL JP for three cells.

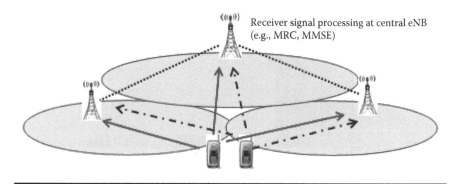

Figure 3.75 Composite MIMO system of UL CoMP.

coefficient per UE per subcarrier. This phase shift and window function value is then multiplied into the MIMO channel matrix of the composite CoMP MIMO channel. This allows use of one single receive combining set of antenna weights for all antennas in the whole CoMP cluster.

However, in uplink CoMP joint processing, one has to cope with multipath propagation and different timing delays to different cells of a CoMP coordination set; thus complexity is increased and there are very tight requirements on the X2 interface capacity and delays. Each UE is transmitting on a time-aligned basis in its serving cell, but the interference signal that arrives at a neighbor cell is not received in an aligned manner (may be delayed generally). If the difference of these timing offsets exceeds the cyclic prefix, the joint processing of the uplink signals of different CoMP reception points in one single receiver is not possible anymore, and intersymbol interference has to be avoided. The related issues will be discussed in the next section.

3.7 CoMP Limitation

As we have discussed, the following issues are foreseen for eNBs to be adequately coordinated in the CoMP approach in LTE-A:

- Configuration and coordination of the CoMP measurement and transmission sets
- Accurate time and frequency synchronization among the collaborating cells and UE
- Accurate intercell channel estimation with high accuracy
- CQI compensation and configuring precoding and beamforming weights
- Processing a larger amount of CQI/PMI/RI feedback compared to Rel-8; the capacity of the UL feedback channel, complex precoder selection, link adaptation, and MAC layer of supercell (i.e., collaborative area) due to an additional spatial dimension is an issue
- Feedback of CSI information; spatial domain-based interference suppression requires very accurate CSI feedback, especially for FDD; meanwhile, a low complexity of the feedback mechanism is highly desired
- Backbone data exchange for CSI and user data sharing among collaborating cells
- HARQ process latency constraints
- Coordinated antenna calibration for CoMP; the complexity of the calibration procedure can be a problem since the coordinated antenna calibration requires calibrations for all the antennas of all the eNBs associated with the CoMP operation

3.7.1 CoMP Delay Spread Analysis

CoMP transmission and reception has been considered an important feature in the LTE-A system to improve the throughput of cell-edge UE. The processing gain from coordinated cells (intersite or intrasite) is based on the assumption that these cells are well synchronized.

Although CoMP is assumed to be carried out in a synchronized network, the signal from cooperating cell sites may arrive at the UE at different times due to different distances between the UE and cell sites (i.e., a mismatch of arrival timing exists). Figure 3.76 illustrates the genesis of the mismatch. If the mismatch is greater than a certain value, the intersymbol interference will counteract the cooperation gain. To overcome this, a threshold can be set up. If the mismatch between a cell site and the reference cell site exceeds the threshold, the cell can then be excluded from the cooperation cell set. For a mismatch within the threshold, a calibration process can be adopted to compensate for the difference. Even after the compensation, there are still residual timing mismatches within the cooperation cell set. The effect of residual timing mismatches depends on the cooperation techniques.

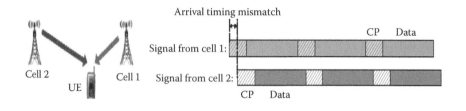

Figure 3.76 Arrival timing mismatch between two cooperating cells.

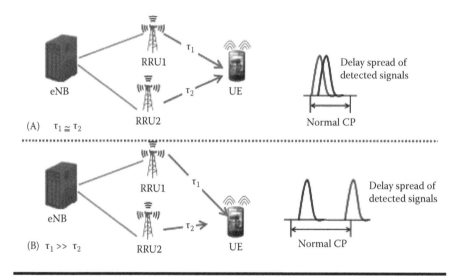

Figure 3.77 Downlink time delay issue in joint transmission.

In an LTE-A CoMP scenario, a smaller delay spread is required. Only the signals whose arrival time lies inside the area [0 μs, 3.9 μs] can be helpful for detection during TTIs with normal CP; otherwise, intersymbol interference (ISI) would occur. When the timing mismatch adjustment is based on the cell with a minimum transmission time delay in the active CoMP set, system performance can be further improved.

3.7.1.1 DL CoMP Delay Spread Analysis

The downlink time delay issue is interpreted in Figure 3.77. For simplicity, it is assumed that up to two cells connected to the eNB serve one UE. Intersite synchronization is accomplished; that is, cells transmit signals at the same time. Define τ_1 and τ_2 as the transmission times from cell 1 and cell 2 to the UE, respectively. The total delay spread at the UE is the sum of the downlink time difference from cell 1 and cell 2 plus the multipath delay, i.e., $(\tau_1 - \tau_1 + \tau_{multipath}) \cdot \tau_{multipath}$, is the multipath delay, as in a centralized cellular system without joint processing, where the CP length is required to satisfy $\tau_{multipath} < CP_{downlink}$, where $CP_{downlink}$ denotes the downlink CP length.

When the UE is located within similar distances from the cells (Case A in Figure 3.77, $\tau_1 \cong \tau_2$), the time delay issue is similar to a centralized cellular system without joint processing. But when the UE is located near cell 2 but far away from cell 1 (Case B in the Figure 3.77, $\tau_1 \gg \tau_2$), the total delay spread may be extended beyond the CP coverage, and thus intersymbol interference would affect the receiving performance.

Therefore, when extended to the scenario where multiple cells serve one UE simultaneously, the requirement for downlink CP is expected to have a CP length larger than the total time delay spread, or

$$\max_{UE\ k}\left\{\max_{RRU\ i}\left[\tau_{UE\ k,\ RRU\ i}\right]-\min_{RRU\ i}\left[\tau_{UE\ k,RRU\ i}\right]\right\}+\tau_{Multipath} < CP_{downlink}$$

where $\tau_{UE\ k,RRU\ i}$ denotes the transmission time delay from the ith serving RRU to the kth UE.

Consequently, the downlink time delay issue is more serious in an LTE-A CoMP system than in a centralized cellular system without joint processing. The issue is more evident as the number of joint processing cells increases.

3.7.1.2 UL CoMP Delay Spread Analysis

In a practical situation, as shown in Figure 3.78, a cell-edge UE transmits UL signals to different cells within a CoMP cooperating set, in which cell 2 is assumed to be the serving cell that manages the timing alignment. Since the delay that spreads to different cells can be significantly different, early and late arrival of transmitted signals (compared to that in the serving cell) will be received in cell 1 and cell 3, respectively. Excessive delay in cyclic-prefix length or even a small amount of advance will introduce interblock interference, degrades the detection performance, and further limits CoMP benefits.

Thus, we can see that the uplink has stricter delay spread requirements than the downlink if multiple RRUs are scheduled for downlink and uplink joint

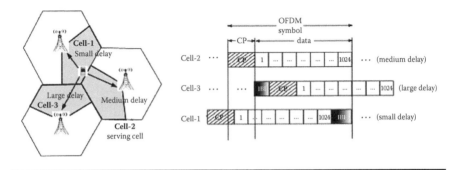

Figure 3.78   An illustration of delay spreads in UL CoMP.

transmission and reception. Moreover, in a multiple RRU joint processing situation, the time delay spread causes more problems in a distributed antenna cell than in the centralized antenna cell without joint processing.

3.7.1.2.1 Uplink Issue: UL Timing Advance

For the uplink in LTE Rel-8, CP is adopted to eliminate ISI and guarantee the orthogonality among signals on all subcarriers. To ensure that uplink signals are covered by the CP window, the UE should transmit the signal in advance, so that the uplink signal arrives at the cell receiver at the expected time, or the UE synchronizes to its serving cell. In a CoMP scenario, the UE UL signal to the nonserving cell (AP) may be out of the receiving CP range of that cell, which results in poor detection performance. In order to achieve coordinated reception, receiving points in the LTE-A network will coordinate TA synchronization aspects of the UE in the CoMP mode.

The uplink time delay issue is interpreted in Figure 3.79. A scenario with 3 UEs and 3 cells (connected to the same eNB) is shown. Assuming the time delays for the kth UE to the ith RRU is $\tau_{UE\,k,RRU\,i}$ and the timing advance is based on the cell with the strongest RSRP, which is referred to as cell TA. Consequently, the timing advance for the UE k is $\tau_{UE\,k,RRU\,TA}$. In the example, we assume that the timing advances for UE 1, UE 2, and UE 3 are based on cell 1, cell 2, and cell 3, respectively. Consequently, the timing advances for UE 1, UE 2, and UE 3 are τ_{11}, τ_{22}, and τ_{33}, respectively.

From the figure, when $\tau_{12} \gg \tau_{11}$ and $\tau_{32} \gg \tau_{33}$, the arriving time difference for signals from UE 1 and UE 3 to cell 2 will be much larger than the normal CP length. This situation can cause serious time delay problems and make it impossible to recover signals from all UEs.

The arriving time of signals from UE k is $\tau_{UE\,k,RRU\,i} - \tau_{UE\,k,RRU\,TA}$. Therefore, the total time delay at the receiver of RRU i is

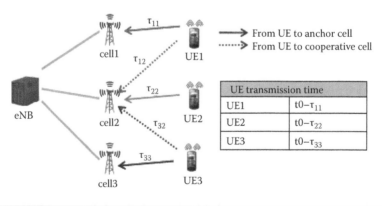

UE transmission time	
UE1	t0−τ_{11}
UE2	t0−τ_{22}
UE3	t0−τ_{33}

Figure 3.79 Uplink time delay issue in joint reception.

$$\max_{RRU\ i} \left\{ \max_{UE\ k} \left[\tau_{UE\ k, RRU\ i} - TA_{UE\ k} \right] - \min_{UE\ j} \left[\tau_{UE\ j, RRU\ i} - TA_{UE\ j} \right] \right\} + \tau_{Multipath}$$

$$= \max_{RRU\ i} \left\{ \max_{UE\ k} \left[\tau_{UE\ k, RRU\ i} - \tau_{UE\ k, RRU\ TA} \right] - \min_{UE\ j} \left[\tau_{UE\ j, RRU\ i} - \tau_{UE\ j, RRU\ TA} \right] \right\}$$

$$+ \tau_{Multipath}$$

When $\tau_{UE\ k, RRU\ i} - \tau_{UE\ k, RRU\ TA}$ varies obviously among multiple UEs, depending on the locations of the served UEs at the same TTI, the total time delay (the sum of the uplink time delay difference and the multipath delay spread) is significantly increased, which is shown in Figure 3.80. The fact requires that the CP length cover all useful signals from all served UEs to avoid ICI and ISI.

The possibility that the uplink total time delay spread at the receiver exceeds the CP range is higher than that in the downlink since the cell would handle the time delays from all its served UEs. Therefore, the uplink time delay spread may cause higher interference than the downlink.

Extended to the scenario that multiple cells serve one UE simultaneously, the requirement for the uplink CP should satisfy the requirement that the CP length is larger than the total time delay spread, or

$$\max_{RRU\ i} \left\{ \max_{UE\ k} \left[\tau_{UE\ k, RRU\ i} - TA_{UE\ k} \right] - \min_{UE\ j} \left[\tau_{UE\ j, RRU\ i} - TA_{UE\ j} \right] \right\} + \tau_{Multipath} < CP_{uplink}$$

where $\tau_{UE\ k, RRU\ i}$ denotes the transmission delay from the kth UE to the ith serving RRU, $TA_{UE\ k}$ denotes the timing advance for the kth UE, and CP_{uplink} denotes the uplink CP length.

Consequently, we must conclude that the uplink time delay issue is more serious in CoMP systems than in the centralized cellular system without joint processing. The time delay issue is more serious in the uplink than that in the downlink. The cumulative density function (CDF) of the uplink signal delay spread is shown in Figures 3.81 and 3.82, where both non-CoMP and CoMP systems are investigated.

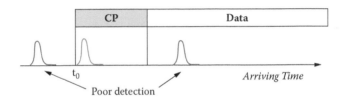

UL signals' arriving time at AP2

Figure 3.80 The total time delay.

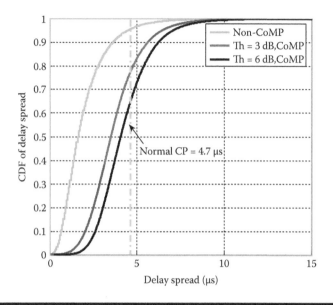

Figure 3.81   Distribution of UL signals' delay spread (case 1).

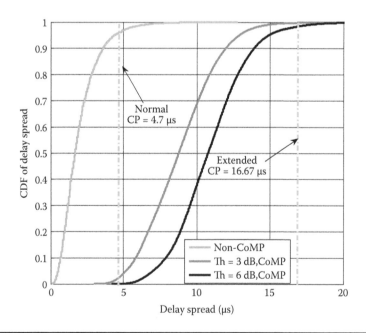

Figure 3.82   Distribution of UL signals' delay spread (case 3).

A higher threshold incurs a lower probability that the uplink signal delay spread is less than the CP length. In case 1 as shown in Figure 3.81 about 1/3 of the cells with normal CP will suffer performance degradation when the threshold for serving cell selection is 6 dB. In case 3 as shown in Figure 3.82, the issue causes almost all cells with normal CP to suffer performance degradation.

In general, the uplink signal delay spread, which is larger than the normal CP length, is high when CoMP is employed. From the previous analysis, extended CP can weaken the delay spread issue. Solutions to the multicell delay spread issue can be applied by adopting extended CP or a new TA adjustment scheme.

For extended CP (or flexible CP, where different TTIs adopt different CP lengths), it can tolerate a larger delay spread at multiple receiving cells, but it incurs high overhead and degrades performance accordingly.

A new TA adjustment scheme could be based on the cell with minimum propagation delay in the active CoMP set. The TA cell is the cell with minimum transmission delay in the active CoMP set (i.e., the closest cell in the set). This can be achieved with a TA adjustment based on the serving cell that receives the UE's signal earliest. With this method, the signal won't arrive at the cell's receiver earlier than the expected time.

3.7.2 Impacts on LTE-A Implementation

DL CoMP operation (relating to JP and/or CBF/CS) generates an additional requirement on LTE-A implementation. Control signaling and procedures on Uu and network internal interfaces need to be enhanced. UE feedback of the downlink channel state information for multiple cells has to be achieved in the CoMP operation. Coordination of sounding resources among cooperating cells is also necessary to avoid intercell interference. All the above requirements imply extra complexity and cost on eNB and UE implementation, and need a strong backhaul to support the complicated X2 interface as well.

With respect to the eNB, a stronger base-band processing capacity and a higher-complexity receiver to support multipoint reception will be needed to support multicell coordination. Furthermore, eNB data transferring capacity should be enhanced to transfer data among cells with low latency and high bandwidth. A more powerful backhaul is also a must if dynamic inter-eNB CoMP is needed.

The UE's complexity will be increased significantly due to neighbor cell channel estimation and interference mitigation ability. UEs also need to support a new feedback mechanism to support different CoMP categories.

From the backhaul perspective, the X2 interface needs new standardization work to bear the additional signaling to support CoMP operation. A strong backhaul mechanism, such as optical fiber, with low latency and high capacity is highly recommended to support JP.

In CoMP operation, higher-layer signaling should be able to support multiple cells in the measurement set and the transmission mode of different CoMP categories. New feedback mechanisms are required for CoMP. A new higher-layer procedure may be introduced for handover enhancement by CoMP.

3.8 CoMP Performance

First of all, we will discuss which UEs should be served under CoMP transmission. In general, pre-CoMP SINR is used as a measure to decide which UEs are served under CoMP transmission. A more targeted approach is to compare the rate with CoMP (R_{comp}) and the rate without CoMP ($R_{non-comp}$), which is not much more complex than SINR calculation. Given the costs of the CoMP transmission approach, a UE may be served under CoMP transmission if $R_{comp} > (1 + k) R_{non-comp}$, where k is the cost factor. It should be noted that the R_{comp} contains a factor of $1/n$, where n is the number of cooperating eNBs. One approach to increasing the system throughput with CoMP transmission is to consider variable-sized eNB clusters for serving different UEs. For example, in a CoMP scenario with three coordinated points, some UEs will be served by only one eNB, some UEs will be served by two eNBs, and the rest will be served by all three coordinated eNBs. The decision on how many eNBs serve a particular UE can be made again based on comparing R_{comp} with $R_{non-comp}$ for different numbers of serving eNBs.

From a theoretical point of view, if an optimal transmission scheme is employed at the coordinated eNBs, which in general requires simultaneous transmission from eNBs to all UEs (i.e., CoMP-MU-MIMO operation mode), there is throughput gain in serving both cell-center and cell-edge UEs under CoMP transmission. However, if the coordinated eNBs serve only one UE at any given time (i.e., CoMP-SU-MIMO operation mode), there can be throughput loss in serving cell-center UEs under CoMP. This is one of the reasons that CoMP has a great potential gain demonstrated at the link level but a much lower gain demonstrated at the system level. Other reasons may include the following:

- Creates overhead for reference signal design and channel status information feedback
- LTE frame structure and HARQ process set limitations to the optimum CoMP concept
- RF performance and synchronization need attention, but there are currently no solutions.

As we know, there are two CoMP modes currently under consideration: CS/CBF and JP. While the former relies on the coordination of neighbor cells to manage and reduce intercell interference, the latter can fully exploit the diversity of multiple transmitting/receiving base stations and thus, can achieve better throughput enhancement. For CS/CBF, each UE has a single serving cell. The throughput improvement of cell-edge UEs is from coordinated interference management. As long as the interference management is effective, the cells can have frequency reuse one and the degrees of freedom offered by neighbor base stations will be used up by the UEs.

3.8.1 CoMP Parameter Impact on Performance

As we discussed, there are two key parameters for a UE-specific CoMP cooperating set: the threshold for UE-specific CoMP cooperating set decisions and the maximum size of the UE-specific CoMP cooperating set. In the following simulation, we evaluate the impact of these key parameters on the CoMP gain of cell-edge user throughput as well as average sector throughput.

At first, we assume the maximum size of the UE-specific CoMP cooperating set as equal to 10. Figure 3.83 gives the probability distribution function (PDF) of the cell number of the CoMP cooperating set with a different threshold (Thr) for the UE-specific CoMP cooperating set decision. As shown in Figure 3.83, when Thr = 3dB, only 2% of the UEs have 3 cells in the CoMP cooperating set and no UE has more than 3 cells. When Thr = 7dB, there 7% of the UEs have 3 cells in the CoMP cooperating set and only 1.2% of the UEs have more than 4 cells in the CoMP cooperating set. Therefore, the maximum size of the UE-specific CoMP cooperating set is no larger than 3. In the following simulation, we assume the maximum size of the CoMP cooperating set equal to 2 or 3 points to evaluate the CoMP gain in terms of cell-edge user throughput and average sector throughput.

Figure 3.84 gives the average sector throughput and cell-edge user throughput as a function of threshold for the UE-specific CoMP cooperating set decision when the maximum size of the CoMP cooperating set is equal to 2 or 3 points.

Because most UEs use no more than two CoMP transmission points, the cell-edge user throughput and average sector throughput of a maximum of 2 cells in

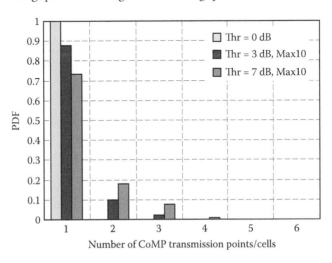

Figure 3.83   PDF of the cell number of the UE-specific CoMP cooperating set.

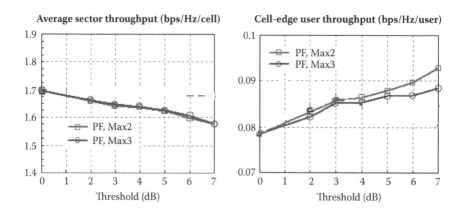

Figure 3.84 Average sector throughput and cell-edge user throughput as a function of threshold for UE-specific CoMP cooperating set decision.

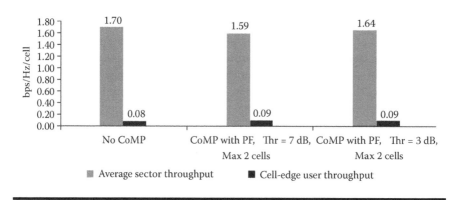

Figure 3.85 Comparison of average sector throughput and cell-edge user throughput.

the UE-specific CoMP cooperating set are similar to those of a maximum of 3 cells. Even when Thr = 7dB, only 7% of the UEs have 3 CoMP transmission points. From Figure 3.84, we can see that at Thr = 7dB, the cell-edge user throughput of a maximum of 2 cells is slightly larger than that of a maximum of 3 cells. This is because the UE with 3 CoMP transmission points has less opportunity to be allocated the same RBs from all the CoMP transmission points than the UE with 2 CoMP transmission points. However, the average sector throughput slightly degrades since the RBs allocated to CoMP UEs in each CoMP transmission point result in fewer remaining RBs for the other UEs, as shown in the Figure 3.85.

According to our evaluation, a maximum of two cells in a CoMP cooperating set is enough to achieve CoMP gain in terms of cell-edge user throughput and average sector throughput for 3GPP case 1. The potential CoMP gain on cell-edge user throughput is as much as 18% at the price of 6% degradation of average sector throughput. More adaptive selection of CoMP transmission points may be required to avoid degradation of average sector throughput.

3.8.2 UL CoMP Evaluation

In this section, system evaluation is carried out when the delay spread issue is neglected to investigate the largest performance improvement introduced by CoMP in the uplink. System-level simulation is carried out to evaluate the benefit from CoMP. According to current research, the maximum number of coordinated cells is around three. The CoMP active set is composed based on a threshold; that is, if $RSRP_{celli} > RSRP_{anchor} - Threshold$, cell i is selected as one of the serving cells, where $RSRP_{celli}$ denotes the RSRP from the ith cell and $RSRP_{anchor}$ denotes the RSRP from the anchor cell. The cell with best channel condition in the active CoMP set is recommended to be the anchor cell, from which the UE detects the system information and its dedicated control signaling of multicell coordination information. We assume the threshold is 3 dB or 6 dB; note that when the threshold is larger, more UEs can enjoy the benefits of CoMP. Throughput improvement over the non-CoMP system for cell average and cell-edge throughput is illustrated in Figure 3.86. The results for both case 1 and case 3 are given.

In conclusion, when the delay spread issue is neglected, CoMP can bring evident improvement to system throughput, especially for cell-edge UE. When more UE is served by multiple cells or more cells could serve the UE, larger throughput improvement can be obtained.

3.8.3 DL CoMP Evaluation

In 3GPP TSG RANWG1 (RAN1), the evaluation scenarios and assumptions for intrasite downlink CoMP were agreed to and multiple companies have provided

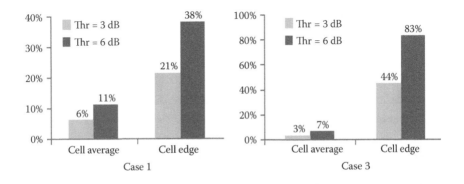

Figure 3.86   Benefits of using CoMP in 3GPP in case 1 and case 3. (Thr—threshold)

CoMP evaluation results for both high-load (10 UEs) and low-load (2 UEs) scenarios. A summary of evaluation results of intrasite CoMP for a full buffer traffic model is described in the following text.

The intrasite CoMP with coordination of up to three colocated cells is the baseline scenario for performance evaluation under configuration 3, which includes antenna installation of both 2 transmit antennas per cell and 4 transmit antennas per cell. For a full buffer traffic model, the evaluations of 2 users per cell and 10 users per cell are done for the low-load scenario and high-load scenario, respectively. The evaluation results are given in Figures 3.87 and 3.88, and show the averaged gain of CoMP (SU-MIMO) over Rel-8 SU-MIMO, CoMP (MU-MIMO) over MU-MIMO, and Rel-8 SU-MIMO.

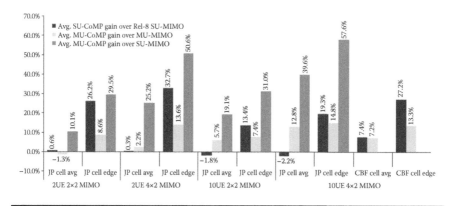

Figure 3.87 Averaged gain of CoMP (SU-MIMO) over Rel-8 SU-MIMO, CoMP (MU-MIMO) over MU-MIMO, and Rel-8 SU-MIMO, respectively, for configuration 3 FDD.

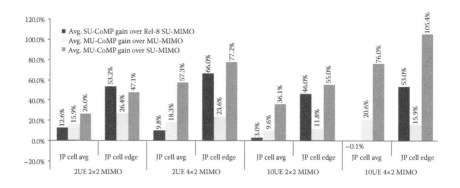

Figure 3.88 Averaged gain of CoMP (SU-MIMO) over Rel-8 SU-MIMO, CoMP (MU-MIMO) over MU-MIMO, and Rel-8 SU-MIMO, respectively, for configuration 3 TDD.

Note: Some companies show quite different results in the CoMP evaluation (far worse than the 3GPP requirement) and 3GPP evaluation (satisfying the 3GPP requirement).

Figures 3.89 and 3.90 show the possible system performance with cell-edge and cell average throughput gain determined from system simulations with the various CoMP techniques in the DL. The simulations adopted a more advanced receiver and interference coordination/cancellation algorithm, quantized feedbacks

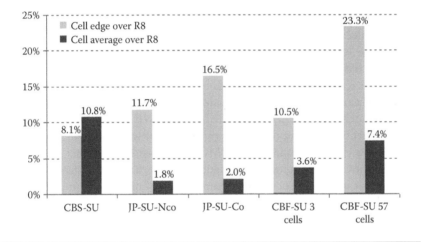

Figure 3.89   Potential gain over Rel-8 (3GPP case 1) of DL CoMP (SU-MIMO).

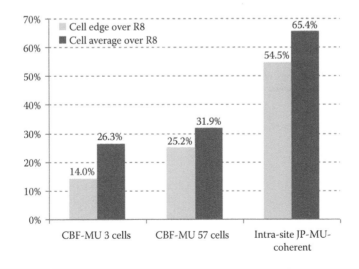

Figure 3.90   Potential gain over Rel-8 (3GPP case 1) of DL CoMP (MU-MIMO).

for CBS with SU-MIMO, intrasite JP with MU-MIMO, and idealized feedbacks for the other schemes. The simulation assumptions were consistent with that agreed to in 3GPP TR 36.814. The CoMP techniques investigated include CBS, noncoherent joint processing (JP-Nco), coherent joint processing (JP-Co), CBF, and intrasite coherent JP.

The previous system-level simulation is presented as an example to show the performance of downlink multipoint joint transmission. The evaluation results support the idea that downlink CoMP would greatly increase the cell-edge user throughput as well as the cell average throughput. Moreover, it is observed that not only do 5% of the cell-edge UEs experience improved throughput, but additional UEs experience improved throughput and cell average throughput by applying CoMP in the whole cell coverage. So CoMP is beneficial to both cell-edge user throughput and cell average throughput. It is a promising technique to achieve the advanced requirement for LTE-A. Simulation results also show that cell-edge user throughput using coordinated scheduling and joint processing are increased by approximately 10% and 20% compared to single-cell transmission, respectively. It is also worth mentioning that CBF as a multipoint technique can improve performance over the single-point MU-MIMO operations by 10–15%.

Finally, the CoMP performance evaluation and comparison with the International Telecommunications Union (ITU) requirement is shown in Figure 3.91, where CoMP exhibits interesting gains with idealized assumptions. LTE Rel-8 already supports simple coordination, so based on long-term channel statistics measured from the uplink, downlink coordinated beamforming can be implemented and shows performance gain in both FDD and TDD systems. We also can see that intra-eNB as well as inter-eNB CoMP will be a tool to improve the coverage of high data rates and cell-edge throughput, and to increase system throughput by spatial domain intercell scheduling and interference coordination, and other cooperation methods. However, CoMP is still an immature technology. The impact of the robustness to

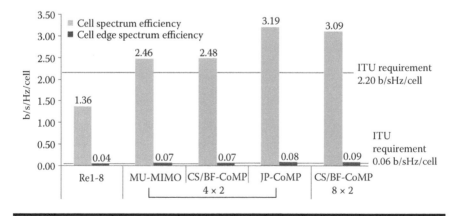

Figure 3.91   ITU requirement and preliminary performance of CoMP.

various channel estimation errors and the implications of feedback delay and quantization errors are not clear; for example, high CoMP gain needs a complex scheduling algorithm; CoMP gain is sensitive to CSI accuracy, which is difficult in a realistic environment, and latency; the data rate of backhual requirements of intersite CoMP is very severe. Currently, it is still not clear which schemes can be used with practical, nonideal assumptions in deployments.

Chapter 4

MIMO

In a Long Term Evolution Advanced (LTE-A) system, the edge throughput and average cell throughput targets are ~10 times higher than the LTE Rel-8 targets. The newly launched collaborative multipoint (CoMP) reception in LTE-A and previously adopted beamforming technology are possible ways to enhance system performance in terms of edge throughput, while the higher-order single-user multiple-input, multiple-output (SU-MIMO) and multiuser MIMO (MU-MIMO) are the possible dominant technologies introduced in LTE-A to enhance the peak and average throughput in a cell.

CoMP was discussed in Chapter 3, and various MIMO-related topics and beamforming technology will be studied in detail in this chapter.

4.1 Wireless Channel Characteristics

MIMO technology has been treated as an emerging technology to meet the demand for higher data rates and better cell coverage without the need to increase the average transmit power or frequency bandwidth. The MIMO structure successfully constructs multiple spatial layers where multiple data streams are delivered on a given frequency-time resource and linearly increases the channel capacity. The performance of a MIMO system is directly related to the received signal-to-interference plus noise ratio (SINR) and the correlation properties that are characteristic of the multipath channel and antenna configuration.

4.1.1 Channel Model in LTE-A

MIMO technology is based on the use of multiple antenna systems, in which data may be transmitted to a user over sets of channels existing between n transmit antennas

Table 4.1 Summary of Delay Profiles for LTE Channel Models

Model	Number of Channel Taps	Delay Spread (r.m.s.)	Maximum Excess Tap Delay (Span)
Extended pedestrian A (EPA)	7	45 ns	410 ns
Extended vehicular A (EVA)	9	357 ns	2510 ns
Extended typical urban (ETU)	9	991 ns	5000 ns

and *m* receive antennas. Multiple simultaneous data streams may be transmitted over the channel set if the channels of the set are sufficiently statistically independent. The ability of the system to successfully transmit multiple simultaneous and different data streams over the MIMO channel set is a function of the channel SINR, and the gain of MIMO transmission is increased for cases of higher SINR and lower channel correlation. That is what we described as "the channel determines the performance."

The radio channel between a transmitter antenna and a receiver antenna is characterized by the following factors: channel frequency response h (f, r, t), interference I, and noise N.

The channel frequency response h (f, r, t) varies with frequency f, position r, and time t. It is further characterized by path loss (including shadowing), delay profile, fading characteristics, co-channel and adjacent channel interference, and Doppler spectrum. The delay profile is a consequence of multipath propagation. It consists of the gain and phase shift per path. In case of multiple antennas, the channel is also characterized by the antenna correlation.

The delay profiles (listed in Table 4.1) are selected to be representative of low, medium, and high delay spread environments. The profiles for low and medium delay spreads are based on the International Telecommunications Union (ITU) pedestrian A and vehicular A channel models, respectively, which were originally defined for the ITU Radiocommunication Sector (ITU-R) evaluation of International Mobile Telecommunications 2000 (IMT-2000). Pedestrian models have a low delay spread, while urban models have a high delay spread. The high delay spread model is based on the typical urban model used for Global System for Mobile Communication (GSM) and in some of the evaluation work for LTE. The models are defined on a 10-ns sampling grid.

As we know, the ITU model and spatial channel model (SCM) (Ped A, Veh-A, Ped-B, etc.) were created for channel bandwidths of less than 5 MHz and may not reflect actual channel conditions when used for wider bandwidths. At wider bandwidths, more paths become resolvable and the channel becomes more Rician (i.e., larger K-factor*). Extended ITU channel models for bandwidths

---

* Open (larger) environments have higher K-factors than smaller environments with close-in reflecting objects (more scattering).

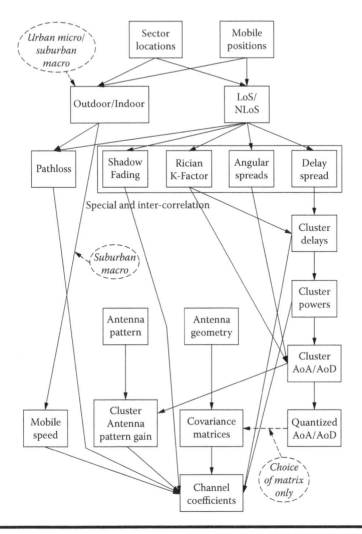

Figure 4.1 Basic characteristics of the channel model.

up to 20 MHz are described in Figure 4.1. In the Third-Generation Partnership Project (3GPP), we usually use two LTE-A channel model approaches:

■ **Ray-based model:** The basic principles of the ray tracing model enable analysis of electric wave propagation through the ray tracing method and obtain the receiving signal field strength through theoretical calculation. Ray-based SCM models explicitly model reflected paths. They are the most accurate, but also the most time consuming to simulate in practice, and are yet to mature.

■ **Covariance-based models:** In order to reduce computational complexity, a covariance matrix can be estimated from sounding reference signal

(SRS) transmissions in cases of time-division duplexing (TDD) and is an efficient method for estimating a covariance matrix for frequency-division duplexing (FDD).

Multipath propagation causes variations in time delay, amplitude, phase, and angles of departure and arrival (AoD & AoA) of the received signals. The received signal power fluctuates in space due to angular spread, or in frequency due to delay spread, or in time due to Doppler spread. This fluctuation in signal level is called *fading* and generally degrades the performance of wireless systems.

Statistical covariance feedback can be classified into long-term and short-term statistical covariance. Long-term statistical covariance is accumulated over an infinite time window; it is the asymptotic covariance matrix that is averaged over fast fading and thus depends on only the propagation environment. Short-term covariance feedback is used when the overhead is not high, but long-term covariance feedback may lose some of its spatial directivity information because antenna cross-correlation reduces when averaging over very long periods and very wide bandwidth.

One of the MIMO channel modeling approaches is cluster modeling, which treats reflection paths as clusters of rays. Each cluster has a power delay profile (power at different delays, i.e., taps in discrete time), which is used in finding MIMO channel tap coefficients. The parameters used to model each cluster are the AoD from the transmitter, the AoA at the receiver, and the angular spread (AS) at both stations (one AS value for each). The number of clusters varies depending on the model, and each cluster can contain many rays. With the knowledge of each tap's power, AS, and AoA (AoD) for a given antenna configuration, the channel matrix **H** can be determined, which **H** fully describes the propagation channel among all transmit and receive antennas.

We assume that $\mathbf{H}_i$ denotes the channel matrix between the ith user equipment (UE) and the evolved Node-B (eNB). The spatial covariance matrix for the the ith UE at the jth frame can be estimated to its statistical mean as follows:

$$\mathbf{R}_{ij} = (1-\alpha)\mathbf{R}_{i(j-1)} + \alpha \frac{1}{N} \sum \mathbf{H}_{ij}^{H} \mathbf{H}_{ij}$$

where $0 < \alpha < 1$, the long-term average of the transmit covariance matrix converges to its statistical mean. N is the total number of subcarriers averaged in the calculation of the transmit covariance matrix; typically this would be several resource blocks (RBs) or even the whole band. Then, the eNB can derive the codebook used by the UE through the transmit spatial covariance matrix.

Regarding statistical channel information, it is well known that channel covariance changes much more slowly than the coherence time and bandwidth of the channel. For this reason, channel covariance has been widely used in TDD systems, and a covariance matrix is also valid on FDD systems for larger FDD duplex distances. Some frequency translation techniques can be used to improve the accuracy, if an eNB measures the spatial channel covariance matrix for a particular UE on uplink (UL) transmissions. The previously discussed frequency translation property and reciprocity states that a similar covariance can be used in downlink (DL) signal processing.

4.1.2 Multipath Propagation Gain

Usually the transmitted signal is reflected on other objects before it reaches the receiver. Each reflection causes a path with a certain delay compared to the shortest path. If there is line of sight, the direct path will be completely dominant and the reflected paths can be ignored. If there is no line of sight, the reflected paths can be comparable. Their amplitudes and relative delays define the delay profile. An example of a delay profile is illustrated in Figure 4.2.

The different paths in the delay profile will add effects constructively or destructively, depending on the phase shift of each path. Since the phase shift depends on the frequency, multipath fading causes frequency selectivity. That is, the channel gain varies with the frequency. The channel gain as a function of frequency is illustrated in Figures 4.3, 4.4, and 4.5 for three different types of shadowing and scattering environments.

If the reflecting objects are located far from the transmitter or receiver, it is referred to as *shadowing* or *slow fading*. This is typical for rural environments where the signal is reflected by hills or mountains. The channel gain varies slowly with position in case of shadowing, since the phase of each path is not much affected if the transmitter or receiver moves a short bit. If the reflecting objects are located close to the transmitter or receiver, it is referred to as *scattering* or *fast fading*. This is typical for urban environments where the signal is reflected by buildings and vehicles. In the DL case, the paths typically arrive at the receiver from all directions, which makes the channel gain sensitive to short movements of the receiver since the phase shift varies quickly with position. If the reflecting objects are moving, the channel gain not only depends on frequency and position, but also varies over time. This is also true if the

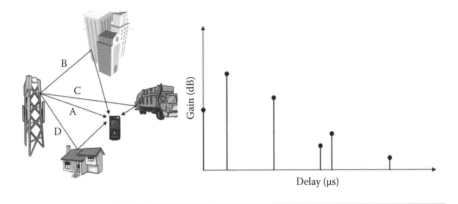

Figure 4.2 **Delay profile with multiple paths. (A, B, C, D are four paths as an example.)**

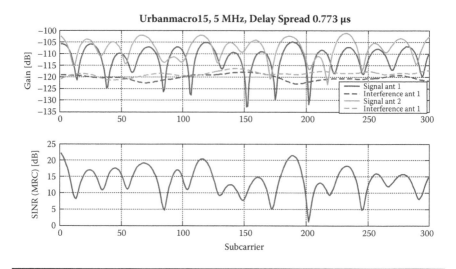

Figure 4.3 Frequency selectivity, urban macro channel model.

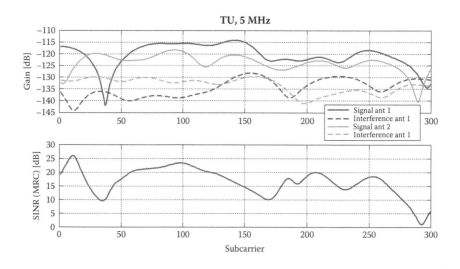

Figure 4.4 Frequency selectivity, typical urban channel model.

transmitter or receiver moves, but in this case the variation over time depends on the position, which changes over time when the transmitter or receiver moves.

The mere distance between transmitter and receiver causes a path loss. This is reflected in the gains of the individual delay profile paths. The path loss is neither time nor frequency dependent. It depends on the position of the receiver compared to the transmitter. There is a large difference in path loss depending on whether there

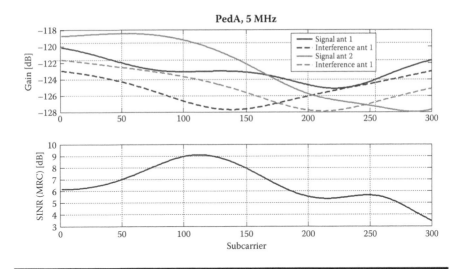

Figure 4.5 Frequency selectivity, pedestrian channel model.

is line of sight between transmitter and receiver. The path loss and multipath fading together result in a channel gain that varies over time and frequency.

4.1.3 Doppler Spread

When the transmitter or receiver moves, an additional effect is the Doppler spread of the signal. With a line of sight, the carrier frequency is offset by the Doppler frequency, which depends on the movement relative to the axis between transmitter and receiver. The Doppler effect is expressed as Doppler shift or Doppler frequency, which can be defined as:

$$f_\mathrm{d} = \frac{1}{2\pi}\frac{\Delta\phi}{\Delta t} = \frac{v}{\lambda}\cos\theta$$

where $\Delta\phi$ is the change of phase, Δt is the change of time, v is the velocity of the receiver (UE) relative to the source (base station), λ is the wavelength of the transmitted signal, θ is the angle between the UE's forward velocity and the line of sight from the UE to the base station. From the equation, we can see that when the relative speed is higher, the Doppler shift can be very high. Thus, the receiver may become unable to detect the transmitted signal frequency if the transmission and reception techniques are very sensitive to carrier frequency offset.

In case of multipath propagation, different paths are affected differently according to the classical Doppler spectrum, as illustrated in Figure 4.6.

The consequence of Doppler spread is that the signal is spread over a wider frequency range and may interfere with signals that are adjacent in frequency.

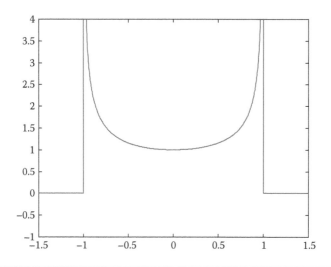

Figure 4.6   Doppler spectrum as a function of normalized Doppler frequency.

4.1.4 Interference

In LTE systems, or in any kind of wireless communication system, there are three kinds of interferences that influence the system performance: noise, intersystem interference, and intrasystem interference. Noise interference inherently exists in wireless communication systems. We cannot avoid it, but can only mitigate the noise effect via advanced signal processing techniques. Intersystem interference exists when other networks operate with LTE systems in the same region. So we also should take into account the frequency bands, coverage, and operational state of the existing networks in LTE network planning. Intrasystem interference exists in the LTE system itself, which is the main factor that influences system performance and should be more seriously considered in network planning.

LTE is a sort of self-inference cellular system. Its intrasystem interference can be categorized into intercell interference and intracell interference. Intercell interference mainly influences the cell-edge users' throughput, which leads to degradation in performance of the whole system. Intracell interference is a more severe problem compared to intercell interference, because it impacts all cell users' throughput.

The following sections describe various kinds of interference that may exist in LTE systems.

4.1.4.1 Downlink Interference

4.1.4.1.1 Intracell Interference

First, there is interference from the eNB transmission on different layers to its own UE. This is applicable in SU-MIMO. The interference depends on antenna

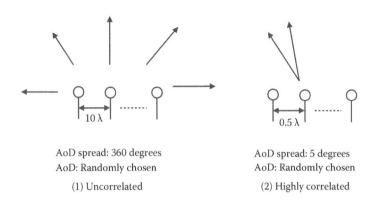

AoD spread: 360 degrees   AoD spread: 5 degrees
AoD: Randomly chosen   AoD: Randomly chosen
(1) Uncorrelated   (2) Highly correlated

Figure 4.7 Spatial correlation models between transmitter antennas.

correlation, which is characterized by antenna separation, AoD, and AoD spread, as shown in Figure 4.7. Second, there is interference from the eNB transmission to other UE on the same resource blocks (RBs). This is applicable in MU-MIMO. The interference also depends on antenna correlation. Third, there is interference from the eNB transmission to other UE on different RBs. This kind of interference depends on the UE's performance, when transmissions on adjacent subcarriers and/ or mirror subcarriers may interfere with its own transmission. As long as the eNB does not use UE-specific power offsets, this interference will be small.

Interference can also be caused by Doppler spread. Interference from the end of one UE's transmission to the start of another UE's reception should not occur if the TDD gaps are dimensioned to handle maximum round-trip time (RTT) between the eNB and the UE.

4.1.4.1.2 Intercell Interference

One type of intercell interference is the interference from neighboring cell eNB transmissions. This is a major source of interference. For a UE at the cell edge, the signal from a neighbor cell eNB is at least as high as the signal from its own cell eNB, provided that the neighbor cell eNB transmits in the same RBs. However, if the neighbor cell eNB does not schedule any transmission in the same RBs, there will be no interference. The interference varies enormously depending on the number of overlapping RBs. It can be noted that LTE does not have any spreading gain as Wideband Code Division Multiple Access (WCDMA). Thus, the signal-to-interference ratio in the decoder in the receiver may be below 0 dB.

Another type of intercell interference is the interference from neighbor cell UE transmissions. This is applicable for TDD only. It will be a problem if the cells are not time synchronized with a common DL/UL pattern. Due to the magnitude of the problem, TDD time synchronization should always be applied and thus this interference should be an abnormal occurrence.

Since the distance between a cell's UE and the UE in a neighbor cell may exceed the distance corresponding to the maximum RTT, it is possible that the end of the neighbor UE's transmission may interfere with the start of the cell's own UE reception. However, since the UEs will be separated by a large distance, the resulting interference should be negligible.

4.1.4.1.3 Intercarrier Interference

Interference from eNB transmissions on adjacent carriers may also exist. This is applicable for any combination of FDD and TDD in a carrier and an adjacent carrier. The adjacent carrier can be LTE or something else, but the transmitter is referred to as the eNB for simplicity.

If the signal from an adjacent carrier's eNB is much stronger than the signal from the home eNB, the adjacent carrier may cause blocking in the UE receiver. This should not be a problem if the carriers belong to the same operator, since handover mechanisms should ensure that the UE is connected to the strongest carrier. Also, the carriers will probably be colocated on the same sites and thus have comparable signal strength.

The problem should be avoided via appropriate guard bands between operators, but there may still be degradation in reception sensitivity due to adjacent carrier leakage.

Interference from adjacent carrier UE transmissions is another case. This is applicable for TDD–TDD and TDD–FDD coexistence.

4.1.4.1.4 Intersystem Interference

Other systems could cause interference due to spurious emissions, but this is an abnormal occurrence.

4.1.4.2 Uplink Interference

4.1.4.2.1 Intracell Interference

Similar to downlink scenario there is interference within a cell from its own UE transmission on other layers. This is applicable in SU-MIMO. The interference depends on antenna correlation. There is also interference from other UE transmissions within a cell to its own eNB on the same RBs. This is applicable in MU-MIMO. The interference also depends on antenna correlation.

Finally, there is interference from other UE transmissions in the same cell on different RBs. This will depend on UE performance, when transmissions on adjacent subcarriers and/or mirror subcarriers may interfere with the transmission from the cell's own UE. The interference level will depend on UL scheduling strategy.

As in the downlink case, interference can also be caused by Doppler spread.

4.1.4.2.2 Intercell Interference

Similar to the downlink case, interference may be generated from a neighboring cell's eNB transmission. This is applicable for TDD only. It will be a problem if the cells are not time synchronized with a common DL/UL pattern. Due to the magnitude of the problem, TDD time synchronization should always be applied and thus this interference should be an abnormal occurrence.

Since the distance between a cell's own eNB and the neighbor cell's eNB may exceed the distance corresponding to the maximum RTT, the end of the neighbor eNB's transmission may interfere with the start of the cell's own eNB reception. This could possibly be a problem when using umbrella cells with large output power together with indoor cells with low power. This potential problem can probably be mitigated via increased guard periods in the indoor cells. It is thus assumed that this interference source will be small.

Interference can also occur when a neighbor cell's UE transmission also exists in the uplink. This is a major source of interference. If a neighbor cell's UE is at its cell border, close to the home cell's eNB, it will cause received signal strength in the home cell eNB being at least as high as the targeted signal strength for its own UE.

In most cases, the UEs will be power limited at the cell edge. Therefore, they will not be able to transmit on many resource blocks. Neighbor cell UEs therefore become typically narrowband interferers.

The interference from neighbor cell UEs varies a lot depending on whether there are any UEs at the neighbor cell's edge and when they are scheduled.

4.1.4.2.3 Intercarrier Interference

Interference from eNB transmissions on adjacent carriers is applicable for TDD–TDD and TDD–FDD coexistence.

Interference from UE transmissions on adjacent carriers is applicable for any combination of FDD and TDD in the home carrier and adjacent carrier. The adjacent carrier can be LTE or something else, but the transmitter is referred to as UE for simplicity. If the signal from the adjacent carrier's UE is much stronger than the signal from the home carrier's UE, the adjacent carrier may cause blocking in the eNB receiver. This should not be a problem if the carriers belong to the same operator, since handover mechanisms should ensure that the UE is connected to the strongest carrier and thus does not transmit with unnecessarily high power. Also, the carriers will probably be colocated on the same sites and thus have comparable signal strength. The problem should be avoided via appropriate guard bands between operators, but there may still be degradation in reception sensitivity due to adjacent carrier leakage.

4.1.4.2.4 Intersystem Interference

Other wireless communication systems could cause interference due to spurious emissions, but this is an abnormal occurrence.

4.1.5 Antenna Correlation

When there is one transmitter antenna and multiple receiver antennas, the channel is referred to as single-input, multiple-output (SIMO). When there are multiple transmitter antennas and one receiver antenna, the channel is referred to as *multiple-input, single-output (MISO)*. When there are multiple antennas on the transmitter and receiver sides, the channel is referred to as MIMO.

A MIMO channel is referred to as N_{TX} x N_{RX}, where N_{TX} is the number of transmitter antennas and N_{RX} is the number of receiver antennas. A 2x2 MIMO channel is illustrated in Figure 4.8.

Each path from a transmitter antenna to a receiver antenna is subject to path loss, fading, and interference.

The paths h_{mn} are in the general case not independent of each other. The transmit antennas (or receive antennas) are considered to be perfectly correlated if h_{2n} can be predicted from h_{1n} for all n. The spatial correlation matrix is defined as the transmit antenna correlation observed at the UE and computed by the UE. The correlation between two antennas can, however, be anything between 0 and 1. Correlation 0 means that the antennas are uncorrelated while values larger than 0 means that they are correlated. For spatial multiplexing, the antennas need to have a low enough correlation. MIMO communication over uncorrelated channel dimensions by transmitting over two polarizations are much better suited for SU-MIMO, and each UE can effectively suppress the interstream interference over the uncorrelated dimensions by means of minimum mean squared error (MMSE) equalization or more advanced successive interference cancellation/maximum likelihood (SIC/ML) receivers. MU-MIMO gains are much smaller in uncorrelated scenarios, and multiple UEs can be multiplexed over the remaining spatially correlated channel dimensions by nonoverlapping beams with low multiuser interference building on long-term channel correlation properties, as shown in Figure 4.9.

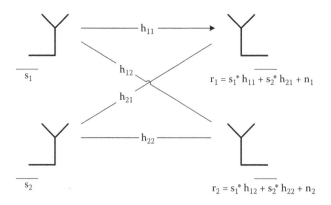

Figure 4.8   2x2 MIMO channel.

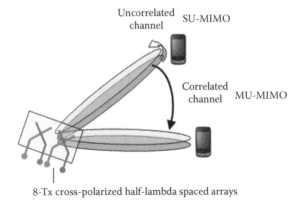

8-Tx cross-polarized half-lambda spaced arrays

Figure 4.9 Correlation domain for separating UEs and the uncorrelated domain for SU-MIMO.

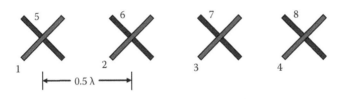

Figure 4.10 8Tx correlated DP antenna configuration.

This is achieved if the antennas have different polarization or if they are spatially separated by a sufficient distance. For a UE that is close to a scattering environment (many reflectors close by), the spatial separation needs to be more than about $\lambda/2$. For an eNB that is usually far from a scattering environment (few reflectors close by), a rule of thumb is that the spatial separation needs to be about $10*\lambda$. With a carrier frequency of 2.6 GHz, $\lambda = c/f = 3e^8 / 2.6e^9 = 0.12$ m. In other words, the antennas need to be separated with at least 6 cm on the UE side and at least 1.2 m on the eNB side to avoid correlation.

Currently some operators prefer to prioritize correlated dual-polarized antennas in an 8Tx antenna configuration. Prioritized scenarios for 8Tx include dual-polarized (DP) arrays with 4 dual-polarized elements with $\lambda/2$ spacing, and uniform linear arrays with 8 co-polarized elements with $\lambda/2$ spacing. As shown in Figure 4.10, with 8Tx DP antennas, the 8Tx can be divided into two antenna groups, each group with four antennas. The antennas in the first group {1,2,3,4} are correlated, and the antennas in the second group {5,6,7,8} are correlated, while the antennas in the different groups are roughly independent due to different polarizations.

As we have discussed, antenna correlation is characterized by antenna separation and angle of departure spread, and so on. Angle spread in the urban propagation

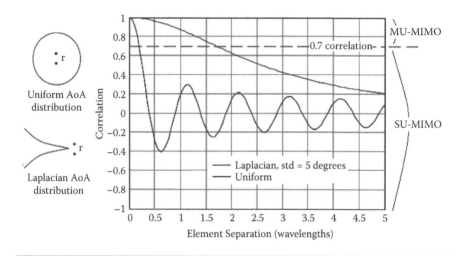

Figure 4.11 Effect of angle spread.

environment adversely affects beamforming systems because it increases the effective beam width, but MIMO systems will benefit from scattering in the propagation environment. With a low channel angle of departure spread, the antennas imply a highly spatially correlated channel; however, in case of dual-polarized arrays, the two polarization branches may still be considered uncorrelated. In any case, such highly correlated channels are ideal for MU-MIMO operation. With a higher angle of departure spread, the channel becomes less correlated spatially even with closely spaced antennas, and SU-MIMO may become more attractive than MU-MIMO. High correlation will reduce SU-MIMO capacity, and correlation of 0.7 or less is acceptable for SU-MIMO, as shown in Figure 4.11.

4.1.6 Consideration of UE Imperfections

Practical limitations in UE performance need to be taken into account when considering power control in the DL and UL. The dynamic range (power spectral density [PSD] difference in dB between highest and lowest UE measured at the eNB) must not be too high.

In the UL, the UE will generate adjacent channel leakage and spurious emissions at the mirror frequency (mirrored in the direct current [DC] subcarrier). In practical receiver implementations, the DC subcarrier or the zero frequency will suffer from high interference/distortion due to local oscillator leakage. The interference levels will be specified by 3GPP as UE performance requirements. Realistic levels are about –30 dBc for both adjacent channel leakage and spurious emission at the mirror frequency.

This means that the received PSD of other UEs must not be too close to those levels in the radio base station (RBS), because then the leakage and emission will

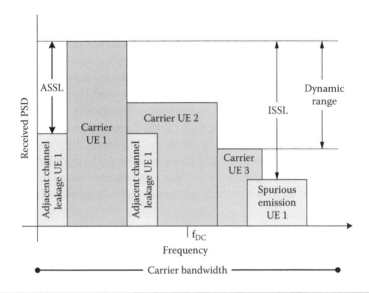

Figure 4.12   Impact on dynamic range due to UE imperfections.

result in lower SINR for UEs scheduled at the disturbed frequencies. This can be mitigated via scheduling and/or power control.

The same problem exists in the DL, this time due to imperfections in the UE receiver (as shown in Figure 4.12). 3GPP will need to specify which DL dynamic range the UE will be able to handle with a specified sensitivity performance.

In Figure 4.12, adjacent channel leakage and spurious emissions from UE 1 causes a disturbance for UE 2 and UE 3 when the PSD levels for those UEs are too far below the PSD of UE 1. This is applicable for both the DL and UL. It can be noted that the RBS can and will be designed in such a way that the corresponding phenomena are negligible. The UE will always have worse performance since it must be relatively cheap. It can also be noted that the mirroring phenomenon will also exist when a UE is scheduled over the whole carrier bandwidth, or at least on both sides of the DC subcarrier. However, in this case the UE will just interfere with itself. When the PSD level is constant over the bandwidth (except for frequency-selective fading), the impact of an interferer at –30 dBc will be negligible.

4.2 Overview of MIMO

In LTE, MIMO technologies have been widely used to improve downlink peak rate, cell coverage, and average cell throughput. To achieve this diverse set of objectives, LTE adopted various MIMO technologies, which have been evaluated since the first day the LTE specification was introduced.

4.2.1 MIMO Schemes

The LTE system has several types of defined multiple antenna transmission schemes, such as transmit diversity (TxD), open loop spatial multiplexing (SM), closed loop spatial multiplexing (precoding based), MU-MIMO, and beamforming. Each of those schemes has already been standardized in LTE Rel-8, works properly according to the channel and system environment, and provides a noticeable gain in each scenario. MIMO schemes (except beamforming) in LTE Rel-8 are specified for a configuration with two or four transmit antennas in the downlink, which supports transmission of multiple spatial layers with up to four layers to a given UE. Beamforming in LTE Rel-8 is specified for a configuration with eight transmit antennas in the downlink.

It is definite that all MIMO schemes defined in LTE Rel-8 should be supported in LTE-A.

4.2.1.1 Scheme Design Principles

The purpose of transmit diversity is to make the transmission more robust. There is no increase in the data rate, through using transmit diversity, which uses redundant data on different paths.

Receive (RX) diversity uses more antennas on the receiver side than on the transmitter side. The simplest scenario consists of two RX and one TX antenna (SIMO, 1x2). Because of the different transmission paths, the receiver sees two differently faded signals. By using the appropriate method in the receiver, the signal-to-noise ratio can be increased.

TX diversity uses more TX antennas than the RX side. The simplest scenario uses two TX and one RX antenna (MISO, 2x1). The same data is transmitted redundantly over two antennas so that the multiple antennas and redundancy coding is moved from the mobile UE to the eNB. To generate a redundant signal, space–time codes are used, which additionally improve the performance and make spatial diversity usable.

Spatial multiplexing is not intended to make the transmission more robust; rather it increases the data rate. To do this, data is divided into separate streams; the streams are transmitted independently via separate antennas. In the open loop method, the transmission includes special sections that are also known to the receiver. The receiver can perform channel estimation accordingly. In LTE, open loop SM is based on cyclic delay diversity (CDD), in which the signals are transmitted by the individual antennas with a time delay. This introduces virtual echoes into orthogonal frequency-division multiplexing (OFDM)-based systems, so as to increase the frequency selectivity at the receiver. In the closed loop SM, the UE reports the channel status to the eNB via a feedback channel. This makes it possible to respond to changing circumstances.

Beamforming is the method used to create the radiation pattern of an antenna array. The transmitter can form a user-specific beam through beamforming technology and has the optimum signal strength only along the center of the beam. The adaptive beamformer adjusts the beam in real time to the moving receiver. Beamforming is

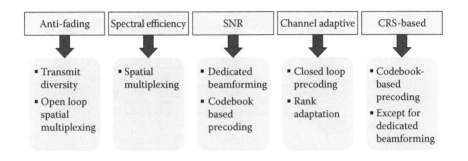

Figure 4.13   MIMO design principles.

well-suited for cell-edge users that operate on the lower signal-to-noise ratio (SNR) condition, and it improves the throughput on the cell edge but not sufficiently close to the cell center.

The usage and design principles of different MIMO schemes are shown in Figure 4.13.

4.2.1.2 Transmit Diversity

The principle of transmit diversity relies on the simultaneous transmission of the signal via two or more independent antennas to get independent signal replicas. In this way there is a high probability that both signals will not fade simultaneously and the deepest fades can be avoided. Therefore, transmit diversity improves the signal quality and achieves a better signal-to-interference ratio (SIR) at the receiver side with proper combining of the different paths. Transmit diversity in LTE is based on space-frequency block coding (SFBC) in the case of two transmit antennas, and combined SFBC and frequency-switched transmit diversity (FSTD) for the case of more than two antennas (Figure 4.14). In general, transmit diversity may provide significant gain by exploiting spatial diversity when channel state information is not available or is unreliable due to channel variation. Transmit diversity mode is one of the most important factors in MIMO support to keep the system coverage as large as possible and provide reliable data transmission for high-mobility UE as well.

In LTE Rel-8, the maximum number of transmit antennas used to transmit common control channels such as physical DL control channel (PDCCH) and physical broadcast channel (PBCH) is four. That means 2Tx and 4Tx transmit diversity (Figure 4.15) could be used for common channels in Rel-8.

4.2.1.3 Spatial Multiplexing

Spatial multiplexing allows the transmission of different streams of data simultaneously on the same downlink resource block(s). These data streams may belong to one single user (SU-MIMO) or to different users (MU-MIMO). While SU-MIMO

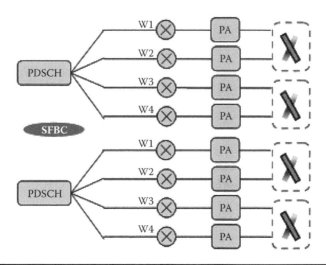

Figure 4.14   Transmit diversity with multiple antennas.

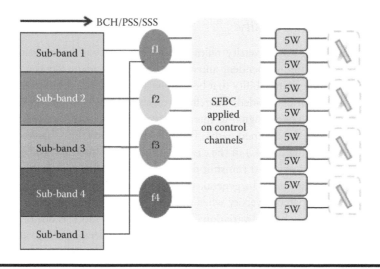

Figure 4.15   Transmit diversity scheme for control channels.

increases the data rate of one user, MU-MIMO allows the overall system capacity to increase. Spatial multiplexing is only possible if the mobile radio channel allows it. Decorrelated channels are required for spatial multiplexing for highest performance.

There are two operation modes in SU-MIMO spatial multiplexing: the closed loop spatial multiplexing mode and the open loop spatial multiplexing mode.

In the closed loop spatial multiplexing mode, the eNB applies the spatial domain precoding on the transmitted signal, taking into account the precoding matrix indicator (PMI) reported by the UE so that the transmitted signal matches the spatial channel experienced by the UE. In other words, the closed loop techniques imply the use of a codebook for MIMO transmission.

Open loop spatial multiplexing differs from its closed loop counterpart mainly in the selection of the precoding matrix. Open loop transmission is used when no accurate feedback can be provided by the UE, for example, due to high velocity or in the case of common channels where no feedback is naturally available. Open loop SM may provide higher throughput performance for UE with high geometry with medium to high mobility.

Based on SU-MIMO, multiple UEs can be co-scheduled on the same time–frequency resources by the eNB. MU-MIMO is a technique that utilizes the fact that MIMO transmission is organized as parallel transmissions of several information streams that are transmitted on separate antenna layers. The separate streams can be directed to separate users by beamforming or channel decoding or both. MU-MIMO is of particular interest for LTE-A because of its ability to increase total cell throughput rather than simply providing very high peak data rates to a small proportion of the UEs in the cell.

The MU-MIMO scheme can provide higher system throughput by exploiting multiuser scheduling gain. It can be implemented by two different modes: transparent mode and nontransparent mode. In transparent mode, the scheduled UEs are not explicitly aware of how many other UEs are co-scheduled or which precoding vectors are used for any co-scheduled UEs. On the other hand in nontransparent mode, scheduled UEs will receive explicit signals to inform them of how other UEs are co-scheduled in pairs with them.

4.2.1.4 Beamforming

Beamforming is supported for improving data coverage when the UE supports data demodulation using the UE-specific reference signal. The eNB generates a beam using the array of antenna elements (e.g., array of 8 antenna elements) and then applies the same precoding to both the data payload and the UE-specific reference signal with this beam. It is noted that the UE-specific reference signal is transmitted in a way such that its time–frequency location does not overlap with the cell-specific reference signal.

Beamforming can be categorized as one-stream beamforming and multiple-stream beamforming. Multiple-stream is a natural evolution of the TDD system, following the one-stream beamforming already adopted in LTE Rel-8. In multiple-stream beamforming, multiple spatial channels are utilized to transmit multiple symbols simultaneously. While one stream beamforming can be applied by either exploiting the channel reciprocity (singular value decomposition [SVD]-based beamforming) or performing direction of arrival (DoA) beamforming, multistream beamforming (BF) builds solely on reciprocity. When considering the number of streams to be deployed, one has to consider the UL resources needed to sustain the DL process, mainly the UL sounding effort. From this perspective, two streams are believed to be able to provide a good trade-off between data rate increase and complexity of UL transmission, and two-stream beamforming has been defined in LTE Rel-9 as transmission mode 8.

4.2.1.5 MIMO Scheme Adaptation

The features of various MIMO schemes mentioned previously are summarized in Table 4.2.

Differentiation is required among transmit diversity, spatial multiplexing, and beamforming. The ability to dynamically adapt to the channel optimal MIMO scheme as channel conditions change is a key focus of LTE systems. So it is required that the eNB scheduler has the capability to optimally select the MIMO scheme that suits the channel conditions of the mobile device.

Antenna configurations at the eNB and the overall channel environment between the eNB and the UE have an impact on the types of MIMO schemes that

Table 4.2   Features of MIMO Schemes

Features of Transmit Diversity	*Features of Open loop Spatial Multiplexing*
• Cell-specific transmit diversity scheme • One scheme for all control channels but primary synchronization signals (PSS) and secondary synchronization signals (SSS) • Support for fallback operation • SFBC (2TxAnt), SFBC + FSTD (4TxAnt)	• Large-delay CDD (cyclic delay diversity) • Support for rank adaptation • Rank 1 open loop spatial multiplexing (OL-SM) = Transmit diversity • Up to 2 codewords transmissions • No precoding (2TxAnt); precoder cycling (4TxAnt)
Features of Closed Loop Spatial Multiplexing	*Features of MU-MIMO*
• Codebook-based precoding due to CRS-based transmissions • Codebook subset restriction • Support for rank adaptation • Up to 2 codeword transmissions	• Codebook-based precoding • Developed under the assumption of highly correlated TX antennas
Features of Dedicated Beamforming	
Non-codebook-based precoding relying on dedicated reference signals (RS)	

are available. The antenna propagation environment affects the properties of the MIMO channel via many factors including UE mobility, path loss, shadow fading, and the polarization of the transmitted signal. This, in combination with the particular antenna configuration at the transmitter and receiver, determines the overall channel characteristics.

Narrowly spaced antennas are ideal for supporting beamforming, while widely spaced or cross-pole antennas are ideal for spatial multiplexing and transmit diversity. Closed loop SM is appropriate in high-SINR areas with rich scattering environments, in combination with suitable antenna configurations. Measurements at the base station receiver and feedback signals by the mobile device help the base station determine the number of streams that can be supported across single or multiple users. When the channel conditions become less favorable to spatial multiplexing, beamforming and transmit diversity can be used.

On the other hand, to yield good performance over a broad range of scenarios, an adaptive multistream transmission scheme is provided in which the number of parallel streams can be continuously adjusted to match the instantaneous channel conditions. When channel conditions are very good, up to four streams can be transmitted in parallel, yielding data rates up to 300 Mbps in a 20-MHz bandwidth. When channel

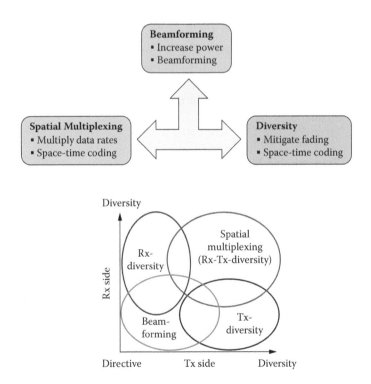

Figure 4.16   MIMO adaptation.

conditions are less favorable, fewer parallel streams are used. The multiple antennas can be used under this condition with a beamforming transmission scheme that improves overall reception quality and, as a consequence, system capacity and coverage.

To achieve good coverage in large cells or to support higher data rates at cell borders, one can employ single-stream beamforming transmission as well as transmit diversity for common channels.

The scenarios of MIMO adaptation are summarized in Figure 4.16.

4.2.2 MIMO Evolution in LTE

4.2.2.1 MIMO in Rel-8

All the previously discussed MIMO schemes are already utilized in LTE Rel-8 in downlink transmission, defined in different downlink transmission modes. There are a total of seven transmission modes defined for LTE Rel-8 downlink transmission, as listed in Table 4.3.

For LTE Rel-8 uplink transmission, the UE supports only one transmit antenna and multiple receive antennas. As such, SU-MIMO cannot be supported on the UL but the UL can support MU-MIMO transparently.

4.2.2.2 Dual-Layer Beamforming in Rel-9

While downlink single-layer beamforming is already supported in LTE Rel-8, for LTE Rel-9 it is considered to be extended with the support of dual-layer beamforming. This enhancement combines the benefits of beamforming and spatial multiplexing, which will primarily improve the throughput of users that are experiencing good channel conditions.

Table 4.3 Transmission Modes in LTE Rel-8

Transmission Mode	Transmission Scheme of PDSCH
1	Single-antenna port, port 0
2	Transmit diversity
3	Transmit diversity if the associated rank indicator is 1, otherwise open loop spatial multiplexing
4	Closed loop spatial multiplexing
5	Multiuser MIMO
6	Closed loop spatial multiplexing with a single transmission layer
7	Single-antenna port, port 5

Dual-layer beamforming simultaneously transmits two beamforming data streams for spatial multiplexing. In addition, it can retain the features of single-stream beamforming technology in order to expand coverage, increase cell capacity, and minimize interference to improve the reliability of cell-edge users as well as the throughput of cell-center users.

Depending on user scheduling, this technology can be categorized into single-user and multiuser dual-stream beamforming.

In single-user (SU) dual-stream beamforming technology, the eNB measures the upstream channels to obtain their status information. It then calculates two beamforming vectors according to this upstream channel information and uses the vectors for downlink beamforming of the two data streams to be transmitted. In SU dual-stream beamforming technology, a UE can simultaneously receive two data streams and thereby benefit from beamforming gains and spatial multiplexing gains at the same time. In this way, it can achieve a higher data rate than single-stream beamforming technology, which increases the capacity of the system.

In multiple user (MU) dual-stream beamforming technology, the eNB matches multiple users according to upstream channel information or UE feedback. It then, according to certain criteria, generates beamforming vectors and uses the vectors for the beamforming of each UE and each stream. MU dual-stream beamforming technology uses the smart antenna beam direction to enable multiuser spatial division multiple access (SDMA).

LTE Rel-9 supports dual-layer beamforming using UE-specific reference signals for both TDD and FDD. The goal of the design of the UE-specific demodulation reference signals and the mapping of the physical data channel to resource elements is forward compatibility with LTE-A demodulation reference signals.

In LTE Rel-9, a new transmission mode 8 has been introduced for dual-layer beamforming scheduling, and a corresponding new channel quality indicator (CQI), PMI, and rank indicator (RI) feedback scheme is also specified. Transmission mode 8 supports two orthogonal streams of UE-specific RS for multiuser MIMO and quasi-orthogonal MU-MIMO with the aid of scrambling IDs.

4.2.2.3 MIMO in LTE-A

The LTE-A target is to support downlink peak spectrum efficiency of 30 bps/Hz and uplink peak spectrum efficiency of 15 bps/Hz. The assumed antenna configuration is 8x8 or less for the DL and 4x4 or less for the UL.

One of the advanced features in LTE-A is support for high-order MIMO. It has been shown that the capacity of a wireless system grows linearly with the number of independent antennas. A higher number of transmit antennas also enables much narrower beamforming to extend the coverage.

LTE Rel-8 is designed to support {1,2,4} transmit antennas at the eNB and {1,2,4} receive antennas at the UE. Downlink transmission at the eNB can be

Table 4.4   Target of Average Spectrum Efficiencies and Cell-Edge Throughput

Radio env. Antenna Configuration		Case 1 [bps/Hz/ cell/user]	Radio env. Antenna Configuration		Case 1 [bps/Hz/ cell]
	2 × 2	0.07		2 × 2	2.4
DL	4 × 2	0.09	**DL**	4 × 2	2.6
	4 × 4	0.12		4 × 4	3.7
UL	1 × 2	0.04	**UL**	1 × 2	1.2
	2 × 4	0.07		2 × 4	2

mapped to a maximum of four layers. Uplink transmission at the UE supports only single antenna mode, while the eNB supports a maximum of four receiving antennas. The LTE-A goal is an instantaneous downlink peak data rate of 1 Gbps. In addition to increased total bandwidth, this target can only be achieved by increasing the number of spatial multiplexed data streams toward a single user.

LTE-A should target the average spectrum efficiencies and the cell-edge user throughput in different environments as listed in Table 4.4.

In LTE-A, up to 8 cell-specific antenna ports (transmit antennas) are supported in the downlink. Up to 4 antenna ports (transmit antennas) are supported in the uplink. The number of receive antennas are receiver implementation specific. At least two receiver antennas are assumed on the UE side. The antenna configuration (coordinated/uncoordinated antennas, co-polar/cross-polar configuration, etc.) is implementation specific.

One more downlink transmission mode 9 has been specified in LTE-A for 8-layer spatial multiplexing. UL spatial multiplexing and UL transmit diversity with multiple TX antennas are also standardized. Enhanced MU-MIMO and CoMP transmission/reception are some of highlights in LTE-A.

The MIMO evolution from Rel-8 to LTE-A is summarized in Figure 4.17.

4.2.2.4 Resource Multiplexing for PDSCH

To support both Rel-8 and Rel-9 legacy UEs and LTE-A UEs in an LTE-A system with more than 4 transmit antennas at the eNB, physical DL shared channel (PDSCH) resources should be multiplexed properly. In general, there are two ways to achieve this multiplexing, as shown in Figure 4.18.

The first alternative is time-division multiplexing (TDM), in which one subframe is assigned either for transmitting for Rel-8 legacy UEs or for transmitting

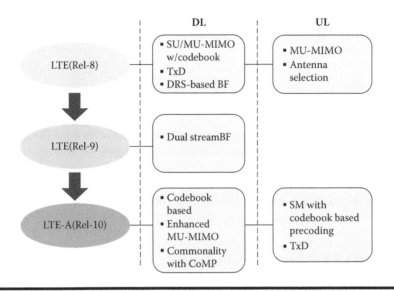

Figure 4.17   MIMO evolution.

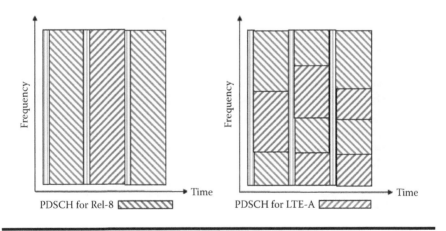

Figure 4.18   Resource multiplexing. (Left) TDM. (Right) FDM.

for LTE-A UE. The other way is frequency-division multiplexing (FDM), in which, within the same subframe, different resource blocks (RBs) could be assigned for Rel-8 legacy UE and LTE-A UE simultaneously. For FDM, additional RSs (> 4Tx) to support high-order MIMO are limited in the RBs assigned to LTE-A UE.

TDM and FDM multiplexing should both be considered at network implementation. In some scenarios, FDM multiplexing is more preferable because there is no impact to legacy UE. The impact-free to legacy UE can be achieved because additional RSs (> 4Tx) to support high-order MIMO are limited in the resources assigned to LTE-A UE, and FDM may support both legacy UE and LTE-A at the same time with low latency.

4.3 DL MIMO

In LTE Rel-8, the maximum number of transmit antennas used for the downlink is four. That leads to the use of 2Tx and 4Tx transmit diversity and up to 4 layers of spatial multiplexing for the downlink. In order to achieve peak spectrum efficiency in LTE-A, a larger number of antennas can be employed in the downlink so that higher-rank transmission is possible as compared with LTE. Therefore, 8x8 downlink MIMO transmission is introduced in LTE-A.

4.3.1 MIMO Scheme Enhancement

In LTE Rel-8, transmit diversity is a main MIMO scheme that has been widely used in live networks. In LTE-A, the original intention of introducing up to 8-layer transmission in the downlink is only to increase the peak data rate for data channels such as PDSCH. High-order transmit diversity with 8 transmit antennas could bring some diversity gain over transmit diversity using 4Tx. However, it is well known that the spatial diversity gain achieved by transmit diversity schemes is quickly saturated as the diversity order goes higher. This implies that if the diversity order is higher than a certain level (e.g., diversity order 4), the gain from the transmit diversity scheme is insignificant and it is more worthwhile to keep the implementation complexity at a minimum. Since the typical channel is frequency selective and therefore sufficiently rich in frequency diversity, the benefit of 8Tx transmit diversity over 4Tx transmit diversity is not apparent.

In addition, high-order diversity gain may not be that significant if the channel estimation error is taken into account. In order to achieve acceptable channel estimation accuracy, orthogonal reference signals are needed for each antenna, which could bring additional RS overhead on top of those defined in Rel-8, in which 4Tx reference signal overhead is already higher than 14%. Considering limited resources in control regions, such an increase in RS overhead may not be tolerable. On the other hand, because Rel-8 UEs and LTE-A UEs share the same control regions, it would be preferable in LTE-A systems to maintain the original number of common RS ports (for example 4) of LTE Rel-8 and reuse the 4Tx TxD in LTE Rel-8 with UE-transparent virtualization so that both LTE and LTE-A UEs can receive the control channel at the same time.

In summary, in LTE-A it is preferable to reuse the 4Tx TxD in LTE Rel-8 for the eNB having 8Tx antennas. The most obvious choice for 8 TxD in LTE-A is SFBC-FSTD, since it has been adopted as a 4 TxD scheme in LTE Rel-8.

$$X_{SFBC-FSTD} = \begin{bmatrix} s_1 & -s_2^* & 0 & 0 & & & & \\ s_2 & s_1^* & 0 & 0 & & & 0_{4\times4} & \\ 0 & 0 & s_3 & -s_4^* & & & & \\ 0 & 0 & s_4 & s_3^* & & & & \\ & & & & s_5 & -s_6^* & 0 & 0 \\ & & 0_{4\times4} & & s_6 & s_5^* & 0 & 0 \\ & & & & 0 & 0 & s_7 & -s_8^* \\ & & & & 0 & 0 & s_8 & s_7^* \end{bmatrix}$$

Open loop spatial multiplexing will also be supported in LTE-A for simpler MIMO operation and for terminals with less accurate channel state information. Generally, an open loop SM is used when no accurate feedback can be provided by the UE. It is beneficial for a UE having high geometry with high mobility and is employed to provide reasonable UE throughput even for medium to high mobility. Therefore, open loop SM should be supported in LTE-A systems as well. In the LTE system, full-rank open loop SM transmission is supported even for a 4Tx antenna system. However, for LTE-A, it should be investigated whether rank 8 open loop SM transmission is needed; full-rank transmission in 8Tx might be unrealistic since the poor channel estimation performance may significantly degrade rank 8 transmission performance in a high mobility scenario. Therefore, it is preferable that the maximum rank for 8Tx open loop SM is smaller than 8 so that implementation complexity is not unnecessarily increased.

LTE-A specifications will support 8 transmit antennas and 8x8 MIMO. The supported MIMO modes will follow LTE Rel-8. Closed loop spatial multiplexing operation is likely the main MIMO mode for 8Tx antenna usage; precoding gain is available also for more practical setups such as 8x2. Closed loop techniques imply the use of a codebook for DL transmission. A codebook-based precoding is employed for a LTE Rel-8 system with 2Tx and 4Tx antennas and each transmit antenna has its own codebook, such as a discrete Fourier transform (DFT)-based codebook (for 2Tx) and a Householder-based codebook (for 4Tx). In order to support 8x8 DL MIMO transmission, it is necessary to define the 8Tx codebook to achieve the peak data rate.

In brief, the main features of DL MIMO introduced in LTE-A include the following:

- Supports up to 8-layer SU-MIMO transmission, 4-layer MU-MIMO transmission, and up to 2-layer transmission per co-scheduled UE; supports dynamic SU/MU-MIMO switching (transmission mode 9). The feedback specified for SU-MIMO can be applied for MU-MIMO operation.
- Supports demodulation reference signal (DMRS)-based PDSCH demodulation; precoding flexibility and signaling overhead reduction is the main reason for adopting precoded RSs.

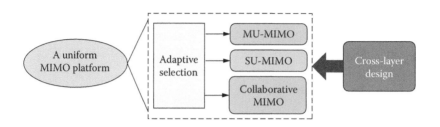

Figure 4.19 Multimode adaptive MIMO for DL/UL.

- Supports channel state information (CSI)-RS-based measurement and thus can reduce overhead due to the sparse configuration in time domain. It is a simplified implementation due to the nested structure.
- Supports a double codebook-based feedback framework.

Moreover, LTE-A supports multimode adaptive MIMO in DL/UL transmission. LTE-A uses adaptive MIMO to accommodate the demand for higher data rates and wider coverage in next-generation broadband wireless access. Generally, SU-MIMO is used for peak user data rate improvement, MU-MIMO is used for average data rate enhancement, and collaborative/network MIMO is deployed for cell-edge user data rate boost (see Figure 4.19).

4.3.2 8Tx MIMO Design

One of the most significant improvements in LTE-A is the support for 8-antenna transmission, which in theory may lead to some technical benefits.

First of all, larger beamforming gain to a UE in SU-MIMO under the same total transmit power can be achieved with 8 antennas. That implies that the lower SNR is required to attain the same block error rate (BLER), and the lowered average interference leaks to other cells. As a result, sector and cell-edge throughput will be improved. In addition, in an 8-antenna system, up to 8 spatial streams of data can be supported in theory, whether they are targeted for one UE or more typically multiple UEs. So an enhanced MU-MIMO scheme with 8 antennas at the eNB will support MU-MIMO with better performance than only 4 antennas.

Furthermore, with more transmit antennas, reduced individual power amplifier (PA) size is achieved when the number of PAs is doubled and the total output power is kept the same. Whether there is a cost advantage is debatable, but the implementation can choose to halve the individual PA size or otherwise double the total TX power.

However, the challenge of the 8-antenna eNB lies in the support of legacy Rel-8 UEs that were designed to support up to 4Tx only. In other words, an 8-antenna eNB needs to support a mixture of LTE-A and Rel-8 UEs. Accordingly, a few guidelines on 8Tx support have been taken into account in LTE-A development.

An 8-antenna eNB should support Rel-8 UEs with minimal, if any, performance impact and with improved performance compared to 4Tx downlink transmission if possible. Rel-8 UE is not capable of estimating 8 channels in PDCCH decoding or PMI-based closed loop operation for PDSCH, so an 8-antenna eNB must be able to function as a legacy eNB with up to 4 antennas. An antenna virtualization scheme can be exploited as one of the ways to cause 8 antennas to be seen as 4 at the Rel-8 UE.

4.3.2.1 Reference Signals for 8Tx

In order to support LTE-A components, such as up to 8-layer MIMO and CoMP, additional RSs have to be defined.

A straightforward design philosophy that supports 8-antenna operation for LTE-A UEs can be easily understood to simply extend the existing 4Tx cell-specific reference signal (CRS) design and 4Tx precoding codebook to 8Tx. This is a CRS-centric design philosophy where CRSs are provided with sufficient density such that a UE can derive, for data demodulation, the effective channel based on the CRS and also the applied transmission scheme such as precoding weights in the form of PMI. The same CRSs are also used for measurements and link adaptation (e.g., CQI, PMI, etc.).

In the CRS-centric design, the MIMO receiver should be implemented according to the number of antenna ports and transmission mode. The precoding information should be available to a UE in each transmission (codebook-based precoding). All physical antenna port channels should be estimated for demodulation, and channel estimation performance is the same irrespective of the rank.

However, provisioning of 4Tx cell-specific RSs in LTE Rel-8 has about 14% overhead. This is a fixed CRS overhead with possible additional overhead incurred if demodulation reference signal (DMRS) is also used. Therefore, the provisioning of 8Tx CRSs will come with much increased system overhead if a simple extension of the same CRS-centric philosophy is adopted.

In the DMRS-centric design philosophy, DMRSs with sufficient density are provided for data demodulation, but low-density CRSs may still be present for measurement purposes only for decoding of common control traffic. Cell-specific RSs are used for two purposes: receiver channel estimation for data demodulation (if not using user-specific RSs) and for measurement and reporting (e.g., CQI, PMI, etc.) and common control messages.

In the DMRS-centric design, a single MIMO receiver can be used regardless of the number of antenna ports and transmission mode (except TxD). Non-codebook-based precoding is used at the eNB side. Only the virtual antenna port channel needs to be estimated for demodulation in a localized manner, and channel estimation performance could be different according to the rank.

Table 4.5 Pros and Cons of CRS and DMRS Design Philosophies

	Pros	*Cons*
CRS-based	• Enable "decent" channel estimation to eNB antennas, covering entire bands (two band edges) and subframe • One set of RS for both purposes: demodulation and measurement • Fixed RS overhead regardless of rank and number of UEs	• Overhead when signaling the transmission scheme (e.g., precoding) such that UE can construct its effective channel • Limitation of allowed precoding weights to a codebook, thus performance degradation • RS cannot benefit from the increased SNR seen by the UE on its data burst, which enjoys a higher total transmit power and a precoding gain
DMRS-based	• RSs also benefit from the increased SNR seen by the UE on its data burst • eNB can tailor precoding weights to individual UE or groups of UEs more flexibly without any constraint • Overhead for low-rank transmission is generally low • Allows greater flexibility in eNB implementation. The eNB may potentially use different antenna ports, antenna array techniques, and coordinated MIMO modes using parameters and methods blind to the UE.	• FDD operation still requires CRS for UE to assist in precoding, even though pilot density can be much lower if CRS is not used for demodulation • Channel estimation quality suffers from the sub-band-edge effect • RS overhead increases with the transmission rank and it can lead to more total overhead than CRS at high transmission rank.

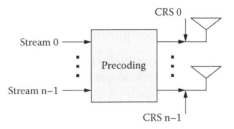

Figure 4.20 CRS-based precoding.

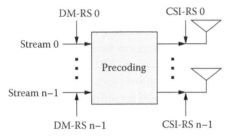

Figure 4.21 DMRS-based precoding.

The above two different design philosophies have pros and cons (see Table 4.5). It is useful, at least for LTE-A operation, to reassess the two design philosophies in the 8-antenna context, even though the backward-compatibility requirement mandates, at least partially, are a fallback to the existing CRS-based design.

When designing the new reference signals, in addition to the CRS defined in LTE Rel-8 and Rel-9 and inband channel estimation, other measurements need to be considered in order to enable adaptive multiantenna transmission. Two additional reference signals have been specified by 3GPP. They are depicted in Figures 4.20 and 4.21, and are explained as follows:

- Channel state information reference signal (CSI-RS): CSI-RS is used for channel sounding; that is, estimation of the channel quality in different frequencies to those assigned to the specific UE. The signals are located in a sparse grid and require low overhead.
- UE-specific demodulation reference signal (DMRS): DMRS is precoded in the same way as the data when non-codebook-based precoding is applied.

The grid pattern for CSI-RS should be extended from the dual-stream beamforming mode defined in Rel-9 where code division multiplexing (CDM) between the RSs of two layers is utilized.

4.3.2.2 8Tx Precoding

4.3.2.2.1 Codebook Design Requirement

The current Rel-8 codebook design is mainly targeted at SU-MIMO, from which the Rel-8 MU-MIMO codebook design is adopted. It has been realized that codebooks should be designed to match the underlying channel characteristics. If optimum performance is desired, the precoding matrix should be matched to certain transmission bandwidths over the air interface. Otherwise, the matching characteristic of the precoding matrix would be degraded, which might result in much decreased performance in practice. For example, a Grassmannian line packing (GLP)-based codebook is shown to achieve near optimal performance for independent identically distributed Rayleigh fading channels. However, those GLP codebooks do not perform as well in correlated fading channels or block diagonal fading channels where dual-polarized antennas are used. An all-weather codebook is desired that performs well in different fading scenarios (e.g., suburban macro fading, urban micro fading, urban macro fading), for different antenna spacing (e.g., 0.5λ, 10λ), and for different antenna polarization profiles. Other attractive features, such as constant modulus, finite alphabet, and a nesting property, are also desired for different usage. Together, these items make designing the codebook difficult. As a direct result, the current LTE Rel-8 codebook designs for SU-MIMO are a compromise of many factors, and the live network can prove that the current codebook has many excellent properties to reduce the codeword search process. As mentioned above, the unitary property, a finite alphabet, nesting structure, and constant modulus are all adopted in LTE-A codebook design.

Because we need to support up to 8Tx antennas and have better support for MU-MIMO in LTE-A, it is straightforward to argue qualitatively that more feedback bits per user are needed. However, a 4-bit, 8-antenna codebook has shown very limited performance improvement over a Rel-8 4-bit, 4-antenna codebook, especially for MU-MIMO. There are some major reasons behind this. First, since the spatial dimension of the channel is increased from 4 to 8, it will expect higher requirements on feedback bits to better represent the higher-dimensional channel. Second, for MU-MIMO, both the desired signal power and the undesired interference power are impacted by the codebook design, and it is natural to see requirements on feedback accuracy rising, while in comparison to SU-MIMO, only the desired signal power is impacted by the codebook design. Third, the number of feedback bits must increase linearly with the SNR in order to achieve capacity. Therefore, we need new generalization methods for the LTE-A codebook.

4.3.2.2.2 Codebook Design Guidelines

The LTE-A 8-Tx codebook is designed for various antenna setups and spatial channel conditions, and priority is given to the following three 8Tx setups:

a. Uniform linear array (ULA) with λ/2 (half wavelength) spacing
b. Four dual-polarized elements with λ/2 spacing between two elements

c. Four dual-polarized elements with 4λ (larger) spacing between two elements

In a practical scenario mentioned previously, a 10λ dimension is usually for the whole array at the eNB, that is, for an 8Tx system, spacing between antennas is less than 1.5λ. This means that the channel is highly correlated at the eNB side. In this correlated channel scenario, it is well known that the smaller size of the codebook can still provide sufficient spectral efficiency. One solution to decrease the necessary physical space for building 8Tx antennas is to employ dual-polarized antennas. A colocated dual-polarized antenna system provides a cost- and space-efficient alternative to other MIMO antenna systems. Thus, the LTE-A 8Tx codebook should provide reasonable spectral efficiency, not only with single polarized antennas, but also with dual-polarized antennas.

The guidelines discussed in the following text are enforced in LTE-A for precoder codebook composition.

4.3.2.2.2.1 Constraint Alphabet

— Low-complexity codebook design can be attained by choosing the elements of each matrix/vector from a small set. An example in LTE Rel-8 of the small set is the 4-alphabet size $\{\pm 1, \pm j\}$. The use of quadrature phase-shift keying (QPSK) and 8PSK $\left\{ \pm 1, \pm j, \pm \dfrac{(1+j)}{\sqrt{2}}, \pm \dfrac{(-1+j)}{\sqrt{2}} \right\}$ alphabet sets avoids the need for computing matrix/vector multiplication.

All the codebooks with a constraint alphabet have the same complexity to calculate the CQI for a given precoder for a given rank and may eliminate all multiplication for CQI calculation. Indeed, the operations $\mathbf{HW}$ (with $\mathbf{H}$ the channel matrix and $\mathbf{W}$ a precoder) used in the MMSE filter do not require any complex multiplication when the entries of $\mathbf{W}$ are constrained to the alphabet $\{0, 1, -1, j, -j\}$. The only complex multiplication comes from the product $\mathbf{H}^H\mathbf{H}$.

4.3.2.2.2.2 Constant Modulus

— Codebook design with constant modulus property is beneficial for avoiding unnecessary increase in peak-to-average power ratio (PAPR), which makes UE PMI selection less complex and avoids power amplifier imbalance at eNB side, thereby ensuring power amplifier balance and guaranteeing equal power transmission from each antenna (especially important for rank 1 transmissions).

4.3.2.2.2.3 Chordal Distance Maximization

— The codebook design should consider maintaining certain minimum chordal distances between codewords. For two random matrixes U, V, the chordal distance's definition is $d(U,V) = \dfrac{1}{\sqrt{2}} \left\| UU^H - VV^H \right\|_F$. The design achieves high chordal distances of (0.7071, 1.0, 0.7071) for rank 1, rank 2, and rank 3, respectively.

For uncorrelated channels, chordal distance is known to be an important performance measure. To be specific, the performance of the codebook is governed by the minimum chordal distance between the codewords in the codebook, which is represented by $G_1 = \min\limits_{\mathbf{f} \in \tilde{F}, 1 \le n < m \le N} \mathrm{dist}(\mathbf{f}_n, \mathbf{f}_m)$, where $\tilde{F}$ is the codebook, $\mathbf{f}_i$ is the ith codeword of the codebook, and $\mathrm{dist}(\mathbf{f}_n, \mathbf{f}_m)$ stands for the chordal distance between the two codewords. Therefore, one of the design objectives is to maximize G_1 for uncorrelated channels. This objective can be represented by the following equation: $F = \arg\max\limits_{\tilde{F}} \min\limits_{\mathbf{f} \in \tilde{F}, 1 \le n < m \le N} \mathrm{dist}(\mathbf{f}_n, \mathbf{f}_m)$.

For correlated channels, the performance of the codebook is partly reflected in the null-direction gain of the codebook projected to the array antennas subspace, which can be expressed as $G_2 = \min\limits_{\mathbf{f} \in \tilde{F}, 0 \le \theta \le \pi} \left| \mathbf{a}(\theta)^* \mathbf{f} \right|^2$, where $\mathbf{f}$ is a codeword of the codebook $\tilde{F}$ and $\mathbf{a}(\theta)$ is the array antenna response vector. Therefore, a second design objective is to maximize G_2 for the codebook. This objective can be represented in the following equation: $F = \arg\max\limits_{\tilde{F}} \min\limits_{\mathbf{f} \in \tilde{F}, 0 \le \theta \le \pi} \left| \mathbf{a}(\theta)^* \mathbf{f} \right|^2$.

It is worth mentioning that in the MU-MIMO scenario, chordal distance threshold is used to avoid pairing two UEs with channels correlated with each other, and this also can reduce the computation of comparing all the combinations.

4.3.2.2.2.4 Nested Property — A nested structure (i.e., lower rank codewords are a subset of higher rank codewords) aims for simplicity of storage and computation. The nested property across ranks is beneficial to accommodate rank override. For each precoder matrix in a certain rank, there is at least one corresponding column subset in all the codebooks of the lower ranks. That implies that the rank 3 codebook is a subset (the first three columns) of the rank 4 codebook, the rank 2 codebook is a subset (the first two columns) of the rank 3 codebook, and the rank 1 codebook is a subset (the first column) of the rank 2 codebook. The nested property can possibly help reduce CQI calculation complexity.

4.3.2.2.2.5 Unitary Precoder — In a unitary precoder, the column vectors of a precoder matrix must be pair-wise orthogonal to one another. While not necessary, it is a sufficient condition for maintaining constant average transmitted power. This constraint is also used in designing the codebook at least for some relevant ranks.

4.3.2.2.2.6 Effective Codebook Size — The effect of codebook size on transmit nulling performance is characterized in terms of the interference suppression ratio metric. A larger codebook can provide higher-resolution feedback and implies more accurate knowledge of the channel at the transmitter, resulting in improved throughput. An increase in the codebook size will improve the codebook performance. However, a big codebook will increase the computational complexity of precoder selection and feedback overhead. So, the balance of performance and complexity should be evaluated.

In a real deployment, there are various propagation environments and antenna configurations at the eNB, such as ULA, split antenna array, cross-polarized

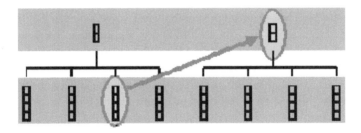

Figure 4.22 **Hierarchical structure.**

antenna, large antenna spacing, and small antenna spacing. It is impossible to have a constant codebook optimized for all of the configurations.

4.3.2.2.3 Codebook Design Details

The previous guidelines and the experience in the LTE Rel-8 codebook design form the foundation for LTE-A 8Tx codebook design. Considering many contributors' proposals, 3GPP adopted a hierarchical structure–based dual-index codebook for 8Tx antenna precoding. The dual codebook structure was introduced in LTE-A to improve feedback accuracy without excessive overhead increases; that is, a precoder **W** for a sub-band is obtained as a matrix multiplication of the two matrices, and each of the two matrices belongs to a separate codebook.

In the hierarchical structure (Figure 4.22) adopted in LTE-A, the codebook entries are organized in hierarchical levels. Such codebook structures can enable the UE feedback to be aggregated over multiple reports, giving more accurate knowledge of the MIMO channel at eNB and resulting in improved throughput. By aggregating the feedback over multiple reports, the codewords can be arranged in a hierarchical structure so that the feedback in one report can use the channel estimates from the previous feedback.

Hierarchical structure–based codebooks exploit the time/frequency correlation of the channel to enable successive refinement of the quantization. It relies on the fact that MIMO/CoMP typically works at low speed where time correlation between two feedback instances is large.

A precoder with hierarchical structure for a sub-band is composed of two matrices; each of the two matrices belongs to a separate codebook. The two codebook matrices together determine the precoder. One of the two matrices targets wideband and/ or long-term channel properties; the other matrix targets frequency-selective and/or short-term channel properties. The first precoder $\mathbf{W}_1$ takes care of correlation properties of the channel and compresses the channel into fewer dimensions. Correlation properties remain roughly constant over bandwidth and change slowly over time, so $\mathbf{W}_1$ does not need to be reported often or in a frequency-selective manner over bandwidth. The second precoder $\mathbf{W}_2$ tries to match the instantaneous properties of the effective channel, such as phase alignment for constructive combining of transmitted signals on the receive side or orthogonalization of the effective channel. Note that Rel-8 precoder feedback can be deemed as a special case of this structure.

The separation of the PMI into two indices provides a computationally efficient way for the UE to search for the preferred precoder in the codebook. Sequentially searching a first index in an N_1-bit codebook and a second index in an N_2-bit codebook is considerably less complex than searching for a single index in an $(N_1 + N_2)$-bit codebook. So the separation of the codebook into two parts is an efficient way of increasing the codebook size while limiting the increase in computational complexity. Further reductions in the computational complexity should, however, not be ignored. Rank nesting, constant modulus, and finite alphabet codebook elements have also been quoted for efficient computation and could be considered as long as they do not unreasonably impair the performance.

The overall precoder is thus formed as $\mathbf{W} = \mathbf{W}_1 \mathbf{W}_2$. The inner precoder $\mathbf{W}_1$ has a block diagonal structure, and targets wideband and long-term channel properties. The outer precoder, $\mathbf{W}_2$, targets the frequency-selective and short-term time channel property and adjusts the relative phase shift between polarizations, so the quantization codebook $\mathbf{W}$ adaptively changes according to the channel covariance matrix (see Figure 4.23).

The UE first selects the first precoder codebook $\mathbf{W}_1$ based on the long-term channel properties such as the spatial covariance matrix. This is done on a long-term basis, which is consistent with the fact that the spatial covariance matrix needs to be estimated over a long period of time and in a wideband manner. Conditioned upon $\mathbf{W}_1$, the UE selects $\mathbf{W}_2$ based on the short-term channel. This selection may also be conditioned upon the selected rank indicator.

Today, the dual-polarized antenna is the most likely antenna configuration for a large number of TX antennas. Therefore, performing wideband and/or long-term

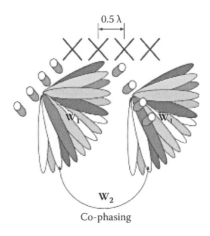

Figure 4.23 LTE-A dual-index precoding codebook.

precoding separately for each of the two polarizations and then co-phasing the two polarizations is a promising approach. This motivates a block diagonal structure of $\mathbf{W}_1$, where each block targets each polarization. $\mathbf{W}_1$, the block diagonal matrix of X, is denoted by $\mathbf{W}_1 = \begin{bmatrix} \mathbf{X} & \mathbf{0} \\ \mathbf{0} & \mathbf{X} \end{bmatrix}$.

From a cross-polarized array (CLA) point of view, the antennas can be divided into two groups. Antennas in each group are on the same polarization and thus can be treated as co-polarized ULA antennas. Theoretically, antennas in different groups are spatially uncorrelated. Therefore, the spatial correlation matrix X could be approximated by a block diagonal matrix.

When all possible X matrices are combined, the goal is to generate 16 4-DFT beams, which represents 4 times oversampling in order to provide sufficient spatial resolution. The beams presented by the DFT vectors offer maximum array gain and represent long-term properties of the channel. X will now present a group of beams rather than just one beam as a long-term channel property. The long-term properties can reflect the antenna configurations and possibly the propagation environment. The UE can estimate the long-term (e.g., 100 ms) wideband spatial correlation matrix based on downlink CSI-RSs and feed back a long-term wideband PMI corresponding to a quantized version of the spatial correlation property with a relatively long feedback interval.

Furthermore, to simplify implementation, the wideband precoding could be selected individually over each polarization of four antennas, and the linear phase shift induced by time misalignment between the two polarizations is captured in the short-term sub-band feedback $\mathbf{W}_2$. $\mathbf{W}_2$ targets the frequency-selective short-term time channel property and is composed of a combination of beam selection and co-phasing operations. To address the higher angle of departure spread scenarios, $\mathbf{W}_1$ should include multiple DFT-4 vectors (per block) from which the selection can be done separately for each sub-band with $\mathbf{W}_2$. This gives the freedom to address a larger angular space with $\mathbf{W}_2$ per sub-band, which is needed in case of a higher angle of departure spread (i.e., less-correlated scenarios).

The $\mathbf{W}_2$ precoder can take the form of one of at least 16 beams in a co-polarized ULA, or one of at least 8 beams for each group of 4 antennas in a closely spaced dual-polarized setup. Such a grid of beams is well known to provide good performance in channels with high spatial correlation. In this case, the feedback/codebook should provide a means of capturing the angle of departure spread. Since the channel becomes less correlated, frequency-selective PMI feedback needs to be supported. This especially becomes an issue for higher ranks, which typically occur in quite uncorrelated channels when the angle of departure spread is very large.

So, block diagonal designs of $\mathbf{W}_1$ are highly efficient for the common case of cross-polarized antennas but can also be made efficient for other antenna configurations. $\mathbf{W}_1$ matches the spatial covariance of a dual-polarized antenna setup, with at least 16 8Tx DFT vectors (different beams for the co-polarized ULA,

8 different beams for each group of 4 co-polarized antennas in the closely spaced antenna setup) generated from $\mathbf{W}_1$ and co-phasing $\mathbf{W}_2$. The beams will fully utilize all PAs and each beam achieves the maximum possible array gain. The 8Tx codebooks adopted by 3GPP in LTE-A are described in the following section.

As mentioned above, $\mathbf{W}_1$ is a block diagonal matrix of X. $\mathbf{W}_1$ matches the spatial covariance of a dual-polarized antenna setup with any spacing (e.g., $\lambda/2$ or 4λ) and provides good performance for both high and low spatial correlation. At least sixteen 8Tx DFT vectors are generated from $\mathbf{W}_1$ and co-phasing via $\mathbf{W}_2$ to match the spatial covariance of a ULA antenna setup.

4.3.2.2.3.1 Rank 1–4 Codebook

— For ranks 1 to 4, X in $\mathbf{W}_1$ is a 4xN_b matrix, consisting of N_b column vectors that correspond to adjacent beams in an angular domain extracted from an oversampled DFT-4 matrix, or in other words, subsets comprising N_b DFT-4 vectors (beams) are included into X. N_b denotes the number of adjacent beams contained in X. For each $\mathbf{W}_1$, adjacent overlapping beams are used to reduce the edge effect in frequency-selective precoding.

For ranks 1 and 2, there are 32 4Tx DFT beams defined for X (oversampled 8x), with beam index 0,1,2...,31, denoted as follows:

$$\mathbf{B} = \begin{bmatrix} \mathbf{b}_0 & \mathbf{b}_1 & \cdots & \mathbf{b}_{31} \end{bmatrix}, \quad [\mathbf{B}]_{1+m,1+n} = e^{j\frac{2\pi mn}{32}}, \quad m = 0,1,2,3, \, n = 0,1,\cdots,31$$

$$\mathbf{B} = \begin{bmatrix} 1 & 1 & \cdots & 1 \\ 1 & e^{j\frac{2\pi}{32}} & \cdots & e^{j\frac{2\pi*31}{32}} \\ 1 & e^{j\frac{4\pi}{32}} & \cdots & e^{j\frac{2\pi*2*31}{32}} \\ 1 & e^{j\frac{6\pi}{32}} & \cdots & e^{j\frac{2\pi*3*31}{32}} \end{bmatrix} \quad \mathbf{b}_n = \begin{bmatrix} 1 \\ e^{j\frac{2\pi*n}{32}} \\ e^{j\frac{2\pi*2n}{32}} \\ e^{j\frac{2\pi*3n}{32}} \end{bmatrix}, \quad n = 0,1,\cdots,31$$

For $\mathbf{W}_1$ ranks 1 and 2, X is formed by $N_b = 4$ adjacent overlapping beams, so there are a total of 16 $\mathbf{W}_1$ matrices per rank with beam indices in each matrix: {0,1,2,3}, {2,3,4,5}, {4,5,6,7},..., {28,29,30,31}, {30,31,0,1}. The $\mathbf{W}_1$ codewords for ranks 1 and 2 are denoted as follows:

$$\mathbf{X}^{(k)} \in \left\{ \begin{bmatrix} \mathbf{b}_{2k \bmod 32} & \mathbf{b}_{(2k+1) \bmod 32} & \mathbf{b}_{(2k+2) \bmod 32} & \mathbf{b}_{(2k+3) \bmod 32} \end{bmatrix} : k = 0,1,\cdots,15 \right\}$$

$$\mathbf{W}_1^{(k)} = \begin{bmatrix} \mathbf{X}^{(k)} & \mathbf{0} \\ \mathbf{0} & \mathbf{X}^{(k)} \end{bmatrix}$$

Codebook 1: $C_1 = \left\{ \mathbf{W}_1^{(0)}, \mathbf{W}_1^{(1)}, \mathbf{W}_1^{(2)}, \cdots, \mathbf{W}_1^{(15)} \right\}$

For $\mathbf{W}_2$ for rank 1, 4 selection hypotheses and 4 QPSK co-phasing hypotheses are defined as follows and compose 16 8x1 $\mathbf{W}_2$ matrices.

$$\mathbf{W}_2 \in C_2 = \left\{ \frac{1}{\sqrt{2}} \begin{bmatrix} \mathbf{Y} \\ \mathbf{Y} \end{bmatrix}, \frac{1}{\sqrt{2}} \begin{bmatrix} \mathbf{Y} \\ j\mathbf{Y} \end{bmatrix}, \frac{1}{\sqrt{2}} \begin{bmatrix} \mathbf{Y} \\ -\mathbf{Y} \end{bmatrix}, \frac{1}{\sqrt{2}} \begin{bmatrix} \mathbf{Y} \\ -j\mathbf{Y} \end{bmatrix} \right\}$$

$$\mathbf{Y} \in \left\{ \tilde{\mathbf{e}}_1, \tilde{\mathbf{e}}_2, \tilde{\mathbf{e}}_3, \tilde{\mathbf{e}}_4 \right\} = \left\{ \begin{bmatrix} 1 \\ 0 \\ 0 \\ 0 \end{bmatrix}, \begin{bmatrix} 0 \\ 1 \\ 0 \\ 0 \end{bmatrix}, \begin{bmatrix} 0 \\ 0 \\ 1 \\ 0 \end{bmatrix}, \begin{bmatrix} 0 \\ 0 \\ 0 \\ 1 \end{bmatrix} \right\}$$

$\tilde{\mathbf{e}}_n$ is a 4x1 selection vector with all zeros except for the nth element with value 1.

For $\mathbf{W}_2$ for rank 2, 8 selection hypotheses and 2 QPSK co-phasing hypotheses compose sixteen 8x2 $\mathbf{W}_2$ matrices.

$$\mathbf{W}_2 \in C_2 = \left\{ \frac{1}{\sqrt{2}} \begin{bmatrix} \mathbf{Y}_1 & \mathbf{Y}_2 \\ \mathbf{Y}_1 & -\mathbf{Y}_2 \end{bmatrix}, \frac{1}{\sqrt{2}} \begin{bmatrix} \mathbf{Y}_1 & \mathbf{Y}_2 \\ j\mathbf{Y}_1 & -j\mathbf{Y}_2 \end{bmatrix} \right\}$$

$$(\mathbf{Y}_1, \mathbf{Y}_2) \in \left\{ (\tilde{\mathbf{e}}_1, \tilde{\mathbf{e}}_1), (\tilde{\mathbf{e}}_2, \tilde{\mathbf{e}}_2), (\tilde{\mathbf{e}}_3, \tilde{\mathbf{e}}_3), (\tilde{\mathbf{e}}_4, \tilde{\mathbf{e}}_4), (\tilde{\mathbf{e}}_1, \tilde{\mathbf{e}}_2), (\tilde{\mathbf{e}}_2, \tilde{\mathbf{e}}_3), (\tilde{\mathbf{e}}_1, \tilde{\mathbf{e}}_4), (\tilde{\mathbf{e}}_2, \tilde{\mathbf{e}}_4) \right\}$$

For ranks 3 and 4, there are 16 4Tx DFT beams defined for X (oversampled 4x), with beam index 0,1,2...,15, denoted as follows:

$$\mathbf{B} = \begin{bmatrix} \mathbf{b}_0 & \mathbf{b}_1 & \cdots & \mathbf{b}_{15} \end{bmatrix}, \qquad [\mathbf{B}]_{1+m,1+n} = e^{j\frac{2\pi mn}{16}}, \qquad m = 0,1,2,3, \; n = 0,1,\cdots,15$$

$$\mathbf{B} = \begin{bmatrix} 1 & 1 & \cdots & 1 \\ 1 & e^{j\frac{2\pi}{16}} & \cdots & e^{j\frac{2\pi*16}{32}} \\ 1 & e^{j\frac{4\pi}{16}} & \cdots & e^{j\frac{2\pi*2*16}{32}} \\ 1 & e^{j\frac{6\pi}{16}} & \cdots & e^{j\frac{2\pi*3*16}{32}} \end{bmatrix} \qquad \mathbf{b}_n = \begin{bmatrix} 1 \\ e^{j\frac{2\pi*n}{16}} \\ e^{j\frac{2\pi*2n}{16}} \\ e^{j\frac{2\pi*3n}{16}} \end{bmatrix}, \qquad n = 0,1,\cdots,16$$

For $\mathbf{W}_1$ ranks 3 and 4, X is formed by $N_b = 8$ adjacent overlapping beams, so there are a total of 16 $\mathbf{W}_1$ matrices per rank with beam indices in each matrix as

$\{0,1,2,\ldots,7\}$, $\{4,5,6,\ldots,11\}$, $\{8,9,10,\ldots,15\}$, $\{12,\ldots,15,0,\ldots,3\}$. The $\mathbf{W}_1$ codewords for ranks 3 and 4 are denoted as follows:

$$\mathbf{X}^{(k)} \in \left\{ \begin{bmatrix} \mathbf{b}_{4k \bmod 16} & \mathbf{b}_{(4k+1) \bmod 16} & \cdots & \mathbf{b}_{(4k+7) \bmod 16} \end{bmatrix} : k = 0,1,2,3 \right\}$$

$$\mathbf{W}_1^{(k)} = \begin{bmatrix} \mathbf{X}^{(k)} & \mathbf{0} \\ \mathbf{0} & \mathbf{X}^{(k)} \end{bmatrix}$$

Codebook 1: $\quad C_1 = \left\{ \mathbf{W}_1^{(0)}, \mathbf{W}_1^{(1)}, \mathbf{W}_1^{(2)}, \mathbf{W}_1^{(3)} \right\}$

For $\mathbf{W}_2$ for rank 2, 16 selection hypotheses and 1 QPSK co-phasing hypothesis are defined as follows and compose 16 8x3 $\mathbf{W}_2$ matrices.

$$\mathbf{W}_2 \in C_2 = \left\{ \frac{1}{\sqrt{2}} \begin{bmatrix} \mathbf{Y}_1 & \mathbf{Y}_2 \\ \mathbf{Y}_1 & -\mathbf{Y}_2 \end{bmatrix} \right\}$$

$$(\mathbf{Y}_1, \mathbf{Y}_2) \in \left\{ \begin{array}{l} \left(\mathbf{e}_1, \begin{bmatrix}\mathbf{e}_1 & \mathbf{e}_5\end{bmatrix}\right), \left(\mathbf{e}_2, \begin{bmatrix}\mathbf{e}_2 & \mathbf{e}_6\end{bmatrix}\right), \left(\mathbf{e}_3, \begin{bmatrix}\mathbf{e}_3 & \mathbf{e}_7\end{bmatrix}\right), \left(\mathbf{e}_4, \begin{bmatrix}\mathbf{e}_4 & \mathbf{e}_8\end{bmatrix}\right), \\ \left(\mathbf{e}_5, \begin{bmatrix}\mathbf{e}_1 & \mathbf{e}_5\end{bmatrix}\right), \left(\mathbf{e}_6, \begin{bmatrix}\mathbf{e}_2 & \mathbf{e}_6\end{bmatrix}\right), \left(\mathbf{e}_7, \begin{bmatrix}\mathbf{e}_3 & \mathbf{e}_7\end{bmatrix}\right), \left(\mathbf{e}_8, \begin{bmatrix}\mathbf{e}_4 & \mathbf{e}_8\end{bmatrix}\right), \\ \left(\begin{bmatrix}\mathbf{e}_1 & \mathbf{e}_5\end{bmatrix}, \mathbf{e}_5\right), \left(\begin{bmatrix}\mathbf{e}_2 & \mathbf{e}_6\end{bmatrix}, \mathbf{e}_6\right), \left(\begin{bmatrix}\mathbf{e}_3 & \mathbf{e}_7\end{bmatrix}, \mathbf{e}_7\right), \left(\begin{bmatrix}\mathbf{e}_4 & \mathbf{e}_8\end{bmatrix}, \mathbf{e}_8\right), \\ \left(\begin{bmatrix}\mathbf{e}_5 & \mathbf{e}_1\end{bmatrix}, \mathbf{e}_1\right), \left(\begin{bmatrix}\mathbf{e}_6 & \mathbf{e}_2\end{bmatrix}, \mathbf{e}_2\right), \left(\begin{bmatrix}\mathbf{e}_7 & \mathbf{e}_3\end{bmatrix}, \mathbf{e}_3\right), \left(\begin{bmatrix}\mathbf{e}_8 & \mathbf{e}_4\end{bmatrix}, \mathbf{e}_4\right) \end{array} \right\}$$

For $\mathbf{W}_2$ for rank 4, 4 selection hypotheses and 2 QPSK co-phasing hypotheses are defined as follows and compose eight 8x4 $\mathbf{W}_2$ matrices.

$$\mathbf{W}_2 \in C_2 = \left\{ \frac{1}{\sqrt{2}} \begin{bmatrix} \mathbf{Y} & \mathbf{Y} \\ \mathbf{Y} & -\mathbf{Y} \end{bmatrix}, \frac{1}{\sqrt{2}} \begin{bmatrix} \mathbf{Y} & \mathbf{Y} \\ j\mathbf{Y} & -j\mathbf{Y} \end{bmatrix} \right\}$$

$$\mathbf{Y} \in \left\{ \begin{bmatrix}\mathbf{e}_1 & \mathbf{e}_5\end{bmatrix}, \begin{bmatrix}\mathbf{e}_2 & \mathbf{e}_6\end{bmatrix}, \begin{bmatrix}\mathbf{e}_3 & \mathbf{e}_7\end{bmatrix}, \begin{bmatrix}\mathbf{e}_4 & \mathbf{e}_8\end{bmatrix} \right\}$$

$\mathbf{e}_n$ is an 8x1 selection vector with all zeros except for the nth element with value 1.

4.3.2.2.3.2 Rank 5–8 Codebook — For ranks 5 to 8, X in $\mathbf{W}_1$ is a 4x4, 4Tx DFT matrix, denoted as follows. There are 4 $\mathbf{W}_1$ matrices for ranks 5 to 7 and one $\mathbf{W}_1$ matrix for rank 8.

$$\mathbf{X}^{(0)} = \frac{1}{2} \times \begin{bmatrix} 1 & 1 & 1 & 1 \\ 1 & j & -1 & -j \\ 1 & -1 & 1 & -1 \\ 1 & -j & -1 & j \end{bmatrix}, \mathbf{X}^{(1)} = diag\{1, e^{j\pi/4}, j, e^{i3\pi/4}\}\mathbf{X}^{(0)},$$

$$\mathbf{X}^{(2)} = diag\{1, e^{j\pi/8}, e^{i2\pi/8}, e^{i3\pi/8}\}\mathbf{X}^{(0)}, \mathbf{X}^{(3)} = diag\{1, e^{j3\pi/8}, e^{i6\pi/8}, e^{i9\pi/8}\}\mathbf{X}^{(0)}$$

For ranks 5 to 8, $\mathbf{W}_2$ is composed of the product of $\begin{bmatrix} I & I \\ I & -I \end{bmatrix}$, and a fixed 8 x rank column selection matrix. $\begin{bmatrix} I & I \\ I & -I \end{bmatrix}$ is introduced to ensure equal usage of both polarization groups for each transmission layer, which will bring good performance for higher rank transmission of spatial channel with richer scattering.

There is one hypothesis, which implies one $\mathbf{W}_2$ matrix per rank. The $\mathbf{W}_1$ and $\mathbf{W}_2$ matrices for ranks 5 to 8 are defined as follows:

Rank 5:

$$\mathbf{W}_1 \in C_1 = \left\{ \begin{bmatrix} \mathbf{X}^{(0)} & \mathbf{0} \\ \mathbf{0} & \mathbf{X}^{(0)} \end{bmatrix} \begin{bmatrix} \mathbf{X}^{(1)} & \mathbf{0} \\ \mathbf{0} & \mathbf{X}^{(1)} \end{bmatrix} \begin{bmatrix} \mathbf{X}^{(2)} & \mathbf{0} \\ \mathbf{0} & \mathbf{X}^{(2)} \end{bmatrix} \begin{bmatrix} \mathbf{X}^{(3)} & \mathbf{0} \\ \mathbf{0} & \mathbf{X}^{(3)} \end{bmatrix} \right\},$$

$$\mathbf{W}_2 = \frac{1}{\sqrt{2}} \begin{bmatrix} \tilde{\mathbf{e}}_1 & \tilde{\mathbf{e}}_1 & \tilde{\mathbf{e}}_2 & \tilde{\mathbf{e}}_2 & \tilde{\mathbf{e}}_3 \\ \tilde{\mathbf{e}}_1 & -\tilde{\mathbf{e}}_1 & \tilde{\mathbf{e}}_2 & -\tilde{\mathbf{e}}_2 & \tilde{\mathbf{e}}_3 \end{bmatrix}$$

Rank 6:

$$\mathbf{W}_1 \in C_1 = \left\{ \begin{bmatrix} \mathbf{X}^{(0)} & \mathbf{0} \\ \mathbf{0} & \mathbf{X}^{(0)} \end{bmatrix} \begin{bmatrix} \mathbf{X}^{(1)} & \mathbf{0} \\ \mathbf{0} & \mathbf{X}^{(1)} \end{bmatrix} \begin{bmatrix} \mathbf{X}^{(2)} & \mathbf{0} \\ \mathbf{0} & \mathbf{X}^{(2)} \end{bmatrix} \begin{bmatrix} \mathbf{X}^{(3)} & \mathbf{0} \\ \mathbf{0} & \mathbf{X}^{(3)} \end{bmatrix} \right\},$$

$$\mathbf{W}_2 = \frac{1}{\sqrt{2}} \begin{bmatrix} \tilde{\mathbf{e}}_1 & \tilde{\mathbf{e}}_1 & \tilde{\mathbf{e}}_2 & \tilde{\mathbf{e}}_2 & \tilde{\mathbf{e}}_3 & \tilde{\mathbf{e}}_3 \\ \tilde{\mathbf{e}}_1 & -\tilde{\mathbf{e}}_1 & \tilde{\mathbf{e}}_2 & -\tilde{\mathbf{e}}_2 & \tilde{\mathbf{e}}_3 & -\tilde{\mathbf{e}}_3 \end{bmatrix}$$

Rank 7:

$$\mathbf{W}_1 \in C_1 = \left\{ \begin{bmatrix} \mathbf{X}^{(0)} & \mathbf{0} \\ \mathbf{0} & \mathbf{X}^{(0)} \end{bmatrix} \begin{bmatrix} \mathbf{X}^{(1)} & \mathbf{0} \\ \mathbf{0} & \mathbf{X}^{(1)} \end{bmatrix} \begin{bmatrix} \mathbf{X}^{(2)} & \mathbf{0} \\ \mathbf{0} & \mathbf{X}^{(2)} \end{bmatrix} \begin{bmatrix} \mathbf{X}^{(3)} & \mathbf{0} \\ \mathbf{0} & \mathbf{X}^{(3)} \end{bmatrix} \right\},$$

$$\mathbf{W}_2 = \frac{1}{\sqrt{2}} \begin{bmatrix} \tilde{\mathbf{e}}_1 & \tilde{\mathbf{e}}_1 & \tilde{\mathbf{e}}_2 & \tilde{\mathbf{e}}_2 & \tilde{\mathbf{e}}_3 & \tilde{\mathbf{e}}_3 & \tilde{\mathbf{e}}_4 \\ \tilde{\mathbf{e}}_1 & -\tilde{\mathbf{e}}_1 & \tilde{\mathbf{e}}_2 & -\tilde{\mathbf{e}}_2 & \tilde{\mathbf{e}}_3 & -\tilde{\mathbf{e}}_3 & \tilde{\mathbf{e}}_4 \end{bmatrix}$$

Rank 8:

$$\mathbf{W}_1 \in C_1 = \left\{ \begin{bmatrix} \mathbf{X}^{(0)} & \mathbf{0} \\ \mathbf{0} & \mathbf{X}^{(0)} \end{bmatrix} \right\},$$

$$\mathbf{W}_2 = \frac{1}{\sqrt{2}} \begin{bmatrix} \tilde{\mathbf{e}}_1 & \tilde{\mathbf{e}}_1 & \tilde{\mathbf{e}}_2 & \tilde{\mathbf{e}}_2 & \tilde{\mathbf{e}}_3 & \tilde{\mathbf{e}}_3 & \tilde{\mathbf{e}}_4 & \tilde{\mathbf{e}}_4 \\ \tilde{\mathbf{e}}_1 & -\tilde{\mathbf{e}}_1 & \tilde{\mathbf{e}}_2 & -\tilde{\mathbf{e}}_2 & \tilde{\mathbf{e}}_3 & -\tilde{\mathbf{e}}_3 & \tilde{\mathbf{e}}_4 & -\tilde{\mathbf{e}}_4 \end{bmatrix}$$

Based on the previous description, the total number of codewords in codebooks C_1 and C_2 are summarized in Table 4.6. Any one codeword within the 109 codewords can be disabled through the codebook subset restriction bitmap in radio resource control (RRC) signaling. The codebook is configured by the eNB through higher-layer control signaling depending on the antenna configuration, UE position, channel condition, and so on. The UE will search in the restricted codebook subset for PMI reporting.

4.3.2.3 Codewords and Codeword-to-Layer Mapping

The maximum number of codewords in spatial multiplexing directly affects the control overhead (number of CQI reports, number of hybrid automatic repeat request [HARQ] processes) and the UE complexity since spatial multiplexing with multiple codewords need to apply adaptive modulation and coding (AMC) and error control on a per-codeword basis in order to achieve the peak performance in a low-mobility scenario. In Rel-8, the number of codewords was limited to two in view of the small performance advantage of larger numbers of codewords in 4x4 configurations. With LTE-A, the effect of this number needs to be evaluated to determine the optimal trade-off between data throughputs, signaling overhead, and complexity. For a higher number of layers, up to 8 in LTE-A, it was finally agreed to use the same layer mapping rule as in Rel-8, and simply extend the Rel-8 mapping so that two codewords are as evenly distributed as possible over each layer.

With spatial multiplexing, eNB may send multiple data streams (or layers) to UEs in downlink transmission over the same frequency. The number of layers or streams is defined as the *rank*. The number of codewords can be one or two depending on the number of layers (see Figure 4.24). The number of layers v is less than

Table 4.6 8Tx Codebook Size

Codebook Size									Sum
Rank	1	2	3	4	5	6	7	8	
C1	16	16	4	4	4	4	4	1	53
C2	16	16	16	8	—	—	—	—	56
Required bitmap size									109

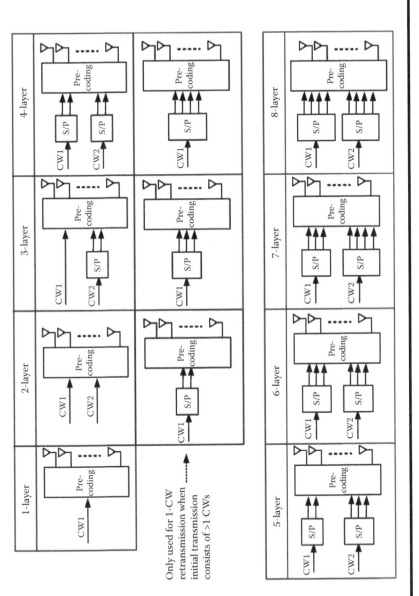

Figure 4.24 DL codeword (CW)-to-layer mapping in LTE-A. (S/P—serial/parallel)

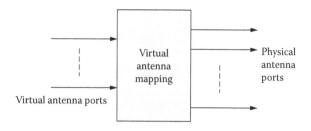

Figure 4.25   Virtual antenna mapping.

or equal to min (N_{TX}, N_{RX}), where N_{TX} and N_{RX} are the number of the eNB TX antenna ports and the number of UE RX antenna ports, respectively (i.e., corresponding to N_{TX}x N_{RX} spatial multiplexing).

4.3.2.4 Antenna Mapping and Virtualization

In an LTE-A system with a large number of antennas (more than 4), in order to support Rel-8 UE, whose capability is limited to 4Tx transmission at the eNB, some solutions should be worked out. An easy and straightforward way is to use virtual antenna mapping, as shown in Figure 4.25, where transmitted streams for Rel-8 UE could be seen as transmitted from virtual antenna ports (four or fewer). As there are more than four physical transmit antennas, these signals transmitted on virtual antenna ports should be mapped onto physical antennas before transmission. This operation should be transparent to Rel-8 UE.

An antenna mapping is required for Rel-8 antenna ports when there are eight antennas at the eNB and all available transmit power shall be utilized for LTE. It may also be needed for Rel-8 antenna ports in a Rel-10 eNB with fewer than eight antennas, if fewer Rel-8 antenna ports than antennas are configured. Although the mapping is an implementation issue, it should have the following desired properties:

■ Antenna virtualization can maintain full power utilization while employing multiple antenna setups. Each power amplifier should be utilized equally (i.e., balanced input) and all amplifiers work at the same operating point and transmit with the same power so the antenna ports should have uniform sector coverage.
■ The impact on PDCCH and Rel-8 PDSCH performance by using the mapping should be minimized.
■ The mapping of the antenna ports must be identical irrespective of whether the LTE-A ports are present in the resource block/subframe.

An antenna virtualization scheme is preferred over the alternative of turning off the other four antennas, which creates inefficient power utilization. Utilizing all PA power for nonprecoding transmission is important, such as in PDCCH and physical multicast channel (PMCH) reception. Legacy UEs will estimate channels from the four virtual antenna ports for both data demodulation (if user-specific RS is not used) and link adaptation support (CQI, PMI, etc.)

It is obvious that utilizing virtual mapping could make the LTE-A system with a large number of antennas (>4) flexible in supporting both LTE-A UE and Rel-8 UE. Virtual antenna mapping can be configured by the eNB on a semistatic or static basis, depending on the deployment scenario. To support LTE-A UE with high-layer transmission, additional antenna ports (>4) can be configured by the eNB depending on channel condition. The virtual antenna mapping can be realized by using fixed precoding, CDD, or radio frequency (RF) switching, and the benefit of virtual antenna mapping is that the system could maintain a minimum number of common RS ports (for example, 4) for decoding PDCCH for both LTE-A UE and Rel-8 UE and PDSCH for Rel-8 legacy UE.

4.3.3 Transmission Mode 9

Transmission mode (TM) 9 was introduced in LTE-A and supports SU-MIMO up to rank 8 and SU/MU dynamic switching. Transmission mode 9 performs precoded DMRS-based transmission, and DMRS will be used for demodulation at the UE side. It is not necessary to inform the precoder index. The DL precoder is not specified, so flexible precoder usage is possible in eNB design.

Downlink control information (DCI) format 2C is defined in LTE-A to support transmission mode 9. TxD is not supported under DCI format 2C. Transmission mode 9 uses one unified signaling system (shown in Table 4.7), regardless of the number of antenna ports or UE capability; that is, it always uses the signaling table with a maximum of eight layers. Joint coding of antenna port(s), scrambling identity (SCID), and number of layers is supported in DCI format 2C by 3 bits. The SCID bit from DCI format 2B is reused to support the joint coding. The number of additional bits added to the current DCI format 2B is 2 bits according to Table 4.7.

New CSI reporting is also defined in LTE-A for transmission mode 9 and introduces a new two-stage precoding scheme.

The UE selects its preferred precoding vector or matrix from the $\mathbf{W}_2$ codebook such that the CQI is maximized for the corresponding vector/matrix according to the quantized spatial correlation matrix. PMI_1 and PMI_2 are reported to the eNB at different rates and different frequency resolutions. The UE feeds back the corresponding short-term narrowband PMI to the eNB with a feedback interval that is shorter than the feedback period of the long-term wideband PMI (as illustrated in Figure 4.26). Different types of feedback can also be configured together for every feedback instance and the differential feedback exploits the correlation between precoding matrices adjacent in time or frequency. In practice, for low-speed UEs, the differential feedback could improve the feedback accuracy.

In particular, let the channel be H. The PMI codebook is adapted by the long-term $\mathbf{W}_1$ first. The adapted codebook is then used to quantize the channel H. The quantization operation may be simply written as

Table 4.7   DMRS Port/SCID Indication for Maximum of Eight Layers

One Codeword		Two Codewords	
Codeword 0 Enabled, Codeword 1 Disabled		Codeword 0 Enabled, Codeword 1 Enabled	
Message		Message	
0	1 layer, port 7, SCID = 0	0	2 layers, ports 7–8, SCID = 0
1	1 layer, port 7, SCID = 1	1	2 layers, ports 7–8, SCID = 1
2	1 layer, port 8, SCID = 0	2	3 layers, ports 7–9
3	1 layer, port 8, SCID = 1	3	4 layers, ports 7–10
4	2 layers, ports 7–8	4	5 layers, ports 7–11
5	3 layers, ports 7–9	5	6 layers, ports 7–12
6	4 layers, ports 7–10	6	7 layers, ports 7–13
7	Reserved	7	8 layers, ports 7–14

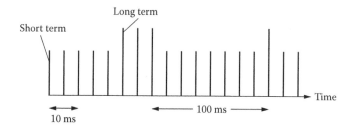

Figure 4.26   Short-term narrowband and long-term wideband PMI report.

$$\arg \max_{w_i \in W_2} \left| HR^{1/2} w_i \right|$$

where $\mathbf{W}_2$ is the matrix representing the baseline codebook matrix. This process is illustrated in Figure 4.27.

In conclusion, with the long-term spatial correlation matrix and the short-term PMI feedback from a UE, the eNB will reconstruct the precoding matrix, which improves feedback accuracy and benefits both SU-MIMO and MU-MIMO performance. This way of splitting the reporting of the recommended precoder into two parts offers a way to increase the spatial resolution of the precoding while still limiting the feedback overhead.

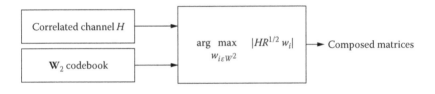

Figure 4.27 Adaptive codebook.

Implicit feedback (PMI/RI/CQI) is used for LTE-A. CQI is computed based on the assumption that the eNB uses a specific precoder (or precoders), as given by the feedback, on each sub-band within the CQI reference resource.

For periodic physical uplink control channel (PUCCH) reporting, a natural extension of the CQI/PMI/RI modes from Rel-8 and Rel-9 is deployed. The $\mathbf{W}_1/\mathbf{W}_2$ reporting procedure has two modes as follows:

- CSI Mode 1: $\mathbf{W}_1$ and $\mathbf{W}_2$ are signaled in separate subframes.
- CSI Mode 2: W is determined by a single report confined to a single subframe.

For aperiodic physical uplink shared channel (PUSCH) reporting, a natural extension of the CQI/PMI/RI modes from Rel-8 and Rel-9 is also deployed. The report in the aperiodic PUSCH is self-contained in the same subframe. One report can contain both $\mathbf{W}_1$ and $\mathbf{W}_2$. If either $\mathbf{W}_1$ or $\mathbf{W}_2$ is fixed, one report can contain $\mathbf{W}_1$ only or $\mathbf{W}_2$ only, regardless of whether the precoder W is derived from $\mathbf{W}_1$ or $\mathbf{W}_2$. The same report also contains a rank indicator (RI) and a CQI.

4.3.3.1 Periodic CSI on PUCCH

For transmission mode 9, periodic CSI reporting modes 1-1 or 2-1 are used if the UE is configured with PMI/RI reporting and number of CSI-RS ports > 1; modes 1-0 or 2-0 are used if the UE is configured without PMI/RI reporting or number of CSI-RS ports = 1. The design details of periodic CSI reporting for transmission mode 9 are listed below.

1. For the 2Tx and 4Tx cases: $\mathbf{W}_1$ is the identity matrix and is therefore not explicitly signaled; $\mathbf{W}_2$ uses the Rel-8 codebook.
2. For the 8Tx case: Three different reporting modes have been agreed upon; further down selection is not precluded yet.
3. Extension of Rel-8 PUCCH mode 1-1 with RI and $\mathbf{W}_1$ signaled in the same subframe:
   a. Codebook subsampling may be performed depending on the final codebook design (to ensure that the total payload is sufficiently small).
   b. W is determined from the two-subframe report (in CSI mode 1, $\mathbf{W}_1$ and $\mathbf{W}_2$ are signaled in separate subframes) conditioned upon the latest RI report.

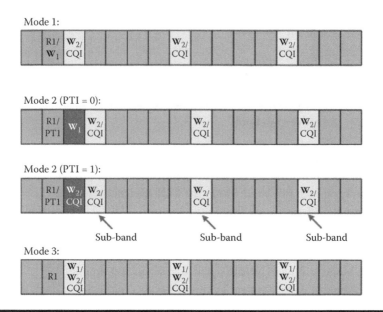

Figure 4.28   Extension of PUCCH CSI report mode 2-1.

  c. Reporting format: Report 1 includes the RI and $\mathbf{W}_1$, jointly encoded; Report 2 includes the wideband CQI and wideband $\mathbf{W}_2$.

4. Extension of Rel-8 PUCCH mode 2-1 (see Figure 4.28) for transmission mode 9, eight CSI-RS ports are configured.

  a. W is determined from the three-subframe report conditioned upon the latest RI report.

  b. Reporting format:

   i. Format 1: RI and 1-bit precoder type indicator (PTI)

   ii. Format 2: PTI = 0: $\mathbf{W}_1$ will be reported; PTI = 1: wideband CQI and wideband $\mathbf{W}_2$ will be reported.

   iii. Format 3: PTI = 0: wideband CQI and wideband $\mathbf{W}_2$ will be reported; PTI = 1: sub-band CQI, sub-band $\mathbf{W}_2$; transmission of sub-band selection indicator versus predefined cycling is for further study (FFS).

   iv. For 2Tx and 4Tx, PTI is assumed to be set to 1 and is not signaled.

5. Extension of Rel-8 PUCCH mode 1-1 with W determined from a single subframe report conditioned upon the latest RI report in a previous subframe

  a. For each rank, a subset of codebook C_1 and/or a subset of codebook C_2 are used to ensure a total payload size ($\mathbf{W}_1$ and $\mathbf{W}_2$ and CQI[s]) of at most 11 bits.

   i. For each rank, the subset of C_1 and subset of C_2 are fixed and therefore not configurable.

   ii. For each rank, the subset of C_1 and the subset of C_2 are designed either separately or jointly.

Table 4.8 PMI Report Formats

	Mode 1-1 with CSI Mode 1	Mode 1-1 with CSI Mode 2	Mode 2-1
Issue	RI reliability issue	Possibly cause performance degradation	More standard effort
Format 1	RI + $\mathbf{W}_1$ (subsampling)	RI	RI, PTI
Format 2	CQI, $\mathbf{W}_2$	CQI, $\mathbf{W}_1$ + $\mathbf{W}_2$ (subsampling)	PTI = 0:$\mathbf{W}_1$ PTI = 1:wideband (WB) $\mathbf{W}_2$, WB CQI
Format 3			PTI = 0:WB $\mathbf{W}_2$, and WB CQI PTI = 1:SB $\mathbf{W}_2$, and SB CQI (subsampling if sub-band index is reported)

b. Different subsets of possible co-phasing are used for different groups of beam angles.

The PMI report variations with PUCCH CSI report mode 1-1 and 2-1 are listed in Table 4.8.

4.3.3.2 Aperiodic CSI on PUSCH

For transmission mode 9, aperiodic CSI reporting modes 1-2, 2-2, 3-1 can be used if the UE is configured with PMI/RI reporting and number of CSI-RS ports > 1; modes 2-0, 3-0 can be used if the UE is configured without PMI/RI reporting or number of CSI-RS ports = 1. A new CSI reporting mode 3-2 has ever been discussed by 3GPP but not put in specification yet. The design details of aperiodic CSI reporting for transmission mode 9 are listed below.

1. Natural extension of the Rel-8 aperiodic PUSCH CQI modes are supported in LTE-A Rel-10 (see Table 4.9).
   a. Support for PUSCH mode 3-2 with sub-band PMI + sub-band CQI targeting feedback accuracy improvements for MU and SU in Rel-10 is under discussion.
   b. PUSCH reporting is self-contained where $\mathbf{W}_1$ and $\mathbf{W}_2$ are always reported in the same subframe.
      i. For 2Tx and 4Tx, $\mathbf{W}_1$ (identity matrix) is not reported.
2. The possibility of reporting multiple CQIs, and if possible PMIs/RIs, (for example, one targeting SU-MIMO and the other targeting MU-MIMO)

Table 4.9   Aperiodic PUSCH CQI Modes

CQI/PMI Mode	CQI	W1	W2
1-2	Wideband CQI for entire system bandwidth (BW)	Single W_1: One for the entire system BW (wideband)	Sub-band PMI W_2
2-2	Wideband CQI for entire system BW + "M-preferred" CQI (for UE-selected bands)		Wideband PMI W_2 + "M-preferred" PMI W_2 (for UE-selected sub-bands)
3-1	Sub-band CQI		Wideband PMI W_2

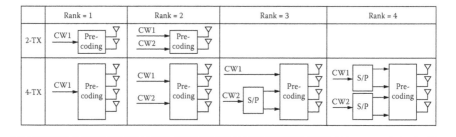

Figure 4.29   UL codeword (CW)-to-layer mapping in LTE-A.

and the frequency granularity of the additional CQIs (and if possible PMIs/ RIs) is FFS.

3. Whether mode 2-2 is finally supported depends on agreements and details of mode 3-2; that is, mode 2-2 and/or mode 3-2 may finally be supported.

4.4 UL MIMO

Uplink spatial multiplexing of up to four layers is being considered for LTE-A for supporting an uplink peak data rate of 500 Mbps and uplink peak spectrum efficiency of 15bps/Hz. In Rel-8, there is only one codeword and one layer in the uplink; in LTE-A, up to two transport blocks and spatial multiplexing with up to four layers can be transmitted from a scheduled UE in a subframe per uplink component carrier. As a baseline, the default UL operation mode is a single-antenna port mode.

With LTE-A, a scheduled UE may transmit up to two transport blocks. Each transport block has its own modulation and coding scheme (MCS). Depending on the number of transmission layers, the modulation symbols associated with each of the transport blocks are mapped onto one or two layers (as Figure 4.29) followed the principle in LTE Rel-8 for downlink spatial multiplexing. The transmission rank can be adapted dynamically. Different codebooks are defined depending on the number of layers that are used. Furthermore, different precoding is used depending on whether 2 or 4 transmit antennas are available. Also the number of bits used for the codebook index is different depending on the 2 and 4 transmit antenna cases, respectively.

TxD is necessary to increase channel diversity for uplink transmission. Multiantenna technology can be deployed to further improve network performance and coverage. Based on the evaluation of all related contributions, the spatial orthogonal resource transmit diversity (SORTD) scheme is determined to provide support for PUCCH formats 1/1a/1b, 2/2a/2b, and 3. In SORTD, the same modulated symbol is transmitted on different orthogonal resources for different antennas. SORTD for PUCCH format 1/1a/1b is deployed in case of no carrier aggregation (CA) for SORTD is the baseline for formats 2/2a/2b. If TxD is supported, SORTD will be chosen for both format 3 and format 1a/1b with channel selection.

For uplink transmission in LTE-A, different antenna ports are defined for transmission of different physical channels. PUCCH antenna ports do not correspond to PUSCH or SRS antenna ports due to the implementation of transparent precoder vector switching on PUCCH. Table 4.10 illustrates the PUSCH, PUCCH, and SRS antenna ports in LTE-A antenna port configuration. Figure 4.30 illustrates the antenna port numbers used for 2Tx.

Two transmission modes are introduced in LTE-A for SU-MIMO PUSCH transmission: a single antenna port mode that is compatible with the LTE Rel-8

Table 4.10 Antenna Ports Used for Different Physical Channels and Signals

Physical Channel or Signal	*Index $\tilde{p}$*	*Antenna Port Number p as a Function of the Number of Antenna Ports Configured for the Respective Physical Channel/Signal*		
		1	*2*	*4*
PUSCH	0	10	20	40
	1	—	21	41
	2	—	—	42
	3	—	—	43
SRS	0	10	20	40
	1	—	21	41
	2	—	—	42
	3	—	—	43
PUCCH	0	100	200	—
	1	—	201	—

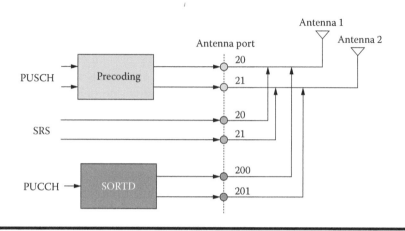

Figure 4.30   Antenna ports mapping in case of 2Tx.

PUSCH transmission and a multiantenna port mode that offers the possibility of two- and four-antenna port transmission. Discussions are ongoing in 3GPP regarding the refinements of PUSCH multiantenna port transmission, such as handling rank 1 transmissions, SRS options, and uplink control information (UCI) multiplexing on PUSCH as well as the precoder design for retransmissions.

The UL reference signal structure in LTE-A will retain the basic structure of that in LTE Rel-8. Two types of reference signals were enhanced: DMRS and SRS. The DMRS is used by the receiver to detect transmissions. In the case of uplink multiantenna transmission, the precoding applied for the DMRS is the same as the one applied for the PUSCH. Cyclic shift (CS) separation is the primary multiplexing scheme of the DMRS. Orthogonal cover code (OCC) separation is also used to separate DMRSs of different virtual transmit antennas. The SRS is used by the receiver to measure the mobile radio channel. Currently, SRS is nonprecoded and antenna specific, and the LTE Rel-8 principles will be reused for multiplexing of the SRS.

4.4.1 Transmit Diversity Schemes for PUCCH

As we have discussed, uplink transmit diversity gains can be exploited to further improve performance in terms of the block error rate or maintaining the coverage requirements while reducing total transmitted power, which is beneficial to intercell interference reduction and eventually also UE power consumption.

PUCCH transmission diversity has been shown to offer substantial performance gains. Such gains are useful in improving reception reliability for the typically interference-limited PUCCH operation. Some transmit diversity schemes were considered as candidates for the PUCCH transmission in LTE-A, including time-switched transmit diversity (TSTD), precoder vector switching (PVS), cyclic delay diversity (CDD), space-time block coding (STBC), FSTD, and space frequency

Table 4.11 Comparison of Two-Antenna Transmit Diversity Schemes for PUCCH

Transmit Diversity Scheme		Diversity Gain	PAPR	Orthogonality with Code Division Multiplexed Rel-8,10 LTE Terminal	Usage of Cyclic Shift or Orthogonal Code (OC) Resources per UE	No. of DMRS Resources per UE
One transmitter	TSTD	Small	Low	Orthogonal	One	One
Two transmitters	PVS	Small	Low	Orthogonal	One	One
	CDD	Medium	Low	Orthogonal	One/Two	One/Two
	STBC	Large	Low	Orthogonal (CQI) / Not orthogonal (ACK/NACK)	One	Two
	FSTD	Medium	Low	Not orthogonal	One	Two
	SFBC	Large	High	Not orthogonal	One	Two

Note: For RS transmission, different antennas will still need to use different orthogonal sequences to make it possible for the eNB to estimate the channels from each TX antenna separately.

block coding (SFBC). Their comparison is shown in Table 4.11 with criteria including transmit diversity gain, a low PAPR/cubic metric (CM), and the orthogonality among code division multiplexed LTE Rel-8 UEs, and so on.

The advantage of these schemes is that both transmit antennas use the same orthogonal sequence for transmission of data symbols. PVS and CDD offer very limited gains relative to single UE antenna transmission, and for this reason they are not being considered further. STBC cannot support format 2a/2b when OCC is used for DMRS. For format 2a/2b, acknowledgment/negative acknowledgment (ACK/ NACK) information is transmitted as the phase difference between the two DMRSs within one slot. To keep the orthogonality of OCC DMRS, the utilization of the phase difference to encode information over the two DMRSs is not allowed with STBC if OCC is used.

Furthermore, service connection time duration (SCTD; or orthogonal resource transmission [ORT]) is proposed for the PUCCH. SCTD obtains the largest transmit diversity gain, similar to STBC and does not need to deal with the unpaired symbol problem as is the case in STBC. Therefore, SCTD seems better than STBC. But the disadvantage of SCTD is that two cyclic shift or orthogonal covering resources per UE are required, thereby doubling the overhead as the UE multiplexing capacity per RB is halved. This implies a potential problem that the system throughput performance will be degraded.

Generally, TxD performance is quite good in most cases, but requires twice the resources compared to no TxD. While considering supporting the transmission of format 2/2a/2b and performance (SORTD provides better performance than STBC, around 1dB), SORTD was adopted in LTE-A for PUCCH transmission. SORTD means that the same modulation symbol from the PUCCH is transmitted from two antenna ports on two separate orthogonal sequences (as illustrated in Figure 4.31). SORTD is defined for the 2Tx antennas, and no explicit TxD scheme is defined for 4Tx antennas, in which scenario the 2Tx TxD scheme is applied as well.

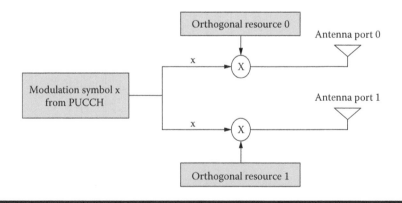

Figure 4.31   PUCCH TxD: SORTD utilizing two PUCCH resources.

The resource allocation scheme using SORTD for different PUCCH formats is as follows:

- **Format 1/1a/1b:** For format 1 and SPS transmission, the two resources for SORTD are semistatically allocated by RRC signaling; for dynamic scheduling, the two resources are implicitly derived from the index of n_{cce} and n_{cce+1} of the corresponding PDCCH.
- **Format 2/2a/2b:** Two resources are semistatically assigned by RRC signaling.
- **Format 3:**
  - When PDSCH scheduling on the primary cell only, use the same resource allocation as format 1/1a/1b depending on dynamic scheduling or SPS transmission.
  - When PDSCH scheduling on multiple cells, the resources are indicated by the ARI (ACK/NACK resource indicator) in the PDCCH corresponding to secondary cell scheduling.

SORTD transmission modes could be configured semistatically or dynamically. If they are configured semistatically, the configuration will be signaled by higher-layer signaling. The eNB could decide when to switch between the single antenna and SORTD modes and then signal this configuration to the UE. The decision could be based on the geometry or mobility of the UE and the availability of the resources. Figure 4.32 shows an example of signal exchange for the configurable transmit diversity for PUCCH.

4.4.2 PUSCH Transmission Mode

In LTE-A, open loop and closed loop spatial multiplexing will be supported by the PUSCH, and in the case of closed loop operation, codebook-based precoding will be used. In contrast to the DL, the transmission schemes in the UL are optimized to preserve the single-carrier property or to reduce PAPR/CM.

Codebook designs also enable precoded transmission with low PAPR/CM. In addition, the codebook will also provide the capability to turn off certain antennas (for power saving) at the instruction of an eNB.

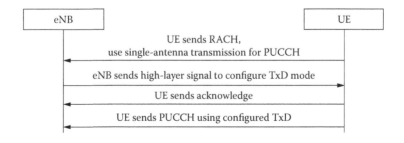

Figure 4.32   Configurable TxD modes for PUCCH. (RACH—random access channel)

Table 4.12 PUSCH Transmission Modes

TX Mode	DCI Format	Search Space	TX Scheme of PUSCH Corresponding to PDCCH
Mode 1	DCI format 0	Common and UE specific by C-RNTI	Single-antenna port, port 10
Mode 2	DCI format 0	Common and UE specific by C-RNTI	Single-antenna port, port 10
	DCI format 4	UE specific by C-RNTI	Closed loop spatial multiplexing

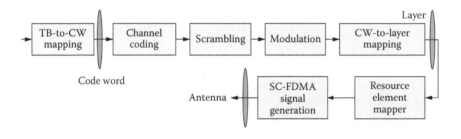

Figure 4.33 PUSCH transmission mode 1. (CW—codeword; TB—transport block; SC-FDMA—single-carrier frequency-division multiple access)

Two PUSCH transmission modes are defined in LTE-A (Table 4.12): PUSCH mode 1 and PUSCH mode 2. A UE is semistatically configured via higher-layer signaling to transmit the PUSCH with one of two PUSCH transmission modes. For PUSCH mode 2, the UE searches the PDCCH for DCI format 0 and format 4. DCI format 0 is used for single-antenna port (SAP) transmission and can be both in the common and UE-specific PDCCH search space. DCI format 4 is used for multiple-antenna port (MAP) with closed loop (CL) spatial multiplexing, only in UE-specific PDCCH search space. For PUSCH mode 1, the UE needs to search only the PDCCH with DCI format 0. PUSCH mode 1 is the default uplink transmission mode for a UE until the UE is assigned a PUSCH transmission mode by higher-layer signaling. The eNB can configure 4Tx antenna UE to use either 4 or 2 antenna ports in PUSCH mode 2. For PUSCH mode 2 and PUSCH mode 1 with the Rel-10 transmission scheme, the Rel-8/9 antenna selection scheme is not used.

The physical layer processing for PUSCH mode 1 and mode 2 is illustrated in Figure 4.33 and Figure 4.34, respectively.

For semipersistant scheduling, the UE needs to search only the PDCCH with DCI format 0, although PUSCH modes 1 and 2 are supported.

The number of SRS transmission ports is also configured according to PUSCH transmission mode. For PUSCH mode 1, the numbers of antenna ports that can

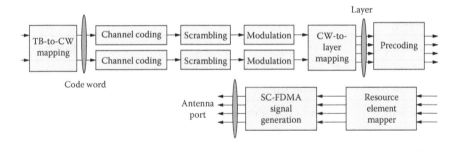

Figure 4.34 PUSCH transmission mode 2.

be used for SRS transmission are 0, 1, 2, or 4. For PUSCH mode 2 with 2 or 4 antenna ports configured, the numbers of antenna ports used for SRS transmission are 0, 1, 2 and 0, 1, 4, respectively.

The PUSCH transmission mode, the PUCCH transmission scheme, and the number of SRS antenna ports are independently configured through RRC signaling.

4.4.3 UL Precoding Technology

As we have already discussed, rank adaptation and channel-dependent precoding is needed to achieve gains from using multiple antennas for UE that faces low SINR environments. Precoding is also an essential component for UL SU-MIMO. Precoding matching of the instantaneous channel properties is required to achieve gains considering that the UE typically operates in a multiscattering environment with a large angular spread of incident waves and, therefore, cannot in general rely on high spatial correlation for directing its transmission. The precoder selection may be based on sounding RS or demodulation RS in the uplink.

4.4.3.1 Codebook versus Non-Codebook

Today, the precoding-based SU-/MU-MIMO is one of the most important technologies to provide high spectrum efficiency gain in the closed loop MIMO system for wireless communications. Precoding should be classified as codebook based or non-codebook based. Both the codebook-based precoding scheme and the non-codebook-based precoding scheme could be implemented, and non-codebook-based precoding should be supported for TDD operation to make full use of channel reciprocity, weighing the possible benefits against the increased complexity of calibrated transmit and receive chains in the UE and the need for additional testing and validation effort. In an FDD system, channel information cannot be obtained through the channel reciprocity property. So, codebook-based precoding should be used to quantize the downlink channel information. A comparison of codebook-based and non-codebook-based precoding is given in Table 4.13. As mentioned,

Table 4.13 Comparison of Codebook-Based versus Non-Codebook-Based Precoding

System Aspect	Codebook-Based	Non-Codebook-Based
Precoder selection	eNB -> lower UE complexity	UE -> higher UE complexity
Performance	Limited by codebook size	Potentially better than codebook-based
Required signaling support	PMI bits on UL grant; additional overhead is small	[FDD] Signaling of UL channel estimate from eNB to UE -> additional DL signaling channel, additional overhead is large
		[TDD] Channel reciprocity can be exploited to avoid signaling the DL; signaling of the UL channel estimate requires precise antenna calibration between the UE and eNB
SRS for link adaptation	Not precoded	Needs to be precoded

the non-codebook-based scheme relies on transmit interference nulling techniques and allows the eNB to have more freedom in selecting rank, MCS, and precoding matrices. Although some benefits and potential gain may be obtained, the additional complication from non-codebook-based precoding may outweigh the potential gain, even for TDD. Therefore, codebook-based precoding may be preferred. In either case, channel estimation for demodulation can be performed via the precoded DMRS.

To see the performance impact of non-codebook-based precoding, link-level simulations comparing it with codebook-based precoding were conducted. The uplink simulation assumptions are given for a 2x2 system with rank adaptation and an MMSE receiver. The scheduling bandwidth is 6 RBs and the channel model is EPA 5 Hz. A single value decomposition (SVD)-based approach employing the downlink channel estimate is used to determine the non-codebook-based precoder and perfect channel reciprocity is assumed, and the precoding vector is the main eigenvector corresponding to the largest eigenvalue of the correlation matrix between the UE and the serving cell. From the simulation result in Figure 4.35, there is less than 1 dB gain using non-codebook-based precoding as compared with codebook-based precoding for 2x2 with rank adaptation. This is attributed to the greater freedom in determining the precoder compared with a codebook approach, thus providing better orthogonalization of the channel at the transmit side. But non-codebook-based precoding increases the cubic metric while the cubic

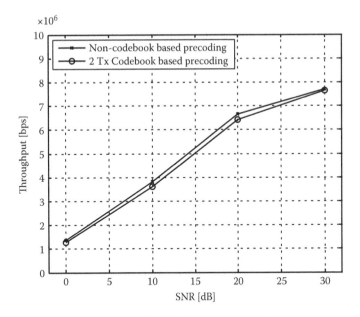

Figure 4.35 Non-codebook compared with codebook.

metric–preserving (CMP) codebook-based precoding does not. It also needs to be kept in mind that the non-codebook-based precoder used in this example does not achieve balanced use of the power amplifiers. Thus, although non-codebook-based precoding exhibits some interesting gains under certain assumptions, several aspects need to be studied further to assess the gains in practice.

4.4.3.2 Considerations in UL Codebook Design

As opposed to downlink codebook design, additional aspects, including the single-carrier property, PAPR requirement, and PUCCH load should be considered in uplink codebook design.

A precoding codebook for a given rank consists of a set of codebook matrices or vectors. The design constraints for DL and UL precoding codebooks are similar, especially concerning the constant modulus constraint. The finite alphabet property is also useful since the eNB may need to perform PMI selection for a host of UEs at a given subframe. These are some general concerns with respct to designing an uplink codebook:

■ **Balanced power amplifiers:** The PAs should be fully and equally utilized; to ensure this, the vector norm of each row of a precoding vector/matrix should be the same (Figure 4.36).

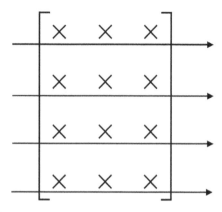

Figure 4.36 **Balanced power amplifiers.**

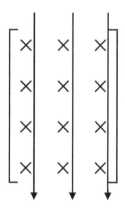

Figure 4.37 **Equal power per layer.**

■ **Equal power per layer:** It is important that each layer is given an equal share of the transmit power in the codebook design. If power control per layer is desirable, it could be obtained by adapting the power of the PUSCH and uplink DMRS accordingly. Therefore, layer power control should be left outside the codebook design to better utilize the limited codebook size. This implies that the column norm of each column of a precoding matrix should be the same (Figure 4.37).

■ **Good distance properties:** The codebook should have good distance properties (e.g., measured by the chordal distance) to ensure good performance in less correlated scenarios.

■ **Good match with transmit array response:** This is important to ensure good performance in highly correlated scenarios.

■ **Constraint alphabet:** Even though the eNB does have larger computational processing power compared to the UE, the eNB will need to support various numbers of UEs in the cell. Any computational complexity reduction in the eNB will reduce the total complexity by a factor of the number of UEs that the eNB can support. As a result, the LTE-A uplink 4Tx codebook vector/matrix will use QPSK values {+1, −1, +*j*, −*j*}.

■ **Nested structure:** The nested structure of the codebook reduces the complexity of downlink SINR calculation per rank and is one of the key design factors in DL codebook design to support efficient rank overriding. Moreover, the nested codebook structure saves memory requirements for UE and eNB. In the case of UL SU-MIMO, since the eNB is always in control of the uplink UE transmission and the UE does not inform the eNB on the preferred rank of the channel, the eNB can always select proper precoding depending on channel conditions. So it is not necessary to maintain the nested structure property for codebook design, since there is no notion of rank overriding for the uplink.

■ **Cubic metric (CM):** The lower CM can translate into higher uplink transmission power. One of several codebook design factors is the PAPR/CM to support the single-carrier property in a power-limited scenario. The cubic metric has been adopted by the 3GPP as a method to determine PA power rating. A low CM (CM preserving) is a reasonable criterion for the codebook design in the power-limited scenario.

For UL transmission, the UE may experience a power-limited situation; thus it is necessary to design a codebook that can maintain a low cubic metric value. A low cubic metric should be maintained in particular in single rank transmission. The main benefit of the CMP codebook is to allow higher power transmission in power-limited situations. As mentioned, the LTE Rel-8 codebook cannot always preserve the low cubic metric.

In addition, it should be noted that physical antenna selection is exempted from the aforementioned concerns. Physical antenna selection is one kind of very special codebook that is useful to save UE power in the scenario of antenna gain imbalance (AGI). Practical constraints of handsets, such as location and orientation of antennas, will cause antenna gain imbalance. In LTE-A, the UE AGI affects receiver performance in DL MIMO, and moves to the transmit side. AGI is defined as the difference in antenna gain between two antennas and leads to less-effective antennas. If the AGI remains near zero, there is a significant increase in the mean channel capacity. As the AGI increases, the benefit of UL multiple antenna transmissions will steadily decrease. When AGI increases to 6 dB or more, the gain that is obtained by using two transmit antennas will be significantly reduced and the performance benefit is observed only in the high SINR regime (as shown in Figure 4.38). This tendency will not only negatively impact the benefits of UL multiple antenna transmission in some very typical use cases, but will actually be harmful as compared to a UL single antenna transmission in terms of performance.

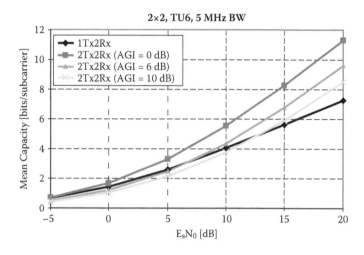

Figure 4.38 E_sN_0 versus mean capacity under different AGIs.

4.4.3.3 Codebook for 2Tx Antennas

For UL rank 2 transmissions, uncorrelated spatial channel is expected, and one identity matrix is enough. Two antenna selection matrixes are necessary for the AGI problem; otherwise it is the same as in the DL. For uplink spatial multiplexing with two transmit antennas, a three-bit precoding codebook is defined as shown in Table 4.14.

The objective of antenna turn-off elements is distinctly different from antenna selection (which provides selection diversity). An eNB may choose to apply an antenna turn-off matrix (or vector) when the perceived uplink channel reflects significant antenna gain imbalance (due to hand gripping, etc.). In the presence of imbalance, the application of antenna turn-off vectors can provide some throughput gain in comparison to a Rel-8 codebook with equal total power transmitted from the UE. In other words, the power from the turned-off antennas is added to the other antennas in this case. Therefore, the application of the antenna turn-off matrices should be left to the discretion of the eNB.

4.4.3.4 Codebook for 4Tx Antennas

It is sometimes claimed that precoding increases the cubic metric and would therefore increase the back-offs used for the power amplifiers. This would make precoding ill-suited to single-carrier transmission where a small cubic metric is desirable. Such a conclusion is, however, only true for certain types of codebooks. It is actually possible to design codebooks with PA back-off properties that are as desirable as 1Tx single-carrier transmission. As long as a single precoder is employed, CM is increased only if the signals for different layers are mixed together onto the same PA. Rank 1 transmission is therefore obviously not a problem at all where any constant modulus codebook is

Table 4.14 Codebook for Uplink 2Tx Antennas

Codebook Index	Number of Layers	
	1	*2*
0	$\frac{1}{\sqrt{2}}\begin{bmatrix}1\\1\end{bmatrix}$	$\frac{1}{\sqrt{2}}\begin{bmatrix}1 & 0\\0 & 1\end{bmatrix}$
1	$\frac{1}{\sqrt{2}}\begin{bmatrix}1\\-1\end{bmatrix}$	
2	$\frac{1}{\sqrt{2}}\begin{bmatrix}1\\j\end{bmatrix}$	
3	$\frac{1}{\sqrt{2}}\begin{bmatrix}1\\-j\end{bmatrix}$	
4	$\frac{1}{\sqrt{2}}\begin{bmatrix}1\\0\end{bmatrix}$	
5	$\frac{1}{\sqrt{2}}\begin{bmatrix}0\\1\end{bmatrix}$	

Antenna turn off vectors, targeted for power saving, e.g., against AGI situation. Antenna selection element.

acceptable from a CM perspective. For higher rank transmissions, it is important that the codebook structure is such that the mixing of layers onto the same PA is limited. One example of such a rank 1 codebook is given in Table 4.15. As seen, there is exactly one nonzero element per row in all precoder matrices and therefore the codebook preserves the CM. Such a CM-preserving codebook structure may seem rather restrictive, and it is therefore reasonable to wonder whether the performance is negatively impacted compared with a codebook whose design is free to increase the CM.

To assess the impact on performance of using a CM-preserving codebook, link-level simulations were conducted comparing the CM-preserving codebook with the rank 2 part of the Householder-based codebook used for the LTE REL-8 downlink. Fixed precoding was also simulated to give an idea of the overall precoding gains. The results of the simulation are displayed in Figure 4.39 and assume the same total transmission power for both codebooks. Channel-dependent precoding is seen to give a gain of around 2 dB relative to channel-independent fixed precoding. Furthermore, the performance for the two channel–dependent precoding codebooks is clearly more or less the same, with a slight benefit for the CM-preserving codebook. Thus, contrary

Table 4.15 4Tx Rank 1 Codebook

Index 0 to 7	$\frac{1}{2}\begin{bmatrix}1\\1\\1\\-1\end{bmatrix}$	$\frac{1}{2}\begin{bmatrix}1\\1\\j\\j\end{bmatrix}$	$\frac{1}{2}\begin{bmatrix}1\\1\\-1\\1\end{bmatrix}$	$\frac{1}{2}\begin{bmatrix}1\\1\\-j\\-j\end{bmatrix}$	$\frac{1}{2}\begin{bmatrix}1\\j\\1\\j\end{bmatrix}$	$\frac{1}{2}\begin{bmatrix}1\\j\\j\\j\end{bmatrix}$	$\frac{1}{2}\begin{bmatrix}1\\j\\-1\\-j\end{bmatrix}$	$\frac{1}{2}\begin{bmatrix}1\\j\\-j\\-1\end{bmatrix}$
Index 8 to 15	$\frac{1}{2}\begin{bmatrix}1\\-1\\1\\1\end{bmatrix}$	$\frac{1}{2}\begin{bmatrix}1\\-1\\j\\-j\end{bmatrix}$	$\frac{1}{2}\begin{bmatrix}1\\-1\\-1\\-1\end{bmatrix}$	$\frac{1}{2}\begin{bmatrix}1\\-1\\-j\\j\end{bmatrix}$	$\frac{1}{2}\begin{bmatrix}1\\-j\\1\\-j\end{bmatrix}$	$\frac{1}{2}\begin{bmatrix}1\\-j\\j\\-1\end{bmatrix}$	$\frac{1}{2}\begin{bmatrix}1\\-j\\-1\\j\end{bmatrix}$	$\frac{1}{2}\begin{bmatrix}1\\-j\\-j\\1\end{bmatrix}$
Index 16 to 23	$\frac{1}{2}\begin{bmatrix}1\\0\\1\\0\end{bmatrix}$	$\frac{1}{2}\begin{bmatrix}1\\0\\-1\\0\end{bmatrix}$	$\frac{1}{2}\begin{bmatrix}1\\0\\j\\0\end{bmatrix}$	$\frac{1}{2}\begin{bmatrix}1\\0\\-j\\0\end{bmatrix}$	$\frac{1}{2}\begin{bmatrix}0\\1\\0\\1\end{bmatrix}$	$\frac{1}{2}\begin{bmatrix}0\\1\\0\\-1\end{bmatrix}$	$\frac{1}{2}\begin{bmatrix}0\\1\\0\\j\end{bmatrix}$	$\frac{1}{2}\begin{bmatrix}0\\1\\0\\-j\end{bmatrix}$

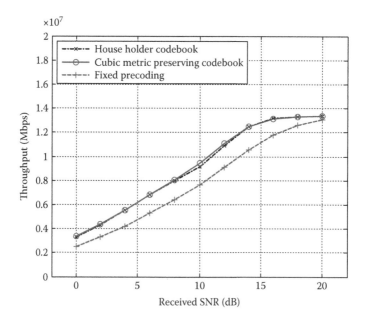

Figure 4.39 Link-level comparison of householder and CM preserving codebook for transmission rank 2.

to initial fears, the CM-preserving structure does not seem to limit the performance. In fact, if we take PA back-offs into account, the CM-preserving codebook would clearly outperform the Householder codebook over the entire SNR range. This indicates that single-carrier transmission and precoding do not necessarily contradict each other and can in fact be supported simultaneously while enjoying a resulting significant performance benefit over codebook designs that do not take CM into account.

Based on the previous discussion, independent codebook design is adopted for different ranks in LTE-A. Antenna-selection codebook elements are designed for rank 1. A CMP codebook is used for rank 2 and rank 3. An identity precoding matrix is utilized for rank 4.

4.4.3.4.1 4Tx Rank 1 Codebook

Since only one layer is transmitted, the single-carrier property will not be impacted by precoding. The only consideration is the precoding performance (i.e., chordal distance, antenna power imbalance, etc.).

- Size 24: 16 constant modulus + 8 antenna-pair turn-off vectors. A few vectors are defined for selected antenna(s) PA turn-off, targeted for power saving in case of, for example, antenna gain imbalance. To determine the codebook size for rank 1, as mentioned previously, the codebook size for the lower ranks may require more vector/matrices to achieve similar relative performance from ideal cases. Precoding gain tends to be larger for lower ranks. At the same time, the gain from increasing the codebook size tends to diminish faster for higher ranks. Therefore, designing larger codebooks for lower ranks is more beneficial.
- QPSK alphabet for rank 1 precoding proposals other than the rank 1 precoding specified in LTE Rel-8.
- The codebook can be designed based on the Grassmannian criterion (i.e., maximize the minimum chordal distance).

4.4.3.4.2 4Tx Rank 2 Codebook

Refer to Table 4.16. For rank 2 and Table 4.18 for rank 3, precoding may increase the CM of the transmitted signal. Therefore, minimizing the resulting CM is beneficial for the system, so a CMP codebook is the main consideration and the rest of the task is finding a set of precoding matrices with maximum/minimum chordal distances.

- Size 16: CM preserving for all the 4x2 precoding matrices; each matrix has one zero and one nonzero $\exp(j*\theta)$ element in each of the 4 rows.
- QPSK alphabet; beneficial for simplicity and computational complexity.

A CM-preserving matrix is a matrix that allows a mixture of only one layer from a single-antenna port point of view. A non-CM-preserving matrix is a matrix that allows a mixture of two layers from a single-antenna port point of view. Table 4.17

Table 4.16 4Tx Rank 2 Codebook

Index 0 to 7							
$\frac{1}{2}\begin{bmatrix}1&0\\1&0\\0&1\\0&-j\end{bmatrix}$	$\frac{1}{2}\begin{bmatrix}1&0\\1&0\\0&1\\0&j\end{bmatrix}$	$\frac{1}{2}\begin{bmatrix}1&0\\-j&0\\0&1\\0&1\end{bmatrix}$	$\frac{1}{2}\begin{bmatrix}1&0\\-j&0\\0&1\\0&-1\end{bmatrix}$	$\frac{1}{2}\begin{bmatrix}1&0\\-1&0\\0&1\\0&-j\end{bmatrix}$	$\frac{1}{2}\begin{bmatrix}1&0\\-1&0\\0&1\\0&j\end{bmatrix}$	$\frac{1}{2}\begin{bmatrix}1&0\\j&0\\0&1\\0&1\end{bmatrix}$	$\frac{1}{2}\begin{bmatrix}1&0\\j&0\\0&1\\0&-1\end{bmatrix}$

Index 8 to 15							
$\frac{1}{2}\begin{bmatrix}1&0\\0&1\\1&0\\0&1\end{bmatrix}$	$\frac{1}{2}\begin{bmatrix}1&0\\0&1\\-1&0\\0&-1\end{bmatrix}$	$\frac{1}{2}\begin{bmatrix}1&0\\0&1\\2&-1\\0&1\end{bmatrix}$	$\frac{1}{2}\begin{bmatrix}1&0\\0&1\\2&-1\\0&1\end{bmatrix}$	$\frac{1}{2}\begin{bmatrix}1&0\\0&1\\2&1\\0&1\end{bmatrix}$	$\frac{1}{2}\begin{bmatrix}1&0\\0&1\\2&1\\0&1\end{bmatrix}$	$\frac{1}{2}\begin{bmatrix}1&0\\0&1\\2&-1\\0&1\end{bmatrix}$	$\frac{1}{2}\begin{bmatrix}1&0\\0&1\\2&-1\\0&1\end{bmatrix}$

Note: For CMP codebook, constant modulus codebook will be acceptable from a CM perspective. For higher-rank transmissions, it is important that the codebook structure is such that there is limited mixing of layers onto the same PA. There is exactly 1 nonzero element per row in all precoder matrices, i.e., only 1 spatial layer is conveyed over each transmit antenna, and therefore the codebook preserves the CM.

Table 4.17 CM Value Analysis of Number of Mixed Layers for a Single-Antenna Port

	CM (dB)
Mix of 1 layer in a single-antenna port	1.22
Mix of 2 layers in a single-antenna port	2.55
Mix of 3 layers in a single-antenna port	3.05
Mix of 4 layers in a single-antenna port	3.30

Table 4.18 4Tx Rank 3 Codebook

Index 0 to 3	$\frac{1}{2}\begin{bmatrix}1&0&0\\1&0&0\\0&1&0\\0&0&1\end{bmatrix}$ $\frac{1}{2}\begin{bmatrix}1&0&0\\-1&0&0\\0&1&0\\0&0&1\end{bmatrix}$ $\frac{1}{2}\begin{bmatrix}1&0&0\\0&1&0\\1&0&0\\0&0&1\end{bmatrix}$ $\frac{1}{2}\begin{bmatrix}1&0&0\\0&1&0\\-1&0&0\\0&0&1\end{bmatrix}$
Index 4 to 7	$\frac{1}{2}\begin{bmatrix}1&0&0\\0&1&0\\0&0&1\\1&0&0\end{bmatrix}$ $\frac{1}{2}\begin{bmatrix}1&0&0\\0&1&0\\0&0&1\\-1&0&0\end{bmatrix}$ $\frac{1}{2}\begin{bmatrix}0&1&0\\1&0&0\\1&0&0\\0&0&1\end{bmatrix}$ $\frac{1}{2}\begin{bmatrix}0&1&0\\1&0&0\\-1&0&0\\0&0&1\end{bmatrix}$
Index 8 to 11	$\frac{1}{2}\begin{bmatrix}0&1&0\\1&0&0\\0&0&1\\1&0&0\end{bmatrix}$ $\frac{1}{2}\begin{bmatrix}0&1&0\\1&0&0\\0&0&1\\-1&0&0\end{bmatrix}$ $\frac{1}{2}\begin{bmatrix}0&1&0\\0&0&1\\1&0&0\\1&0&0\end{bmatrix}$ $\frac{1}{2}\begin{bmatrix}0&1&0\\0&0&1\\1&0&0\\-1&0&0\end{bmatrix}$

shows a CM value analysis of a mixture of signals from different numbers of layers. We can see that when multiple signals from different layers are mixed into a single antenna port, the result is higher CM values. The CM value increase is proportional to the number of layers mixed into a single antenna port.

In the uplink, a CM-preserving codebook should be employed for all ranks in 4Tx codebook designs. For higher-rank codebooks, non-CM-preserving may also be considered as a part of the codebook.

4.4.3.4.3 4Tx Rank 3 Codebook

Refer to Table 4.18. There is only one layer of data of a rank 3 codebook in one antenna port; this is used mainly in power-limited scenarios. The CM property is also important for rank 3 cases because there is still the possibility of suffering

Table 4.19   4Tx Rank 4 Codebook

Codebook Index	Number of Layers $\upsilon=4$
0	$\dfrac{1}{2}\begin{bmatrix} 1 & 0 & 0 & 0 \\ 0 & 1 & 0 & 0 \\ 0 & 0 & 1 & 0 \\ 0 & 0 & 0 & 1 \end{bmatrix}$

a power-limited situation even in a higher rank due to uplink power sharing between PUSCH and PUCCH transmissions when both uplink channels are scheduled at the same time (i.e., each row of the precoding matrix can have at most one nonzero entry for cubic metric preserving). For the trade-off between codebook size and low complexity, 4Tx rank 3 codebook design in LTE-A is featured by size 12 CMP codebook, binary phase shift keying (BPSK) alphabet, minimum chordal distance = 0.2887.

As rank 3 transmission is mainly used for higher-geometry UEs, a situation where power is not limited could be the typical case. So, a rank 3 codebook can relax the CM-preserving criterion. The CM-preserving criterion requires that the structure be unnecessarily restrictive. It is possible to limit the amount of CM increase by utilizing a certain structure where the cubic metric is increased but not as much as in the worst case of mixing the signals from all three layers onto the same PA.

4.4.3.4.4 4Tx Rank 4 Codebook

A single identity matrix (Table 4.19) is used for precoding.

For spatial multiplexing, up to two codewords and four layers are defined and dynamic rank adaptation is supported. For the UE with 2 transmit antennas, the 3-bit precoding codebook for spatial multiplexing with up to 2 layers is predefined, in which six precoding vectors are used for single-layer transmission and an identity matrix is used for two-layer full-rank transmission. For the UE with 4Tx antennas, the 6-bit precoding codebook for spatial multiplexing with up to 4 layers is predefined; the rank 1 codebook is size 24, the rank 2 codebook is size 16, the rank 3 codebook is size 12, and only an identity matrix is used for rank 4 (i.e., full-rank transmission).

4.4.4 Procedure for UCI Multiplexing on PUSCH

In LTE-A, simultaneous transmission of uplink control signaling and data is supported through control signaling being multiplexed with data on the PUSCH according to the same principle as in LTE Rel-8, or control signaling is transmitted on the PUCCH simultaneously with data on the PUSCH (see Figure 4.40).

As multiple transmit antennas are supported in LTE-A for uplink transmission, up to two CWs could be used for the PUSCH with MIMO transmission.

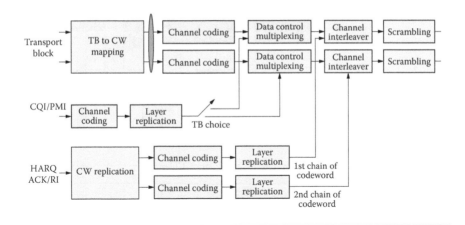

Figure 4.40 UCI multiplexing on PUSCH.

When there is only one CW, the control information could be just multiplexed with data on it. For the UEs with two CWs, the UCI could be multiplexed with data on only one CW or on all the CWs.

In LTE-A, a new DCI format, DCI format 4, is introduced for PUSCH scheduling. For uplink HARQ in the case of UL SU-MIMO, the number of MCS fields in DCI format 4 is two. Separate link adaptation of two codewords is supported. Two new data indicators (NDIs) (one NDI per CW) in the DCI format 4 will be associated with UL SU-MIMO and two HARQ ACK/NACK bits are supported. Spatial HARQ bundling is not supported, and there is one ACK/NACK for each codeword. For a single component carrier (CC) uplink MIMO transmission, the physical HARQ indicator channel (PHICH) resources for CW1 and CW2 are identified by the following: The first PHICH is determined by the Rel-8 equation, the second is determined by replacing the (lowest PRB index) with the (lowest PRB index + 1) in the same Rel-8 equation.

HARQ-ACK and RI bits are replicated over both CWs before channel coding, and TDM is multiplexed with data such that UCI symbols are time-aligned across all layers. Same as in Rel-8, ACK/NACK bits puncture the resources of the PUSCH, while RI information is mapped to resources around those used by HARQ-ACK, as shown in Figure 4.41.

CQI/PMI transmission will reuse Rel-8 multiplexing and channel interleaving mechanisms and transmit only on one codeword. Since CQI/PMI is important for the proper functioning of the system, it needs to be protected more than data so that it can be received by the eNB correctly. CQI is only transmitted on the codeword with the higher MCS on the initial grant, or on the first codeword when MCSs of both codewords are the same.

When there is no data transmitted, the UCI for multiple carriers is jointly coded and only rank 1 transmission for UCI is supported.

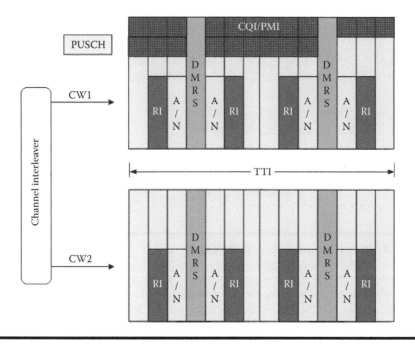

Figure 4.41　Channel interleaver of UCI on PUSCH.

4.5　Multiuser MIMO

MU-MIMO is a set of advanced MIMO technologies that exploit the availability of multiple independent radio terminals in order to enhance the communication capabilities of each individual terminal. In contrast, single-user MIMO considers access to only the multiple antennas that are physically connected to each individual terminal. MU-MIMO entails scheduling multiple UEs on the same time-frequency resources. Instead of maintaining orthogonality only over time and/or frequency, the spatial dimension is used for separating the signals to the co-scheduled UEs by designating appropriate transmit and receive antenna weight vectors. MU-MIMO can be seen as the extended concept of SDMA, which allows a terminal to transmit (or receive) a signal to (or from) multiple users in the same band simultaneously. By jointly taking advantage of multiuser scheduling gain and space frequency diversity, MU-MIMO can achieve significantly higher cell spectrum efficiency than SU-MIMO, in theory. MU-MIMO can provide spatial multiuser diversity and outperform single-user MIMO if designed properly. MU-MIMO's philosophy is to enhance the total cell throughput by allowing users with similar high SNR, and similar data transmission requirements in the same subframe, to share the same time and frequency resource. This is also known as co-scheduling (or pairing) the UEs.

The MU-MIMO scheme is simply described in Figure 4.42. Each UE estimates the MIMO channel matrix for each resource block. Based on the channel estimate,

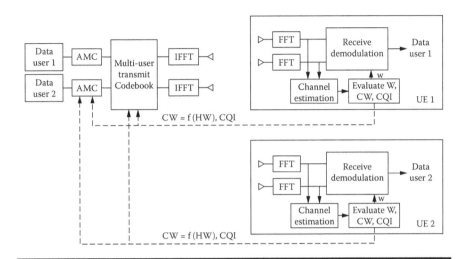

Figure 4.42 MU-MIMO DL system.

each UE designs a receive beamformer, elects an appropriate codeword from a given codebook of a vector, and computes the effective CQI independently for each RB. The codeword **CW** is selected from a codebook for each RB to best represent the vector quantity **HW** where **W** is the adaptive receive channel precoding matrix and **H** is the channel matrix corresponding to the RB (with N_t rows and N_r columns; N_t and N_r are the number of eNB and UE antennas, respectively).

With MU-MIMO, multiple UEs can be scheduled on the same resources and will benefit from multiuser diversity and spatial multiplexing gain in a low mobility scenario. As we have discussed, it is suitable to multiplex multiple UEs over the correlated channel, and the remaining uncorrelated channel dimensions can be utilized by allocating multiple layers to each UE. MU-MIMO transmission with a single layer per UE would be beneficial for a deployment with high system load and high geometry. The scenario with very high geometry, a cross-polarized eNB antenna array, and a light system load can introduce multiple-layer transmission per UE in MU-MIMO operation.

However, in order to obtain reasonable spatial multiuser diversity gain, it is necessary to handle multiuser interference properly. The MU-MIMO scheme in LTE does not provide any interfering channel information to the UE. Therefore, there is a huge CQI mismatch between UE feedback and the downlink channel when the eNB schedules multiple UEs in the same time-frequency resource. Therefore, more sophisticated MU-MIMO operation should be introduced for LTE-A to provide better performance.

It is worth mentioning that the total power in the eNB is fixed in downlink MU-MIMO. The power per resource element is also fixed. From that point of view, it can be expected that downlink MU-MIMO has less potential for capacity gain than its uplink counterpart.

4.5.1 MU-MIMO in LTE

MU-MIMO in LTE Rel-8 specifies a simple version of downlink MU-MIMO and is very closely related to the techniques designed for SU-MIMO, whereas in practice more significant improvements in total system throughput and spectral efficiency can be achieved by means of different feedback and beamforming techniques. In practice, MU-MIMO can provide spatial multiuser diversity and outperform SU-MIMO in terms of bps/Hz, if designed properly. LTE Rel-8 UEs that are capable of receiving PDSCH transmissions in transmission mode 5* (MU-MIMO) do not necessarily have the capability to receive multiple PDSCH codewords in the same subframe. The downlink signaling from the eNB to the scheduled UEs is used to indicate the selected precoding vectors and to implicitly indicate the presence of a co-scheduled UE. This is done by use of DCI format 1D with a PMI and one power offset bit.

The precoding vector for MU-MIMO is restricted within the codebook defined in LTE Rel-8 and only the wideband PMI precoding report is supported. The use of a wideband PMI implies that MU-MIMO works well with correlated channels because the spatial correlation may not change much over frequency. The granularity of the codebook (4 bits for four-antenna codebook), the wideband nature of the PMI, and the use of CRS limit the use of transmitter-side interference suppression such as zero-forcing beamforming.

In LTE Rel-8, both the channel quality estimation and the data demodulation are based on cell-specific reference signals. Since the CRS cannot capture the UE-specific precoding applied to the PDSCH, the precoder index should be signaled on the PDCCH. Meanwhile, LTE Rel-8 defines only one pattern of dedicated RS so that the dedicated beamforming supports rank 1 transmissions only. UE needs to know the precoding matrix for demodulation because the scheduling grant provides no information about the precoding vector of the other co-scheduled UE. Because the UE does not know the precoding vectors of other UEs when it estimates CQI, it will cause a CQI mismatch problem and degrade the effectiveness of interference suppression and system throughput. According to SU CQI feedback from the UE, MU CQI can be calculated based on the following equation:

$$CQI_{MU} = CQI_{SU} - n - \alpha - \Delta$$

where n is a power offset due to power allocation, which depends on the total number of layers in an MU transmission. For example, $n = 3$ dB if two rank 1 UEs are paired up to do MU transmission. α is a back-off factor that depends on the UE geometry. Δ is a CQI adjustment factor based on ACK/NACK.

In LTE Rel-8 MU-MIMO, each UE is restricted to receiving a single layer. Although Rel-8 MU-MIMO does not explicitly forbid scheduling more than two users in a resource block, in practice, only up to two UEs can be well supported

---

* In LTE Rel-8, SU-MIMO and MU-MIMO are two separate transmission modes. Mode switching between these two is semistatically configured and signaled through higher-layer signaling.

due to the single power offset bit. The only difference between SU-MIMO and MU-MIMO is that power sharing information is provided in the PDCCH so that a UE can demodulate data when higher modulation is used. Since the UE is not informed of co-channel interference information, a CQI is calculated based on single-user MIMO transmission.

We assume that the MMSE receiver is used at the UE. Each user reports the SNR (instantaneous rate) on all the precoding vectors/matrices calculated according to the formula:

$$SNR_k^{MMSE} = \frac{\varepsilon_s}{N_0[F^*H^*HF + N_0/\varepsilon_s I_M]^{-1}k,k} - 1$$

where F is the precoding matrix of dimension *number of Tx antennas* × *rank*; allowed F matrices can only be from a standardized codebook. The UE feeds back its best choice of F (known as PMI) and corresponding CQI; H is the channel matrix (includes path loss, fading, etc.).

For the SU-MIMO rank 1 report, the above SNR is calculated without accounting for interference while the rank 2 report considers interstream interference, which is from the interaction of the user's channel matrix with the orthogonal precoding vector. So the rank 2 report is the sum of the spectral efficiencies of the individual streams.

After obtaining the CQI reports from all users, the eNB calculates the metric for every user i – vector/matrix j pair. For SU-MIMO-only scheduling, the scheduler picks the best user–vector pair that provides the highest metric. When considering MU-MIMO, we compute a metric for each of the unitary matrices. This metric is the sum of the highest metric on each column of the matrix. We select the matrix that offers the highest metric and compare this metric with that of the best SU-MIMO metric. If the MU-MIMO metric is larger, we allocate in MU-MIMO fashion (Figure 4.43).

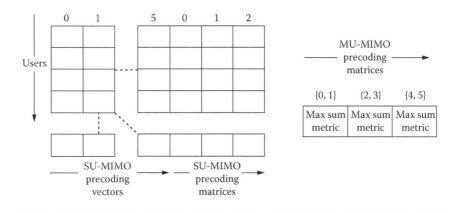

Figure 4.43 MU-MIMO pairing.

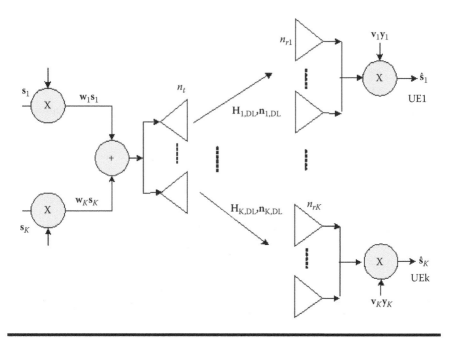

Figure 4.44 A typical downlink MU-MIMO system.

LTE Rel-8 eNBs with UE channel state information may schedule multiple users on the same radio resources.* A typical downlink MU-MIMO system can be illustrated as in Figure 4.44. Consider the scenario where $\mathbf{s}_k$ and $\mathbf{n}_k$ denote the kth transmit symbol vector and the additive white Gaussian noise vector, respectively. The actual transmitted signal vector for user k is then given by $\mathbf{w}_k\mathbf{s}_k$, where $\mathbf{w}_k$ denotes the precoding matrix for the kth user. We assume that service will be provided to a set of K selected uncorrelated users and each user is equipped with $n_{r,k}$ uncorrelated antennas (thus able to receive up to $n_{r,k}$ streams). The received signal vector at the kth user is: $\mathbf{y}_k = \mathbf{H}_k\mathbf{w}_k\mathbf{s}_k + \mathbf{H}_k \sum_{l=1, l\neq k}^{K} \mathbf{w}_l\mathbf{s}_l + \mathbf{n}_k.$

The goal of linear precoding is to design $\mathbf{W} = [\mathbf{w}_1, \mathbf{w}_2, \dots, \mathbf{w}_K]$ based on the full channel matrix $\mathbf{H} = [\mathbf{H}_1^T, \mathbf{H}_2^T, \dots, \mathbf{H}_K^T]^T$, so that $\mathbf{H}\mathbf{W}$ is diagonal, e.g., $\mathbf{H}_i\mathbf{w}_j = 0$ for $i\neq j$.

There are several mature schemes, for example, zero-forcing precoding or block-diagonalization or maximum signal-to-leakage ratio (SLR) beamforming, which can be used to null out the interuser interference.

---

* The bandwidth of multiplexed UEs has to be same because CS cannot keep the orthogonality among the UEs with different bandwidth in Rel-8 MU-MIMO. Further throughput improvement is expected by applying MU-MIMO with different bandwidths.

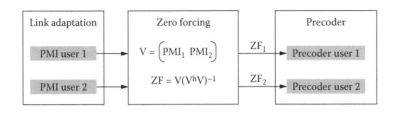

Figure 4.45   Downlink MU-MIMO with zero-forcing.

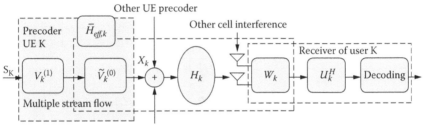

Figure 4.46   Downlink MU-MIMO with BD.

Zero-forcing (ZF) precoding is a potential precoder design technique for downlink MU-MIMO. The main benefit of ZF precoding is that the interference is precanceled at the transmitter side. This implies that the eNB has most of the computational complexity in designing the precoder, and each UE needs only information regarding its own data streams for reception. The ZF precoding process is shown in Figure 4.45.

For a single-cell transmission, consider a two-user scenario where the receive signal is given by

$$y_1 = H_1 F_1 s_1 + H_1 F_2 s_2 + n_1$$

$$y_2 = H_2 F_1 s_1 + H_2 F_2 s_2 + n_2$$

where F_1 and F_2 are the precoding matrices for user 1 and user 2, s_1 and s_2 are the data vectors for user 1 and user 2, H_j is the channel matrix from the eNB to the jth user, and n_1 and n_2 are the interference and noise with covariance matrix $R_{1,n}$ and $R_{2,n}$. The zero-forcing design principle of MU-MIMO is to obtain precoding matrices F_1 and F_2 to minimize or completely precancel the interuser interference to make $H_1 F_2 = 0$ and $H_2 F_1 = 0$, while achieving good system performance for the effective single-user channel $H_1 F_1$ and $H_2 F_2$.

Block diagonalization (BD) is another approach for linear precoding in the MIMO channel that sends multiple interference-free data streams to different users in the same cell. The principle idea of BD (shown in Figure 4.46) is to project the precoding matrix into the null space of a co-scheduled user, so interuser interference is precanceled at the eNB and the UE does not need to perform additional intracell interference mitigation.

The SLR beamforming vectors are designed to maximize the ratio of the desired signal received at the target user and the interference leaks to a co-scheduled user in the same frequency resources. SLR beamforming allows a certain level of residual interference as a trade-off for a more flexible beamforming design, and residual interuser interference could be further mitigated at the mobile terminal using on the UE-specific interference suppression method. The precoder of user 1 with SLR is expressed as:

$$W_1 = \arg\max_{W} \frac{W^H H_1^H H_1 W}{W^H \left(H_2^H H_2 + R_{2,N} \right) W}$$

$$= evec\{(H_2^H H_2 + R_{2,N})^{-1} \times H_1^H H_1\},$$

where *evec*{·} is the right dominant/principle eigenvector, and $R_{j,N}$ is the noise/interference covariance estimate of the *j*th user. Sufficient accuracy of both the transmission and noise/interference covariance need to be known in order to effectively mitigate interuser interference. It is easy to see that W_1 follows the Rayleigh-Ritz quotient theory for maximizing the ratio of user 1's desired signal to the leakage (interference) that user 1 causes to user 2. If the $R_{j,N}$ estimate is considered unreliable, an alternative solution is to report the noise/interference power P_0 for SLR beamforming formulated as $W_1 = evec\{(H_2^H H_2 + P_0 I_{Nt})^{-1} \times H_1^H H_1\}$.

Although the aforementioned schemes exist MU-MIMO support in LTE Rel-8 is suboptimal: Precoding is based on the SU-MIMO codebook (no zero-forcing) and the rank 1 SU-MIMO CQI report is also used for MU-MIMO. Meanwhile, the UE does not know about the precoder of the interferer; the UE is unaware of the interference when performing CQI measurement, and therefore the reported CQI is too optimistic.

Rel-8 channel estimation for demodulation is based on cell-specific reference signals. This implies that downlink signaling from the eNB to the scheduled UEs is required to indicate the selected precoding vectors and to implicitly indicate the presence of a co-scheduled UE. This is done by use of DCI format 1D with a PMI and power offset bit to indicate to the UE working under the MU-MIMO that the power of the data addressing this UE is only half of the energy per resource element (EPRE). Current Rel-8 MU-MIMO design has some limitations; for example, the precoding vector is constrained by the predefined codebook, each of the two co-scheduled UEs is limited to rank 1 transmission that will lead to limited throughput, and the scheduling grant provides no information about the precoding vector of the other co-scheduled UE. In summary, the main features of LTE Rel-8 MU-MIMO are as follows:

■ Semistatic SU–MU MIMO switching: The eNB semistatically configures the UE into SU-MIMO or MU-MIMO operation. Here, the term *semistatic* implies that the switching is performed via an RRC reconfiguration. In closed loop (CL) operation, UE feedback for MU operation is based on the SU rank 1 assumption. The SU-based feedback could be the bottleneck to achieving the higher throughput predicted by advanced MU operation, which becomes more feasible and important as the number of transmit antennas at the eNB increases. One approach to improve MU operation is to amend PMI-based feedback to include information about the null space.
■ Continuous and aligned resource allocation of co-scheduled UEs
■ Inaccurate feedback, no difference in CQI/PMI feedback signaling between SU-MIMO and MU-MIMO
■ Less effective interference suppression: The codebook defined for Rel-8 MU-MIMO is the same as for SU-MIMO. Wideband precoding is used in general. On the other hand, the UE uses the cell-specific reference signals with no additional signaling regarding the co-scheduled UE, so it limits the effectiveness and flexibility of UE-based interference suppression.

In LTE Rel-8 MU-MIMO, transparent MU-MIMO is supported using transmission mode 5, where good spatial separation between scheduled UEs depends on the eNB scheduler. In LTE Rel-9, an extension was adopted to improve the orthogonality between DMRSs using different UE-specific antenna ports with orthogonal patterns and DMRS scrambling initialization ID, which is a form of nontransparent MU-MIMO. The UE-specific MU-MIMO scheme supports dynamic indication that the DMRS port is supported in case of rank 1 transmission, to enable scheduling of two UEs with rank 1 transmission using different orthogonal DMRS ports on the same PDSCH resources. There is no explicit signaling of the presence of co-scheduled UEs in case of rank 1 transmission. In case of rank 1 transmission, the UE cannot assume that the other DMRS antenna port is not associated with the PDSCH assigned to another UE. The main features of MU-MIMO in Rel-9 (TM8) are

■ Dynamic SU-/MU-MIMO switching
■ Transparent SU-/MU-MIMO transmission
■ Up to 4 total layers and a maximum of 2 layers per UE
■ Flexible resource allocation and pairing

LTE Rel-9 MU-MIMO supports one stream transmission of up to 4 UEs simultaneously, and 2 orthogonal DMRS ports and 2 scrambling sequences in a quasi-orthogonal manner are defined, but the UE reports PMI and CQI in the SU-MIMO context only.

In general, MU-MIMO in Rel-8/9 is very closely related to the techniques designed for SU-MIMO, whereas in practice, more significant improvements in total system throughput and spectral efficiency can be achieved by means of different feedback and beamforming techniques. More advanced MU-MIMO techniques should be considered for LTE-A to improve overall system spectral efficiency, and LTE-A should be able to approach the theoretical capacity more closely than LTE Rel-8.

4.5.2 MU-MIMO in LTE-A

In LTE Rel-8, as discussed previously, the MU-MIMO scheme is not fully optimized to minimize co-channel interference. Simply reusing the SU-MIMO codebook and feedback mode may cause serious performance degradation with huge CQI mismatch. A UE cannot foretell what kind of precoder will be used for PDSCH transmissions and how much interference the UE will experience from other users' beams, so it is very difficult for a UE to measure and report accurate and appropriate CQI. The interference cannot be nulled out perfectly in practice; codebook and feedback design play a major role in this problem. Without accurate CSI feedback, the eNB cannot properly schedule multiple users on the same radio resources and use, for example, zero-forcing precoding or block diagonalization to null out the interuser interference. Since both the desired signal power and the undesired interference power are impacted by codebook selection in MU-MIMO, it is natural to see higher requirements on feedback accuracy compared to SU-MIMO.

Since the DL MU-MIMO scheme available in LTE Rel-8 has been simply designed, there is much room for MU-MIMO enhancement in LTE-A. LTE-A adopted many base features that were used in Rel-9, such as transparent SU-/MU-MIMO transmission, dynamic SU-/MU-MIMO switching, up to 4 total layers, a maximum 2 layers per UE, DMRS-based demodulation, two SCID and two orthogonal DMRS per SCID, and support for flexible resource allocation and UE pairing. The main enhancement features of LTE-A MU-MIMO are discussed hereafter.

4.5.2.1 CSI-RS-Based CSI Measurement

The CSI-RS is transmitted in each physical antenna port or virtualized antenna port and is used for measurement purposes only. Its channel estimation accuracy can be relatively lower than DMRS.

CSI feedback enhancement uses a two-matrix ($\mathbf{W}_1$, $\mathbf{W}_2$) feedback framework in which $\mathbf{W}_1$ targets wideband/long-term channel properties and $\mathbf{W}_2$ targets frequency-selective/short-term time channel properties.

4.5.2.2 Dynamic SU/MU Switching

The switching between SU- and MU-MIMO can be done either dynamically or semistatically. However, compared to semistatic switching, dynamic switching has some notable benefits. With dynamic switching, the eNB can be more flexible in adapting to different system conditions, such as type of traffic and number of UEs, and the eNB can better adapt to the time variation of the wireless channel conditions. The eNB can also adaptively optimize its transmission on a subframe-by-subframe basis to maximize system performance. But the issue of SU-/MU-MIMO dynamic switching implies that SU- and MU-MIMO should have the unified CQI feedback frame and CQI calculation method in the UE, which is not based on the transmission mode hypothesis.

In LTE-A, UE-specific DMRS may provide an opportunity for dynamic switching between SU- and MU-MIMO. Switching between SU- and MU-MIMO transmissions is possible without RRC reconfiguration. If the same DMRS for both SU-MIMO and MU-MIMO is used, this implicit MU-MIMO may provide scheduling gain by allowing dynamic switching between SU- and MU-MIMO, and the UE does not need to know the transmission mode. For historic reasons, SU-MIMO and MU-MIMO are the two separate transmission modes in the current LTE Rel-8 and Rel-9, and mode switching semistatically configured and signaled through higher-layer (RRC) signaling, and two different DCI formats are specified to support these two modes. LTE-A supports dynamic SU-/MU-MIMO switching, which indicates that SU- and MU-MIMO transmissions are within the same transmission mode, and the same DCI format is used to indicate both SU- and MU-MIMO transmission. A block diagram of the eNB processing for dynamic SU/MU switching is given in Figure 4.47.

4.5.2.3 Transparent versus Nontransparent MU-MIMO

The MU-MIMO DCI would be transmitted from the eNB to the UE via the PDCCH. The DCI will include an indication of the existence of co-scheduled

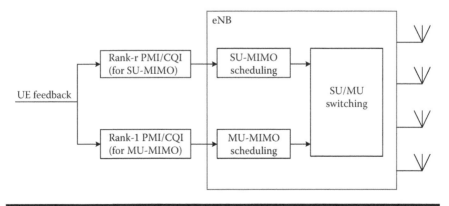

Figure 4.47   eNB scheduling procedure with multirank PMI/CQI feedback.

Table 4.20 Transparent versus Nontransparent MU-MIMO

	Advantages	*Disadvantages*
Transparent	No additional signaling to indicate co-scheduled UE	Inter-UE interference suppression at the receiver side is difficult; solely relying on signal processing at the network side may not be able to fully exploit the MU-MIMO gain
	Common feedback mode for SU-MIMO and MU-MIMO	Must rely on spatial isolation DMRS (nonorthogonal) to some extent if more than two UEs are co-scheduled
Nontransparent	Good co-scheduled UE interference suppression capability by selecting optimized MMSE combining weights in the receiver, or nonlinear interference cancellation techniques	Additional signaling overhead and should have MU-MIMO-specific feedback mode

UE(s), precoding information of the co-scheduled UE(s), DMRS pattern, and so on. The DCI may be transparent or nontransparent (see Table 4.20).

Transparent here means that no DL signaling (DCI) is provided to indicate to a UE whether a DL transmission to another UE is taking place in the same RB. In other words, transparent MU-MIMO means that from the UE perspective there is no difference between SU-MIMO and MU-MIMO transmission and the UE can simply utilize an interference rejection combining receiver for SU-MIMO demodulation. *Nontransparent* DCI is defined such that some information about the co-scheduled UE is explicitly or implicitly signaled to the UE via the DCI, which enables interference suppression between the co-scheduled UEs. Due to precoded UE-specific DMRS, power-sharing information is not needed in LTE-A since the DMRS already contains power-level information so that fully UE transparent MU-MIMO can be implemented. So in LTE-A, no dynamic DL signaling is provided to indicate to a UE whether a DL transmission to another UE is taking place in the same RB in which the UE is scheduled. In the transparent case, knowledge of the interference, such as the spatial signature, cannot be accurately obtained by the desired UE. The performance of transparent MU-MIMO heavily relies on how well spatial isolation can be achieved between co-scheduled UEs. The degradation of channel estimation accuracy due to interference on DMRSs leads to a marginal gain of MU transmission over SU transmission. Inter-UE interference cancellation would be difficult by the UE receiver, and interference cancellation at the transmitter side becomes absolutely necessary.

From the previous discussion, we know that the performance of transparent MU-MIMO will depend on spatial isolation, which depends on channel characteristics, antenna setup, number of users to be scheduled, and knowledge of the channels at the eNB side that can be achieved by transmitter precoding between co-scheduled UEs.

4.5.2.4 Double Codebook-Based Feedback Modes

The new codebook supports finer granularity of channel feedback and interfering vector indication for better support of the CQI calculation. A smart scheduler can select an appropriate precoder to reduce the interuser interference in MU-MIMO operations. LTE-A UE simply follows the Rel-8 feedback framework and reports implicit SU CQI/PMI, and MU-MIMO precoding decision and link adaption is performed at eNB accordingly. The MU precoders can be derived with a transformation on the PMI that is received from multiple UEs by various solutions including block diagonalization, maximum SLR, and zero-forcing.

4.5.2.5 LTE-A MU-MIMO Dimensioning

LTE-A must support the enhanced MU-MIMO dimensioning in terms of the number of co-scheduled UEs and the number of layers per UE. The maximum number of co-schedulable users will be determined by the number of DMRSs. MU-MIMO is based on one layer transmission per UE and only two co-scheduled UEs operating on the cell-specific RS in LTE Rel-8. Two orthogonal DMRS ports and two scrambling sequences are defined and can accommodate up to four layers using a combination of orthogonal and quasi-orthogonal DMRSs in Rel-9. From an implementation perspective, having more UEs multiplexed adds to the UE complexity due to the need for more sophisticated interference cancellation and complicated eNB scheduling. On the other hand, multilayer transmission to scheduled UEs should also be considered to improve the total system throughput in LTE-A. For the design of DL signaling and DMRS, the following principles are assumed for MU-MIMO:

- No more than 4 UEs co-scheduled, and these UEs should have very high geometry and spatial multiplexing gain, which is more significant due to power splitting and increased intracell interference.
- No more than 2 layers per UE with 2 orthogonal DMRS ports, because having a high number of layers per UE in MU-MIMO requires a high SINR and therefore very good spatial separation between the layers to avoid excessive MU interference.

Aligned and continuous resource allocation (Rel-8)   Flexible resource allocation (LTE-A)

Figure 4.48   Flexible pairing and resource allocation.

■ No more than 4 transmission layers in total for MU-MIMO transmission, because supporting more than four layers in MU-MIMO would require a very high SINR and an extremely high level of spatial separation, which seems unrealistic in practice.

4.5.2.6 Resource Allocation in MU-MIMO

LTE-A supports flexible pairing and resource allocation in MU-MIMO (see Figure 4.48), which can provide 10% more performance gain over aligned resource allocation used LTE Rel-8.

Based on all of the previous analysis, we can see that there have been many valuable changes in MU-MIMO in LTE-A. In summary, the MU-MIMO technology evolution from Rel-8 to LTE-A is shown in Table 4.21.

4.5.3 Analysis of Mutual Interference between UEs

It is well known that MU-MIMO can provide a multifold increase in total throughput, if the per-UE performance is not severely degraded due to the increased mutual interference experienced by UEs, which is the biggest technical challenge to overcome to achieve the theoretical potential. In MU-MIMO, there are two types of interference: interference among the layers allocated to the same UE (intra-UE interference), and interference among the layers allocated to difference UEs (inter-UE interference), which is due to the lack of perfect channel state information at the eNB and UE. Given that a typical 2Rx UE has no extra degree of freedom to cancel any interferer, the eNB has to assume most of the burden of reducing the cross-UE interference. This can be challenging if the eNB does not have good channel information from all UEs. The following points should be considered to avoid unnecessary implementation complexity when introducing the newly designed higher-order MIMO schemes or more sophisticated MU-MIMO operation for LTE-A. First, it should be guaranteed that the new schemes or operation can provide significant gain as compared with those in the LTE system. Second, the backward/forward compatibility should be taken into account to support legacy UEs as well as LTE-A UEs. Therefore, in principle, it is possible to enhance MU-MIMO in LTE-A in the following aspects.

The MU-MIMO design in Rel-8 lacks interference suppression capabilities at both transmitter and receiver. With the introduction of precoded UE-specific RSs,

Table 4.21 The Evolution of MU-MIMO Transmission

Release	Main Features
Rel-8	1. Semistatic MU-MIMO transmission mode
	2. One layer of UE-specific reference signals; each of the two co-scheduled UEs is limited to rank 1 transmission
	3. CQI/PMI/RI feedback similar to SU-MIMO
	4. 4-bit codebook-based feedback; precoding is based on the SU-MIMO codebook (no zero-forcing) and the rank 1 SU-MIMO CQI report is also used for MU-MIMO
	5. UE does not know about the precoder of the interferer; UE cannot suppress cross-talk due to MU-MIMO
	6. UE is unaware of the interference when performing CQI measurement; therefore reported CQI is too optimistic
	7. Possible MU-MIMO algorithm: eNB paired users with orthogonal rank 1 vectors; eNB chooses best user pair; predicts MCS; UE cannot suppress cross-talk due to MU-MIMO
	8. Channel estimation in Rel-8 MU-MIMO is based on the use of antenna-specific common reference signals
Rel-9	1. Supports 2 CDM streams of UE-specific reference signals: Same overhead as Rel-8; 1 stream UE-specific RS; enables cross-talk suppression with transparent MU-MIMO
	2. A single transmission mode for rank 1, rank 2, MU rank 1: Dynamic transition between SU and MU and dynamic transition between rank 1 and rank 2
	3. No signaling to indicate the presence of co-scheduled UEs
	4. No PMI feedback; LBUE always feeds back one CQI based on transmit diversity; UE does not feed back the rank indicator; MCS/rank determined by the eNB; UE not aware of SU or MU transmission during feedback
	5. PMI feedback (2Tx): CQI/PMI/RI feedback (same as 2Tx spatial multiplexing mode)
	6. In TDD, SRS transmission can be used for estimating covariance matrix for MU-MIMO; KBZero-forcing-based MU-MIMO for two rank 1 users; in FDD, translation of UL covariance to DL covariance may be possible in some situations
	7. Codebook-based feedback with CRS is inefficient

(Continued)

Table 4.21 (Continued) The Evolution of MU-MIMO Transmission

Release	Main Features
Rel-10	1. UE-specific reference signals extended to 8 streams; supports more than 2 users for MU-MIMO and multistream MU-MIMO
	2. CSI-RS (midamble) for CQI estimation; supports reduced CRS overhead
	3. Novel feedback method for FDD/TDD; may support covariance feedback, high-resolution PMI feedback
	4. Supports distributed antenna systems; transmitting to each mobile from several network antennas; receiving each mobile at many network antennas

more precoding flexibility is allowed at the transmitter side; however, residual spatial interference detection and suppression still need to be performed at the receiver. UE-specific RS designs could allow UEs to track the spatial interference, and channel diagonalization and interference nulling are generally regarded as attractive precoding strategies for single-cell MU-MIMO schemes for boosting system spectral efficiency and user experience. The Rel-9 feedback mechanism along with DMRS already allows the eNB to use precoding strategies such as zero-forcing or its many variants to co-schedule two UEs and minimize their mutual interference. However, practical feasibility and realistic gain of such processing at the UE should be thoroughly investigated.

Intra-UE interference can be mitigated by PMI-based precoding or receiver interference cancellation (IC). Inter-UE interference mitigation can be achieved only by an interference nulling precoder, which sometimes requires heavy CSI feedback.

A smart scheduler may be able to select an appropriate precoder to reduce the interuser interference in MU-MIMO operations. However MU-MIMO, as specified in LTE Rel-8, just relies on the codebook-based approach, so the smart algorithm for determining MU-MIMO precoders is limited. LTE-A can get better interference canceling in precoding (e.g., ZF) with covariance matrix feedback and signaling the interfering precoders so that the UEs can suppress residual interference using, for example, a linear minimum mean square estimation (LMMSE) receiver.

It is also necessary to enhance CQI reporting so that interference is taken into account. In other words, the UE needs to be interference aware when measuring CQI.

A number of MU-MIMO algorithms are possible with the transmit covariance information such as block diagonalization (BD) and maximum signal-to-leakage ratio (SLR) beamforming.

With covariance matrices and CQI information feedback from the UEs, the eNB scheduler may choose to schedule a DL transmission using MU-MIMO. The DL beamforming vectors are chosen to reduce the cross-user interference and to maximize the desired user signal strength simultaneously. Specifically, in the case of MU-MIMO with UE1 and UE2 being a pair, the beamforming vector for UE1 is derived by optimizing the following objective:

$$v_1^* = \arg\max_{v_1} \frac{v_1^H R_1 v_1^H}{v_1^H R_2 v_1^H + ni_2}$$

where R_1 and R_2 are the averaged channel covariance matrices of UE1 and its pair UE2 over a group of frequencies and ni_2 is the noise and interference power level at UE2, which can be directly reported or derived from the UE2 CQI feedback. So the optimal solution for v_1^* is the zero-forcing beamformer denoted as

$$v_1^* = Eig_m(R_{2,NI}^{-1} R_1)$$

or equivalently v_1^* is the eigenvector corresponding to the largest generalized eigenvalue of $(R_1, R_{2,NI})$, where $R_{2,NI} = R_2 + ni_2 \cdot I$, and $ni_2 \cdot I$ accounts for interference plus noise power per receive antenna measured at UE2.

In the case of SU-MIMO with covariance matrix feedback, the beamforming vectors are the eigenvectors corresponding to the largest one or two eigenvalues of R_1 depending on the rank of the transmission scheduled by the eNB scheduler. Meanwhile, as we can see, the ability to configure a UE to compute a covariance matrix over wideband or a specified sub-band is needed. This implies the need to define additional reporting modes.

Considering the interference cancellation from the UE side, if the MU-MIMO transmission mode is known by the UE, the interference from other co-scheduled UEs may be eliminated at the receiver with more signaling support. The necessary signaling may include the following:

■ The number of co-scheduled UEs and the layer number and location of co-scheduled UEs
■ Reference signal sequences of other co-scheduled UEs for DMRS
■ Allocated resource blocks for MU-MIMO transmission

To conclude, MU-MIMO highly relies on accurate channel knowledge at the transmitter and is very sensitive to the quantization error and to the feedback design. As SNR increases, MU-MIMO becomes interference limited due to the intracell interference induced by the quantization error. In an ideal scenario, two DMRSs can be separated with orthogonal MU pairing in which the transmit weight of the interfering user is orthogonal to the UE's own channel. However, because of the inevitable feedback quantization error and delay error, interference exists even though the reported PMIs of the two co-scheduled users are orthogonal. The interference can be reduced by using different scrambling codes. However, the orthogonality provided by the two scrambling codes is weak, especially when the precoding granularity is small (e.g., one RB), and therefore the performance cannot improve much. So, under more practical user distribution, quantized CSI feedback, and nonideal CQI, two co-scheduled UEs seems to be a much more typical and reasonable situation for live network MU-MIMO operation. LTE-A adopts double codebook and flexible feedback schemes that change depending on the user's spatial correlation, deployment scenarios, environments, and so on. When MU-MIMO is scheduled with limited feedback, in TDD, if channel reciprocity is ideal and antenna calibration is applied, multiuser interference in the spatial domain can be reduced easily with respect to FDD. In FDD on the other hand, spatial orthogonality may be hard to guarantee due to limited feedback and feedback delay, and special methods including the interference-suppression capability of the eNB and the UE need further study.

4.6 Cooperative MIMO

Cooperative MIMO (Co-MIMO), also known as network MIMO (Net-MIMO) or Ad-hoc MIMO, utilizes distributed antennas that belong to other users, while conventional MIMO, or single-user MIMO, only employs antennas belonging to the local terminal. Co-MIMO improves the performance of a wireless network by introducing multiple antenna advantages, such as diversity, multiplexing, and beamforming. If the main issue is diversity gain, it is known as *cooperative diversity*. It can be described as a form of macro-diversity, which is used, for example, in soft handover. Cooperative MISO corresponds to transmitter macro-diversity or simulcasting. A simple form that does not require any advanced signal processing is the single-frequency network (SFN), used especially in wireless broadcasting.

Co-MIMO is a technique that is useful for future cellular networks that employ wireless mesh networking or wireless ad-hoc networking. In wireless ad-hoc networks, multiple transmit nodes communicate with multiple receive nodes. To optimize the capacity of ad-hoc channels, MIMO concepts and techniques can be applied to multiple links between transmit and receive node clusters. Contrasted with multiple antennas in a single-user MIMO transceiver, participating nodes and their antennas are located in a distributed manner. So, to achieve the capacity of this network, techniques to manage distributed radio resources are essential. Strategies such as autonomous interference cognition, node cooperation, and network coding with dirty paper coding have been suggested as solutions to optimize wireless network capacity. So the basic concept of Co-MIMO is to perform joint MIMO transmission and reception between multiple coordinated eNBs and a single UE or multiple UEs over the same radio resources.

Despite the benefits Co-MIMO may bring, there are key challenges we have to face to deploy Co-MIMO. Co-MIMO will raise very strict requirements on the backhaul capacity and latency. In the downlink, both data and channel state information need to be available in all cooperating base stations; in the uplink, receiver soft outputs may need to be exchanged between cooperative base stations.

Multicell channel estimation is also needed for channel state feedback purposes, but arranging reference signals with tolerable overhead for channel estimation of a large number of antennas seems extremely challenging, and heavy channel estimation (and precoding calculation) capabilities are required by the terminals (e.g., all antennas of up to 2–5 cells).

Furthermore, UL resources for feedback transmission are very scarce. Naturally, channel state feedback needs to be very limited in order to not fill UL capacity with control. While the cell-edge terminals are exactly the ones that benefit most from Co-MIMO cooperation, but because they may be power-limited, serious challenges are faced especially with cell-edge terminals.

Co-MIMO will face similar problems as single-cell multi-user MIMO. Since channel state information at base stations can never be perfect, CQI reporting becomes tricky as interuser interference nulling cannot be perfect. If the interuser interference is unknown, accurate CQI reporting is difficult to achieve; therefore, for good performance, interferer precoding weights may be needed at the CQI reporting phase.

The most common scenarios in Co-MIMO can be divided into two types. In the first type, each UE can be jointly served by multiple eNBs through eNB coordination over the same radio resource. By doing so, the ICI can be mitigated or even changed into useful signal power. The second type is in some ways similar to MU-MIMO scenarios in which each eNB can serve multiple UEs over the same radio resource. By doing so, the overall sector throughput can be improved.

An example of Co-MIMO on the downlink is shown in Figure 4.49 in which two eNBs jointly serve two UEs through coordination over the backhaul.

Cooperative multicell MIMO can increase throughput by spatial multiplexing from multiple sites, as shown in Figure 4.50. The eNBs use the same resource to the same UEs and transmit multiple streams. For less co-channel interference and better cell-edge performance, independent channels will be used by eNBs for both UL and DL. In multicell MIMO scenarios, UE synchronization to more than one cell and a synchronized network are required, and the increased signaling overhead due to the required feedback information should be considered. It is desired that the multicell

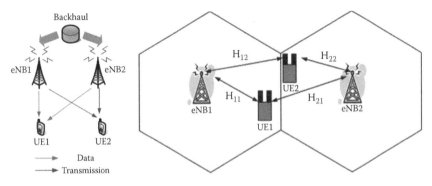

Figure 4.49   Illustration of Co-MIMO on the downlink.

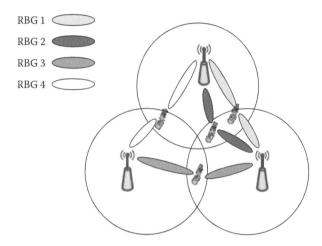

Figure 4.50 Cooperative multicell MIMO.

feedback utilize the single-cell feedback as much as possible while keeping in mind the trade-off between feedback overhead and the corresponding performance.

Co-MIMO involves three basic operations. First, each eNB acquires the channel information of all UEs that it serves. Channel information,* including the channel matrix of each user and channel covariance matrix, should be acquired. Second, eNBs should exchange information needed for coordination. Each eNB may need to share complete or partial channel information and scheduling information between itself and the UEs that it may serve through Co-MIMO. For partial channel information exchange, each eNB may need to share only a part of the channel information, for example, the long-term CQI, received signal strength indicator (RSSI) in conjunction with other DL preamble measurements, or the covariance channel matrices between the UEs and itself. Third, based on the information obtained in the first two operations, multi-eNB transmitter processing is applied on the downlink for the UEs being jointly served by the coordinating eNBs. Different multiuser transmitter and receiver processing techniques, such as block diagonalization, SINR maximization, interference cancellation, and beamforming, may be used. Various technical alternatives can be considered to provide different trade-offs between performance and implementation cost. Co-MIMO can be regarded as an evolution of macro-diversity and multiuser MIMO that allows the benefits of both macro-diversity (MD) and MU-MIMO to be realized across a set of coordinating eNBs (Figure 4.51).

---

* In LTE, the channel measurement mechanism has already been defined for single eNB MIMO in both TDD and FDD mode. Co-MIMO can reuse this mechanism by introducing some enhancements. In particular, in single-eNB MIMO, each eNB only needs the channel information of users associated to it; while in Co-MIMO each eNB needs the channel information of users associated to other eNBs as well. The existing channel measurement mechanism cannot support this function and needs to be improved.

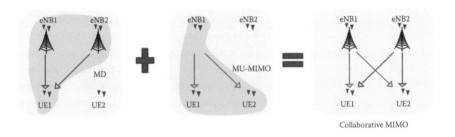

Figure 4.51 Collaborative MIMO viewed as a combination of macro-diversity and MU-MIMO.

A main target for Co-MIMO is that multiple eNBs collaborate to mitigate ICI or even change the interfering signal into desired signals in the downlink. These collaboration levels could be different according to the data and the CSI-sharing scenarios. Therefore, the level of information sharing among the collaborating eNBs should be decided by considering the trade-off between performance and control overhead. For example, multiple PMIs could be reported by the UE to capture per-cell spatial direction or the UE could report one PMI and the intercell information for each coordinated cell. Co-channel interference limits single-cell MIMO gains. To mitigate the interference, transmissions in multiple cells can be done cooperatively. In LTE-A, a set of cells (under one or several base stations) cooperate to improve the SINR perceived by the terminals. In the downlink, cooperative zero-forcing beamforming or block diagonalization can be used to null the interuser interference. In the uplink, some cooperation schemes are also under study, including multicell joint detection, MU-MIMO pairing and grouping over multiple cells, and interference cancellation.

Co-MIMO and CoMP need to work together, feedback and control signaling for MU-MIMO need to be designed in a forward-compatible manner, and CSI-RS design and reuse patterns for single-cell (SU- and MU-MIMO) need to be forward-compatible with Co-MIMO and CoMP.

4.7 Beamforming and DL Dual-Stream Technology

4.7.1 Beamforming

The main idea of beamforming is to address the power problem using multisignal processing techniques. The system sends multiple small signals (instead of one high-powered signal) such that they combine effectively at the end user terminal but cancel each other out in other places. This process is similar to dropping multiple small pebbles in water; all the different waves combine to form a much larger wave at the end user, but do not combine in the same way anywhere else. This process is called *beamforming.* Multiple signals are combined to form an RF beam to the user. The application of smart antennas to cellular networks allows the operator to utilize its available frequency spectrum more efficiently compared to a single-antenna system. This is achieved by ensuring that as little power as possible is wasted by transmitting to, or receiving from, unwanted locations. The result is that cell capacity can be enhanced, or if capacity isn't a problem, the cell size can be increased. Both will lead to significant cost savings for the operator.

In LTE, it is possible to do UE-specific beamforming in the downlink, that is, the precoding for the downlink transmission is chosen by the eNB separately for each user. The UE-specific reference signals are precoded with the same precoding vectors as the payload data and used for demodulation at the UE side. As mentioned previously, beamforming has proven beneficial for increasing cell throughput and especially essential to users at the cell edge in the commercial network of time-division synchronous code division multiple access (TD-SCDMA).

Additionally, it is worth mentioning that beamforming and closed loop spatial multiplexing are two kinds of multiantenna technologies with different key properties. In closed loop spatial multiplexing, the UE decodes the PDSCH by channel estimation based on CRS, and the eNB sends the PDSCH by codebook precoding according to closed loop feedback from the UE. In beamforming, PDSCH channel estimation is based on dedicated UE-specific reference signals at UE side, and the eNB sends PDSCH with non-codebook precoding according to open loop feedback.

4.7.1.1 Basics of Beamforming

Because the uplink in a TDD system can provide enough spatial information about the downlink to do beamforming, MU-MIMO, and CoMP through normal feedback methods, such as covariance feedback or codebook feedback, the system overhead can be significantly lowered because regular uplink traffic (or sounding) can be used to determine the downlink transmit weighs instead of requiring a feedback channel. Actually, the TD-LTE beamforming operation is non-codebook based and relies on UE sounding and channel reciprocity and UE-specific reference signals (RS), also referred to as dedicated RSs (i.e., DRS) as opposed to cell-specific RSs. The eNB generates a beam using the array of antenna elements (e.g., an array of 8 antenna elements) and then applies the same precoding to both the data payload and the UE-specific reference signal with this beam.

The reciprocity property of the TDD system is fully exploited in non-codebook-based precoding. The UE can estimate the channel information utilizing the downlink reference signals. Then uplink precoding is performed based on the downlink channel information and CSI estimation is performed by the eNB derived from the SRS. Precoded demodulation reference signals can be assisted to improve CSI estimation in the eNB. In corresponding CSI estimation of non-codebook-based precoding in the eNB, CSI that is based on the sounding RS is usually decomposed to get the precoding matrix first, and then corresponding SINRs are calculated with the aid of an equivalent channel. The decomposition is expected to use the same arithmetic as the precoding process in the UE, such as SVD. SVD is the main, but not the only, arithmetic to get the precoding matrix from the channel matrix. Other methods such as ZF, MMSE, or QR decomposition* can also be used for precoding and CQI estimation.

Two kinds of beamforming schemes are deployed in LTE downlink transmission: beamforming for the traffic channel (PDSCH) and beamforming for broadcast or control channels (PBCH, physical control format indicator channel [PCFICH], PDCCH, and PHICH). Traffic channel beamforming considers the reciprocity property of the TDD system in low-mobility UE scenarios. An array of beamforming weights are retrieved from uplink channel estimation, and then used as downlink weights assigned to each transmit antenna for downlink PDSCH transmission to get beamforming gain at the UE side. Although beamforming is not a coverage booster, it can increase user data rates with higher MCS by improving the SINR. Compared to the real-time beamforming weight update in traffic channel beamforming, fixed broadcasting weights are used for broadcast and control channel beamforming, in which no beamforming gain exists so that beamforming does not improve the coverage of control channels, which may become a limiting factor. On the other hand, broadcast weights may be configured semistatically according to the coverage requirement, to reduce common channel interference among multicells for network optimization.

Beamforming utilizes channel state information to achieve array-processing SINR gain. Channel state information mainly includes the fast fading channel coefficient (instantaneous or average), direction of arrival (DoA) of the signal, and CQI information. Channel state information can be obtained in different ways, including feedback from the receiver and estimation from the reverse link assuming channel reciprocity.

Any CSI errors at the UE side can cause inappropriate BF matrix computation. In case of fast-changing channel conditions, the received CSI is most probably outdated, which makes it useless for building the channel matrix. Since the short-term spatial correlation matrix is frequency selective, the eNB should gather a lot of CSI in order to get the information about particular sub-bands. This would be possible with SRS hopping; however, it is extremely time-consuming to scan the full bandwidth.

4.7.1.2 Multiantenna Technology

Beamforming technique is based on the antenna array with a small interelement distance. In LTE Rel-8, beamforming is supported on the single-antenna port (port 5).

---

* In linear algebra, a QR decomposition of a matrix is a decomposition of a matrix A into a product A = QR of an orthogonal matrix Q and an upper triangular matrix R. QR decomposition is often used to solve the linear least squares problem, and is the basis for a particular eigenvalue algorithm, the QR algorithm.

Beamforming can make the spatial selectivity of the interference obvious. This is an important issue as it has a decisive impact on the performance. When doing beamforming (SVD or DoA precoding) with a correlated transmit antenna, the intracell interference radiation will depend on the DoA to both the desired and interfered UE. Figure 4.52 illustrates how the spatial interference selectivity for different antenna spacing varies as predicted by the 3GPP-agreed SCM channel mode. In the simulation, the mean interference for different antenna spacing is seen by a user moving around an antenna array of four single polarized antennas with an omnidirectional antenna pattern. An active user is situated at 0 degrees and transmits with eigenbased beamforming. The radio channel is modeled with 3GPP SCM.

Figure 4.52 shows the short-term value of the interference for different azimuth angles to the interfered UE and we observe that the typical arrays used for beamforming (λ/2 spacing) show important spatial interference selectivity. When antenna spacing increases, the selectivity is reduced. To predict the proper performance, one must take this into account in system simulations.

Another simulation (Figure 4.53) shows that relative channel capacity variation decreases with an increasing number of RX antennas.

Beamforming requires that instantaneous CSI be sent via sounding reference signal in order to get instantaneous channel knowledge. The beamforming matrix is built based on eigenvectors of the instantaneous channel correlation matrix. The correlation matrix is frequency selective, which means the beamforming weights should be calculated separately for every subcarrier. In practice, it is applied on a sub-band basis only. The fact that the BF matrix is computed based on the

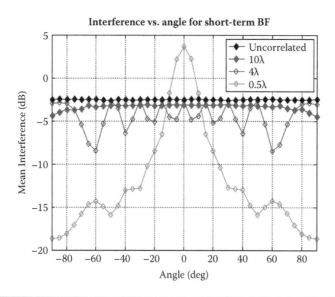

Figure 4.52   Interference for different azimuth angles to the interfered UE.

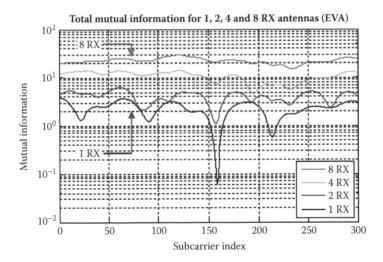

Figure 4.53 **Total mutual information (= capacity) versus frequency.**

instantaneous spatial correlation matrix makes beamforming useful mainly in cases of highly uncorrelated MIMO channels.

4.7.1.3 Beamforming Algorithms

Beamforming for both DL and UL should be a killer feature, making TD-LTE more competitive than FD-LTE for potential TD-LTE operators. DL TX beamforming can provide ~30% gain over the baseline in Rel-8. UL RX beamforming (maximum ratio combining [MRC] and interference rejection combining [IRC]) should, in theory, also enhance the UL performance greatly.

There are many possible ways of choosing the beamforming vector. Two different algorithms for estimating the precoding are considered in this section: the simpler grid of beams (GoB; based on long-term channel information) algorithm and the more complex eigenvalue-based beamforming (EBB, based on short-term channel information). Short-term BF is the best technique for highly uncorrelated eigenvectors. This may happen especially in a dense urban environment with many obstacles, reflections, and a strong multipath that has higher angular spread. There is almost no difference between 15° and 5° angular spread in the case of short-term BF because it uses water-filling power allocation in the case of more than one dominant eigenvector.

The difference of the two algorithms is that the GoB uses fixed beams and the EBB uses an eigenvalue-based weight vector (Figure 4.54). In the GoB algorithm, the direction of arrival of the signal is determined by correlating the received signal to a set of fixed beams. The fixed beam that is most correlated to the received signal is then chosen as the precoding in the downlink. In the EBB algorithm, the precoding is determined from the eigenvectors of the covariance matrix of the channel

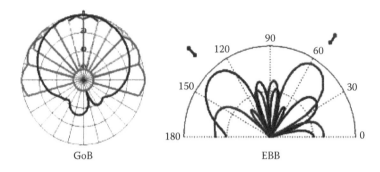

Figure 4.54 GoB and EBB.

estimate; EBB fits the scenarios with increased multipath (high angular spread) and low UE mobility. The gain varies greatly depending on channel variation, angular spread, number of RX antennas, and so on.

For transmission to a certain user, the beamforming vector **v** is chosen from a predefined static set of complex numbered antenna weights. In the case of one spatial layer, the receive vector of each user and subcarrier has the following form:

$$\underbrace{\mathbf{y}}_{M_R \times 1} = \underbrace{\mathbf{H}}_{M_R \times M_T} \underbrace{\mathbf{v}}_{M_T \times 1} \underbrace{p}_{1 \times 1} \underbrace{s}_{1 \times 1} + \underbrace{\mathbf{z}}_{M_R \times 1}$$

The weights are chosen from a fixed codebook $\mathbf{v} \in \{\mathbf{v}_1 \mathbf{v}_2 \dots \mathbf{v}_I\}$ with I denoting the number of available beams. **H** is the MIMO channel matrix, p the square root of the transmit power, s the transmitted symbol, and **z** the vector of interference plus noise. The antenna weights $\mathbf{v}_i$ (with nontapered beams) are calculated from the main beam direction of beam i and from the mth transmit element position for all elements.

4.7.2 DL Dual-Stream Technology

Not only does the beamforming technique increase capacity and coverage, it is also suitable for single multistream transmission and multiuser multistream transmission. Single-layer beamforming has been considered mainly as an enhancement for cell-edge user performance. Dual-stream beamforming can then be seen as a throughput enabler for users experiencing good channel conditions. This evolution from single- to dual-layer beamforming is also favorable due to the usage of cross-polarized antenna arrays because they can typically reduce the antenna array size compared to arrays using a single polarized antenna. Due to the low correlation among different polarizations, it is believed that arrays based on a cross-polarized antenna, under typical radio channel conditions, will have at least two strong MIMO subchannels. To maximize spectral efficiency under these conditions, dual-layer transmission is needed. So enabling dual-layer transmission can be seen as a natural evolutionary step to exploit the full benefits of beamforming with a dual polarized antenna. LTE Rel-9 further extended the single-layer beamforming in Rel-8 to single-user dual-layer beamforming and dual-user single streams (MU-MIMO) to realize a combined beamforming and spatial multiplexing technique (as shown in Figure 4.55). In order to support dual beamforming, Rel-9 defines the new dual ports, 7 and 8, for UE-specific RSs, with new control signaling. The new defined transfer mode 8 can adopt two kinds of feedback mode: PMI or non-PMI.

To experience gain from the dual-layer operation, the two virtual antennas should be uncorrelated, and this can be achieved by using a dual polarized antenna or by having wide spacing between the two virtual antennas. This virtual 2Tx MIMO system can be operated in a manner similar to the existing 2Tx Rel-8 MIMO solution (transmission mode 4 with two active CRS ports).

The baseline configuration for dual-layer beamforming design is two active CRS ports, each mapped to a virtual antenna. Meanwhile, since only one MCS is needed for single-stream scheduling, it requires less indication information compared to dual-stream scheduling. The transmission mode combined with the DCI format determines how the PDSCH is transmitted. In LTE Rel-8 and Rel-9, transmission mode 7 is defined for single-layer beamforming, and transmission mode 8 is defined for dual-layer beamforming. The DCI formats used for transmission modes 7 and 8 are listed in Table 4.22.

In LTE Rel-9, two layers of UE-specific reference signals have been introduced, as shown in Figure 4.56. This enables the eNB to transmit two layers of data to a UE using spatial multiplexing in a closed loop mode by constructing antenna weights using channel reciprocity. Switching between single- and dual-layer transmission to a single UE, and switching between SU-MIMO and MU-MIMO, is supported in a dynamic fashion. The control signaling overhead for supporting a dynamic and transparent MU-MIMO transmission is small because a UE is not explicitly informed of the presence of a co-scheduled UE, either for the purposes of feedback or for demodulation. The two layers of UE-specific reference signals are overlaid on top of each other (separated by length 2 orthogonal cover codes) and a UE, after subtracting its channel estimate, may estimate a covariance matrix that represents the combined interference from a co-scheduled UE

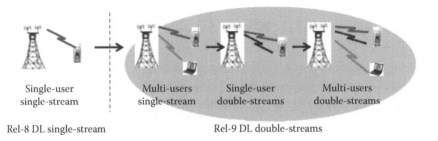

Single-user single-stream

Multi-users single-stream

Single-user double-streams

Multi-users double-streams

Rel-8 DL single-stream

Rel-9 DL double-streams

Figure 4.55   Dual beamforming in Rel-9.

Table 4.22   Transmission Mode 7 and Mode 8

Transmission Mode	DCI Format	Search Space	Transmission Scheme of PDSCH Corresponding to PDCCH
Mode 7	DCI format 1A	Common and UE-specific by cell radio network temporary identifier (C-RNTI)	If the number of PBCH antenna ports is one single-antenna port, port 0 is used, otherwise transmit diversity
	DCI format 1	UE-specific by C-RNTI	Single-antenna port, port 5
Mode 8	DCI format 1A	Common and UE-specific by C-RNTI	If the number of PBCH antenna ports is one single-antenna port, port 0 is used, otherwise transmit diversity
	DCI format 2B	UE-specific by C-RNTI	Dual-layer transmission, ports 7 and 8 or single-antenna port, port 7 or 8

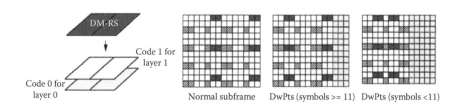

Figure 4.56   DMRS for dual stream beamforming. (DWPTS—downlink pilot time slot)

and outer cell transmissions. The main features of dual-stream beamforming are as follows:

- Extension of DMRS-based beamforming to support spatial multiplexing, forward compatible with LTE-A.
- Introduction of code-domain multiplexing DMRS (as shown in Figure 4.56); the code-domain multiplexing-based DMRS can be a good solution that lets a UE detect co-channel interference with orthogonal DMRS by using energy detection in each scheduled PRB, and the UE can suppress the interfering signal in each scheduled PRB if interfering UE exists.

Two layers of UE-specific reference signals are separated by code in a resource block. The shaded resources in Figure 4.56 indicate cell-specific reference signals for the four antenna ports present in Rel-8.

4.7.2.1 MU-MIMO by Dual Layer

In Rel-8, SU- and MU-MIMO are supported by different transmission modes which are configured semistatically by higher layers. Thus, the SU and MU transmission schemes could be optimized independently. With dual-port DMRSs in Rel-9, two or more UEs can be supported simultaneously, each DMRS supporting one or two layers per UE, without interference between the two DMRSs. The eNB can obtain instantaneous channel information about each UE from SRS, so UE pairing for MU-MIMO transmission can be based on the instantaneous channel information. According to the maximum throughput criterion or other pairing criterion, the eNB selects two suitable UEs for grouping and selects a suitable MCS level for each UE. The interference between two layers can be canceled with an MMSE SIC receiver at the UE. Due to precoding the DMRSs, the PMI is not needed to inform the UE.

The channel state information (e.g., DoA) can be obtained by the eNB from measurements of the uplink channels. According to this information, the eNB pairs the user groups that satisfying a certain constraint condition. The beam corresponding to the weight of a certain user has the following features: the main lobe is formed in the DoA direction of the target user, and the widening null steering is formed in the DoA directions of other users in the same group with the target user. Therefore, the interuser interference is reduced. Moreover, the transmission weights of different users in the same group can also be orthogonalized to further eliminate the interference. Figure 4.57 is a block diagram of multiuser dual-layer BF.

In Rel-9, UEs may be co-scheduled in transmission mode 8. MU-MIMO can be implemented based on SU-MIMO with additional control signaling, such as DMRS port indication and DMRS scrambling initialization ID (shown in Figure 4.58). code-domain multiplexing with length 2 Walsh spreading was adopted to provide two orthogonal DMRS ports and two DMRS sequences by setting the proper scrambling ID n_{SCID} in DCI format 2B, in order to support up to four UEs without large degradation. Actually, the use of the second DMRS sequence is up to the eNB implementation since it is only quasi-orthogonal to the first DMRS sequence. Being aware of the allocation of its own DMRS port and DMRS scrambling initialization ID, the desired layer(s) can be detected regardless the existence of co-scheduled UE.

If the eNB co-schedules two UEs, it is better to use the same DMRS sequence to optimize channel estimation performance. If the eNB co-schedules more than two UEs, both DMRS sequences will have to be used. If the UE can know the DMRS sequence of a co-scheduled UE, the UE could use LMMSE equalization to cancel the inter-UE interference for better demodulation performance. Considering port

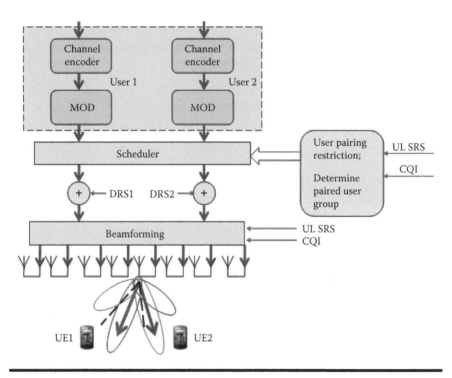

Figure 4.57   Block diagram of MU dual-layer BF scheme. (MOD—modulation)

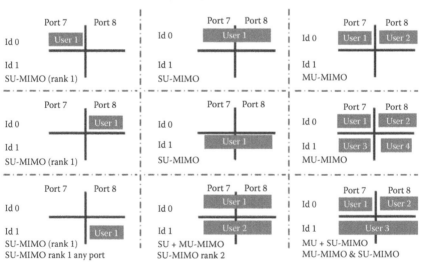

Up to 2 layers per user (SU-MIMO). Up to 4 layers in total (MU-MIMO).

Figure 4.58   DMRS port indication and DMRS scrambling initialization ID.

orthogonality in MU-MIMO, and since port and ID allocations are wideband, it is not always possible to ensure orthogonality even though only two users are multiplexed in MU-MIMO. On the other hand, even when DMRS sequences of major interfering UEs are known, such DMRS sequences are only quasi-orthogonal to the DMRS sequence of desired UE.

In MU-MIMO, a UE does not have the channel information of the other UE (in a matched pair), so the CQI feedback mechanism will be similar to LTE Rel-8 (i.e., the UE feeds back only its own CQI). For a pair of matched UEs, each of rank 1, CQI feedback is the same as in transmission mode 7 in LTE Rel-8, but tuning at the eNB is still necessary. In a TDD system, with channel information obtained from the SRS at the eNB, the CQI for paired UEs can be calculated more precisely and thus a suitable MCS can be selected.

4.7.2.2 Channel Reciprocity and Mismatch Calibration

One aspect related to beamforming applied in TDD mode operation is to utilize channel reciprocity to derive the downlink CSI from uplink channel estimation over sounding signals transmitted by the terminals. In this case the channel is actually made up of the propagation channel (the medium between the transmitter and receiver), the antennas and the transceiver RF, and IF and baseband circuits at both sides of the link.

LTE Rel-8 supports SVD-based beamforming (rank 1 transmission) using channel reciprocity to obtain the CSI. Since there is no codebook for this, UE-specific reference signals are required and supported for one stream. In LTE-A, extension to ranks greater than 1 SVD-precoded transmissions will likely happen. Here, we first estimate the channel covariance per each N PRBs ($N \geq 1$, total N_f subcarriers) as

$$R = \frac{1}{N_f} \sum_{i=f_0}^{f_1} \mathrm{H}^T(i)\mathrm{H}^*(i)$$

Then, the optimal precoding matrix for rank p transmission is obtained from the p eigenvectors corresponding to the p largest eigenvalues of R.

In theory, though the propagation channel can be assumed to be nearly reciprocal if the time interval between UL and DL transmission is much less than the coherence time of the propagation channel, the transceiver circuits are usually not reciprocal (i.e., the TX and RX frequency responses are different), and this may jeopardize the performance of beamforming realized by non-codebook-based precoding that relies on UL/DL channel reciprocity (i.e., derives DL CSI from UL sounding). This is schematically illustrated in Figure 4.59.

In wireless transmission, a sudden change in temperature may vary the properties of the crystal oscillator, filters, and amplifiers of antennas. When a UE changes its state from idle to connected mode, due to the activation of the circuitry, the temperature may rise quickly which significantly impacts the RF properties differently

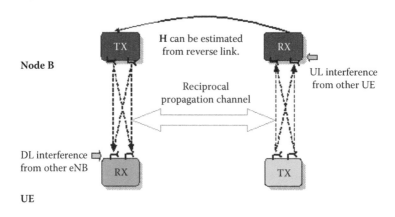

Figure 4.59　Channel reciprocity.

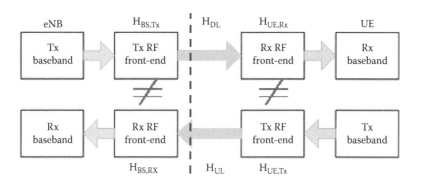

Figure 4.60　Practical imperfections of hardware.

on multiple antennas. The practical imperfections of the RF front-end (as shown in Figure 4.60) include A/D(D/A), low noise amplifier (LNA), mixer, local oscillator (LO), bandpass filter, and I/Q modulator, and many of them are nonlinear and difficult to analyze. Reciprocity-based beamforming needs accurate calibration of the amplitude–phase mismatch between the transmit and receive antennas of the eNB. To support non-codebook-based precoding in the uplink, the multiple antennas at the eNB and terminal should be calibrated so that all of the antennas have similar amplitude and phase properties to maintain good matches of TX and RX RF. UE-specific reference signal demodulation performance, CQI reporting difficulty due to the UE not knowing the precoder at the time of reporting, and TX/RX RF mismatches are the key challenges to channel reciprocity utilization. In this section we discuss the fact that TX/RX RF mismatches will destroy channel reciprocity and why good RF calibration is crucial for performance.

The discussion of TX and RX mismatch calibration uses the following notations:

- T_{BS} and R_{BS} are square diagonal matrices of size m and denote the TX response and RX response of m antenna/transceivers respectively, at the base station (eNB).
- T_{UE} and R_{UE} are square diagonal matrices of size n and denote the TX response and RX response of n antenna/transceivers at the UE, respectively.
- X_D and X_U denote the DL and UL transmitted data symbol vectors, respectively.
- W is the DL precoding matrix, H is the propagation channel from eNB to UE, and N_0 is the Gaussian noise perceived at the receiver.
- The DL received signal is represented by $y_D = H_D \times W \times X_D + N_0$ where $H_D = R_{UE} \times H \times T_{BS}$.
- The UL received signal is represented by $y_U = H_U \times X_U + N_0$ where $H_U = R_{BS} \times H^T \times T_{UE}$.

From above, we can derive $H_U^T = T_{UE}^T \times H \times R_{BS}^T$ and $H = (T_{UE}^T)^{-1} \times H_U^T \times (R_{BS}^T)^{-1}$ and $H_D = R_{UE} \times (T_{UE}^T)^{-1} \times H_U^T \times (R_{BS}^T)^{-1} \times T_{BS}$.

From this equation it is clear that if we do not have $R_{UE} \times (T_{UE}^T)^{-1} = I$ and $(R_{BS}^T)^{-1} \times T_{BS} = I$ (I represents unit matrix), the effective UL and DL channels will be different, so if we use an effective UL channel to derive the DL precoder, the precoder will be suboptimal. To restore channel reciprocity, we should introduce a calibrated channel $H_{D,C}$ and $H_{U,C}$. The calibrated channel is generated from the effective channel by applying precoding in the two transmitters as follows:

$$\text{Downlink calibrated channel: } H_{D,C} = H_D K_{BS}$$
$$\text{Uplink calibrated channel: } H_{U,C} = H_U K_{UE}$$

where $K_{BS} = R_{BS} / T_{BS}$ and $K_{UE} = R_{UE} / T_{UE}$ are square diagonal matrices of size m and n representing the calibration factor at the eNB and UE, respectively. The calibration process is basically to derive K_{BS} and K_{UE}, the delta of amplitude and phase of each Rx/Tx chain.

Therefore, a feasible calibration procedure for TDD LTE-A system is described as follows (see Figure 4.61):

1. The eNB estimates the uplink CSI $\mathbf{H}_{UL}$ from SRS or DMRS.
2. The eNB performs SVD and obtains the precoding unitary matrix $\mathbf{V}_{UL}$.
3. The eNB transmits UE specific reference signal precoded by $\mathbf{V}_{UL}$ to the UE for channel estimation.
4. Channel estimation occurs at the UE. With dedicated UE specific reference signal, the downlink *effective channel* $\mathbf{H}_{eff} = \mathbf{H}_{DL}\mathbf{V}_{UL}$ is estimated.
5. The UE calculates the channel calibration factor. SVD decomposition is performed by the UE so that $[\mathbf{U}_{eff} \mathbf{D}_{eff} \mathbf{V}_{eff}] = svd(\mathbf{H}_{eff})$. The first r columns of $\mathbf{V}_{eff}$ can be exploited as a calibration factor, and that is $\mathbf{E} = (\mathbf{V}_{eff}(:, 1:r))^H$.
6. The UE feeds the calibration factor $\mathbf{E}$ back to the eNB. Note that when the narrowband precoding approach is adopted by the system, the UE will feed back multiple $\mathbf{E}$ matrices with each $\mathbf{E}$ corresponding to a certain sub-band.

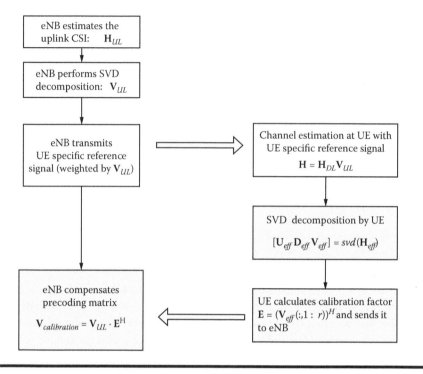

Figure 4.61   Proposed calibration procedure for SVD-based SU-MIMO.

7. The eNB utilizes the feedback calibration factor to compensate for the precoding matrix. The calibrated precoding matrix is, therefore, $\mathbf{V}_{calibration} = \mathbf{V}_{UL} \cdot \mathbf{E}^H$. The same $\mathbf{E}$ will be utilized until the calibration procedure is triggered again.

4.7.2.3 Multilayer Beamforming Schemes

First we illustrate the transmitter and receiver structure of an SVD-based SU-MIMO system (as shown in Figure 4.62). The transposition of effective uplink CSI H_{UL}^T is processed by the SVD technique, which is represented by the following equation:

$$\mathbf{H}_{UL}^T = \mathbf{U}_{UL} \Sigma_{UL} \mathbf{V}_{UL}^H$$

Herein, matrices $\mathbf{U}_{UL}$ and $\mathbf{V}_{UL}$ are both unitary matrices such that $\mathbf{U}_{UL} \mathbf{U}_{UL}^H = \mathbf{I} = \mathbf{V}_{UL} \mathbf{V}_{UL}^H$, and $(\cdot)^H$ is known as the *Hermitian operation* and denotes the transposed complex conjugate of the argument. Suppose there are n_t antennas at the transmitter and n_r antennas at the receiver, and thus $\mathbf{U}_{UL} \in \mathbf{C}^{n_r \times n_r}$, $\mathbf{V}_{UL} \in \mathbf{C}^{n_t \times n_t}$. The rank of CSI matrix $\mathbf{H}_{UL}$ should fulfill $r \le \min(n_t, n_r)$. The diagonal

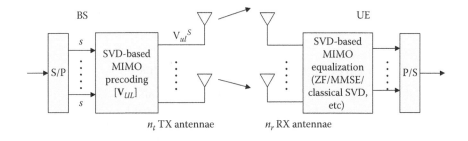

Figure 4.62 Transmitter and receiver structure of an SVD-based MIMO system. (S/P—serial/parallel; P/S—parallel/serial)

matrix Σ_{UL} can be expressed as $\Sigma_{UL} = \begin{bmatrix} \Sigma_{UL}^r & 0 \\ 0 & 0 \end{bmatrix}_{n_r \times n_t}$, where $\Sigma_{UL}^r = diag(\lambda_1, \lambda_2, \cdots \lambda_r)$ with ordered singular values λ_i such that $\lambda_1 > \lambda_2 > \cdots \lambda_r$.

For the SVD-based precoding technique, the $\mathbf{V}_{UL}$ matrix is exploited, where a column of $\mathbf{V}_{UL}$ is called an eigenvector of $\mathbf{H}_{UL}^H \mathbf{H}_{UL}$, which is related to an eigenmode of the communication channel. The precoding technique at the eNB is operated by matrix–vector multiplication given by $\mathbf{c} = \mathbf{V}_{UL}\mathbf{s}$.

Then, the received signal vector $\mathbf{y}$ is determined by $\mathbf{y} = \mathbf{H}_{DL}\mathbf{V}_{UL}\mathbf{s} + \mathbf{n}$. Herein, $\mathbf{s} = [s_1, s_2, \ldots, s_{n_t}]^T$.

Please note, while exploiting the eigenmode selection precoding approach, a number of largest eigenvectors (e.g., K eigenvectors), which are usually chosen to be less than the number of transmit antennas ($K < n_t$), construct the precoding matrix.

If $\mathbf{H}_{UL}^T = \mathbf{H}_{DL}$ and uplink and downlink channels are ideally reciprocal, we have $\mathbf{V}_{DL}^H \mathbf{V}_{UL} = \mathbf{I}$. Therefore, the received signal can be expressed as $\mathbf{y} = \mathbf{U}_{DL}\Sigma_{DL}\mathbf{s} + \mathbf{n}$.

Considering a flat fading channel, the received signal vector at UE is

$$Y = HWS + N$$

where H is the DL MIMO channel, S is the transmission data vector precoded by W at the eNB, and N is additive white Gaussian noise.

In single-layer BF, the optimal precoding vector is

$$W_0 = \arg\max_{W}\left\{W^H H^H HW\right\}$$

That is, the precoding vector W_0 is the eigenvector of matrix $H^H H$ corresponding to the largest eigenvalue.

In multilayer BF, one criteria to select the precoding matrix is

$$W_0 = \arg\max_{W}\left\{trace\left(W^H H^H HW\right)\right\}$$

That is, the precoding matrix W_0 is composed of eigenvectors corresponding to the two largest eigenvalues of matrix $H^H H$.

4.7.2.4 DL Control Signaling and CSI Feedback

LTE Rel-8 supports seven transmission modes that offer various features to improve the performance of the radio link. Single-layer beamforming is supported in transmission mode 7. The new transmission mode 8, new DCI formats, and two new dedicated antenna ports 7 and 8 are defined for dual-layer beamforming in LTE Rel-9. Because DMRS is used for PDSCH demodulation, there is no need to indicate the precoding information in the downlink control signaling. The DMRS port for rank 1 transmission and the DMRS scrambling sequence index that is used should also be indicated. Transmission mode 9 is introduced in LTE-A Rel-10 to support multistream beamforming with CSI-RS. All of the supported transmission modes so far and their corresponding CQI report modes are listed in Table 4.23.

Generally there is no need for feedback in TDD when we can get enough and accurate channel state information from SRS based on channel reciprocity. However, in FDD, CSI feedback should be considered due to much less correlation in channel reciprocity. The CQI estimation can be based on the SRS and all available CRSs. Two different CQI reporting schemes for an antenna setup with cross-polarized UE antennas and an antenna with four columns of cross-polarized antennas can be used in a suburban macro environment. First, UE can determine the CQI according to transmission mode 7 of Rel-8 and the eNB determines beamforming vectors from an eigenvalue decomposition of an 8x8 covariance matrix and compensates the CQI assuming no interference between the two layers. Second, the eNB performs wideband eigen beamforming and determines a set of beamforming weights that is used for both sets of co-polarized antennas from a single 4x4 covariance matrix. The eNB then combines the eigen beamforming weights with the PMI reported by the UE to determine the precoding vectors. The interference between the two layers can be adjusted by the eNB according to the weight vectors; therefore, the eNB can further revise and compensate for the CQI. The CQI feedback mode for SU dual-layer BF can be the same as the one for closed loop spatial multiplexing in Rel-8. For rank 1 MU dual-layer BF, the CQI feedback mode is the same as that for transmission mode 7 in Rel-8.

Because the UE does not know what beamforming weights will be used, a major challenge with non-codebook-based beamforming is link and rank adaptation in addition to channel-dependent scheduling. Meanwhile, with codebook-based precoding, the problem of orthogonalization between the layers does not exist, and the UE can account for the interference between the two layers for the selected precoding matrix.

Table 4.23   DL Transmission Modes Supported in Rel-8, 9, and 10

DL Transmission Mode	Transmission Scheme of PDSCH		CQI Mode	DCI Format	Rel
Mode 1	Single-antenna port; port 0. No precoding.	Only CQI	Aperiodic: 2-0, 3-0 Periodic: 1-0, 2-0	Format 0/1A, 1	R8
Mode 2	Transmit diversity with two or four antenna ports using space frequency block code	Only CQI	Aperiodic: 2-0, 3-0 Periodic: 1-0, 2-0	Format 0/1A, 1	R8
Mode 3	Open loop spatial multiplexing with rank indication feedback	Only CQI	Aperiodic: 2-0, 3-0 Periodic: 1-0, 2-0	Format 0/1A, 2A	R8
Mode 4	Closed loop spatial multiplexing with precoding feedback	CQI, RI, PMI	Aperiodic: 1-2, 2-2, 3-1 Periodic: 1-1, 2-1	Format 0/1A, 2	R8
Mode 5	Multiuser-MIMO	CQI, PMI	Aperiodic: 3-1 Periodic: 1-1, 2-1	Format 0/1A, 1D	R8
Mode 6	Closed loop rank 1 without spatial multiplexing	CQI, PMI	Aperiodic: 1-2, 2-2, 3-1 Periodic: 1-1, 2-1	Format 0/1A, 1B	R8
Mode 7	Single-antenna beamforming with dedicated reference signals	Only CQI	Aperiodic: 2-0, 3-0 Periodic: 1-0, 2-0	Format 0/1A, 1	R8
Mode 8	Dual-stream beamforming with dedicated reference signals	CQI, RI, PMI	Aperiodic: 1-2, 2-0, 2-2, 3-0, 3-1 Periodic: 1-0, 1-1, 2-0, 2-1	Format 0/1A, 2B	R9

(Continued)

Table 4.23 (*Continued*)   **DL Transmission Modes Supported in Rel-8, 9, and 10**

DL Transmission Mode	Transmission Scheme of PDSCH	CQI Mode		DCI Format	Rel
Mode 9	Mutilayer SU/ MU; multistream beamforming with CSI-RS	CQI, RI, PMI	Aperiodic: 1-2, 2-0, 2-2, 3-0, 3-1 Periodic: 1-0, 1-1, 2-0, 2-1	Format 0/1A, 2C	R10

New DCI format 2B is defined in Rel-9 for dual-layer beamforming based on DCI format 2A:

- Add 1 bit for scrambling sequence initialization
- Remove swap flag
- Rank 1 transmission: The NDI bit of disabled transport block (TB) is re-used to indicate port information; 0 – enabled transport block associated with port 7; 1 – enabled transport block associated with port 8
- Rank 2 transmission: TB1 associated with port 7, TB2 associated with port 8

The structure of DCI format 2B is listed in Table 4.24.

4.7.2.5 Performance Analysis

Dual-layer beamforming targets single-user throughput increase. Multiuser beamforming is already supported in LTE Rel-9 because two UEs in transmission mode 8 can be scheduled to the same PRB. The dominant interference source for most of the users will be from intercell interference. In addition to the interference, the other reasons described below will impact performance.

The performance of dual-layer beamforming is impacted by considering different imperfections as follows:

- **Channel estimation errors:** Estimating radio channel parameters from UL sounding transmission in the presence of noise and interference cannot be done perfectly. The channel estimation errors will depend on the signal-to-noise and interference ratio and this parameter is to be estimated.
- **Calibration errors:** These errors model the possible mismatch between DL and UL transceiver chains. These errors need to be accounted for when considering reciprocity in TDD, which is using the UL channel state measurements to represent the DL channel state. This error can be reduced if some form of calibration is implemented to handle the mismatch in the DL and

Table 4.24   DCI Format 2B for Dual-Layer BF in Rel-9

PDCCH Field	Number of Bits	Description
Random access (RA) header	1 / 0 (for 1.4 MHz)	
RB assignment	$\lceil N_{RB}^{DL} / P \rceil$	
TPC	2	
DAI	2 (TDD only)	
HARQ Process ID	3 (FDD), 4 (TDD)	
Scrambling ID	1	New entry to select scrambling ID for DRS sequence Replaces TB-CW switch
Transport Block 1		
MCS	5	
NDI	1	
RV	2	
Transport Block 2		Rank signaled via TB disable mechanism (MCS = 0, RV = 1)
MCS	5	
NDI	1	NDI is used to select port 7 or 8 when TB2 is disabled
RV	2	
Precoding information	0 (number of antenna ports is 2)	
	2 (number of antenna ports is 4)	
Cyclic redundancy check (CRC)	16	

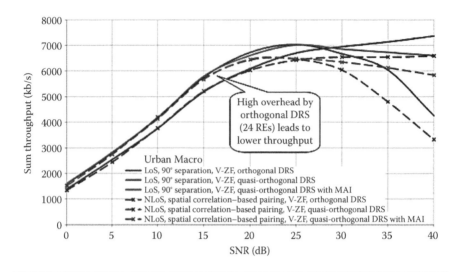

Figure 4.63 Quasi-orthogonal reference signal.

UL transceiver chain. Otherwise this error could be nontrivial. The UE calibration error is not sensitive to the performance of channel reciprocity–based short-term beamforming, and in this case the calibration error at base station due to imperfection is considered.

■ **SRS transmission rate:** Sounding symbols are not transmitted in the whole band of every transmission time interval (TTI) for all UEs for all antennas. If maintaining an estimate of the full channel state, this can only be updated according to the sounding transmission. Under typical conditions it will be difficult to achieve an update period of less than 10 ms.

To accurately estimate the performance of beamforming, it is important to take into account these three types of imperfection.

A simulation of the performance of the quasi-orthogonal reference signal used for MU-MIMO is illustrated in Figure 4.63. In the simulation, two UEs with 8Tx + 2Rx/UE are involved, and rank 2 TX; a total of four layers are simulated.

We can see from the figure that quasi-orthogonal reference signals for MU-MIMO work well.

4.7.3 Beamforming Evolution

The LTE to LTE-A beamforming evolution path is from single-cell single-layer beamforming in LTE Rel-8 to single-cell dual-layer beamforming in Rel-9, followed by multicell, multilayer, multiuser beamforming in the future releases for LTE-A. Therefore, it is important from a standards perspective to have a proper

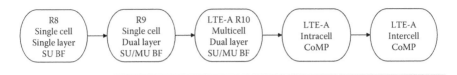

Figure 4.64   Beamforming evolution path.

feature roadmap to ensure a smooth yet backward-compatible evolution of the beamforming techniques in 3GPP.

The specification work for Rel-9 DL dual-layer beamforming and LTE-A multicell, multilayer, multiuser beamforming based on user-specific reference signals will be ongoing in parallel. Ensuring a smooth yet forward-compatible beamforming design based on LTE-A DMRS becomes extremely important. The exact RS pattern combining FDM/TDM multiplexing with a CDM dimension offers a good trade-off with reasonable overhead for all MIMO modes with beamforming operations (for example, two or more layers, single or multicell, single or multiuser) and allows transparent operation to some extent.

Before and beyond LTE-A, the evolution path for beamforming can be summarized in Figure 4.64.

Coordinated multicell beamforming targets interference reduction by coordinated scheduling. It can enhance the signal quality particularly of cell-edge UE and reduce interference caused to/by other UEs. The coordination is achieved considering the DoA of the involved UEs. The implementation of coordinated multicell beamforming will have no impact on radio standardization and just a minor change on the X2 interface.

Combined multicell beamforming aims to strengthen signals by joint transmission. The eNBs jointly schedule data to the UEs using different weights, so UE-specific reference signals must be used. The joint channel state will be reported among eNBs, and UE might actually be unaware of network cooperation.

Chapter 5

Relaying

Long Term Evolution Advanced (LTE-A) considers relaying for cost-effective throughput enhancement and coverage extension. Relaying is deployed in LTE-A as a tool to improve coverage for high data rates, group mobility, temporary network deployment, and cell-edge throughput, and/or to provide coverage in new areas.

The relay node (RN) is wirelessly connected to the radio-access network via a donor cell. Relay transmission can be seen as a kind of collaborative communication in which an RN helps to forward user information from neighboring user equipment (UE) to a local evolved Node-B (eNB). In doing this, an RN can effectively extend the signal and service coverage of an eNB and enhance the overall throughput performance of a wireless communication system. The performance of relay transmission is greatly affected by the collaborative strategy, which includes the selection of relay types and relay partners (i.e., to decide when, how, and with whom to collaborate).

If the network is expanded by adding relays, there is a transition from a pure macro deployment to a mixed macro/pico deployment. The deployment of relays could be a cost-effective way to improve system throughput and coverage for Rel-8 systems and enhance the Rel-8 user experience. The relay node may be applied with different purposes. In a rural area the relay goal is to improve cell coverage with two or more hops; in an urban hot spot the relay goal is to achieve higher spectrum efficiency; and in a dead spot the relay goal is to resolve the coverage problem for UE in coverage holes. The typical relay deployment scenarios are shown in Figure 5.1. The relay system may be categorized into three types associated with L1, L2, and L3 relays with different functions, requiring various complexities in terms of control channel, data processing, and high layer interface.

In LTE-A, the eNB serving the RN is called the donor eNB (DeNB). The terms *backhaul link* and *access link* are often used to refer to the DeNB–RN connection and the RN–UE connection, respectively. The donor cell may, in addition to one or

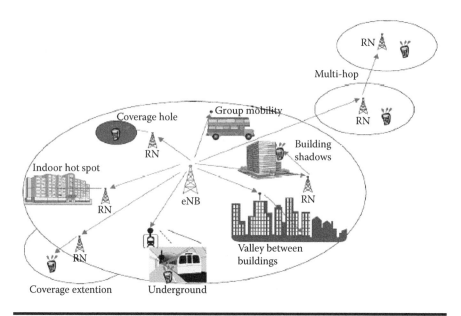

Figure 5.1 Relay deployment scenarios.

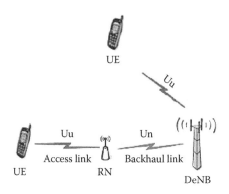

Figure 5.2 Relay air links.

several RNs, also serve UEs not connected via an RN. The backhaul link and access link are illustrated in Figure 5.2.

The radio interface between the DeNB and the RN is called the *Un interface*, which is a modified version of the LTE radio interface. By adding relay nodes that use the Un interface as backhaul, we need to distribute the macro capacity to the relay nodes and the UEs in the cells served by the donor eNB. The addition of relay nodes in a network will only increase the service availability for the users with the worst throughput; however, this network evolution will not increase the network capacity due to the limitation in the backhaul link.

5.1 Relay Technology

5.1.1 Relay Classification

A number of different classifications for relay technology exist. From the aspect of the transmission scheme, there are *amplify-and-forward* (AF) relays and *decode-and-forward* (DF) relays. According to the number of protocol layers that an RN implements, it is distinguished as a layer 0/1 relay, a layer 2 relay, or a layer 3 relay. With respect to the RN's usage of spectrum, its operation can be classified as an *inband* relay or an *outband* relay. With respect to the knowledge in the UE, relays can be classified as *transparent* relays and *nontransparent* relays. Furthermore, there are Type 1 and Type 2 relays introduced in the Third-Generation Partnership Project (3GPP) specification. This section will explain the diverse relay classifications.

5.1.1.1 AF and DF Relay

In general, AF and DF relays are categorized by the way they handle and forward received signals. Amplify-and-forward relay nodes, commonly referred to as *repeaters*, just enhance the received signal strength, as illustrated in Figure 5.3.

The AF relay functions mostly as a simple repeater, by amplifying and retransmitting the received signal, including noise and interference. Since amplification is carried out to an already distorted signal with noise, an AF relay node amplifies the negative effects of the first-hop radio channel toward the second hop. Although they are simple and cheap, because of this simplicity, their potential usage is quite limited. Still, AF relays might be sufficient for simple coverage extension scenarios. They are also fast because there are no decoding delays introduced.

Decode-and-forward relay nodes are more sophisticated and do not blindly repeat the whole signal. Instead, they first decode and then regenerate the original signal, as shown in Figure 5.4. Due to this ability, they can utilize link adaptation and interference control. However, this comes with increased complexity and protocol overhead.

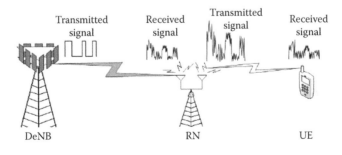

Figure 5.3 Amplify and forward.

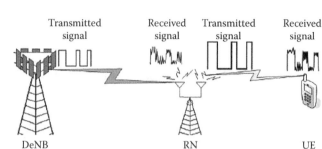

Figure 5.4 Decode and forward.

In AF relaying, the power received by the UE is sent from the DeNB and amplified by the RN. The noise and interference received by the RN are also amplified. So the signal-to-noise ratio (SNR) at a UE connected to RN can be expressed as

$$SNR_{AF-eNB-RN-UE} = \frac{P_{eNB-RN}^{RX} \cdot \beta \cdot PL_{RN-UE}}{N + N\beta \cdot PL_{RN-UE}} = \frac{SNR_{eNB-RN} \cdot \beta \cdot PL_{RN-UE}}{1 + \beta \cdot PL_{RN-UE}}$$

where P_{eNB-RN}^{RX} is the power received at the RN from the eNB transmission, N is the noise power, PL_{RN-UE} is the path loss on the RN–UE link (we consider antenna gains included in this factor), and β is the amplification factor of the RN.

A DF relay node is a smarter repeater that regenerates and amplifies only the relevant parts of the received signal based on the users that are targeted by the relay. Therefore, there is no noise or interference enhancement in the DF relay scheme. A dense relay deployment will not only improve the signal-to-interference plus noise ratio (SINR) for edge UEs, but also offer capacity increases. In DF relaying, the SNR at the UE side is the minimum of the SNR of the backhaul link and the SNR of the access link.

$$SNR_{DF-eNB-RN-UE} = \min(SNR_{eNB-RN}, SNR_{RN-UE})$$

The DF relay node has the ability to make decisions to send only the useful portion of the packets toward the recipient. The DF relay node can also utilize different channel environments by employing different channel-coding parameters and can make higher-level decisions, such as the quality of service, when scheduling the packets.

5.1.1.2 Layered Relay

Based on the protocol layer at which the user data packet is available at the RN, it can be classified as L0, L1, L2, and L3 relays.

An L0 relay node is the conventional repeater that has been deployed in existing systems where the received signal is amplified and forwarded in the analog front end. The L0 relay does not even involve the physical (PHY) layer.

Although an L1 relay is another kind of AF relay, L1 relay nodes can be viewed as smarter or advanced repeaters, where the received signal is processed by the PHY layer. One exemplary PHY process in the L1 relay is frequency domain filtering, by which only the useful signal is forwarded. The L1 relay may incur additional processing delay, which exceeds the cyclic prefix length or even one orthogonal frequency-division multiplexing (OFDM) symbol length.

L1 relays have been used extensively for coverage reinforcement in Global System for Mobile Communication (GSM) and code division multiple access (CDMA) and wideband code division multiple access (WCDMA) networks and are well suited to LTE. An L1 relay node in its simplest form (see Figure 5.5) has two back-to-back radio frequency (RF) amplifiers connected via duplexers with one connected to a transit antenna and the other to an access footprint antenna. It is important that the transit path be relatively unobstructed to reduce fading and enhancement effects, and the total uplink gain should be controlled to avoid elevating the noise floor of the donor eNB receiver and shrinking its coverage. L1 relay nodes retransmit the received signal on the whole system bandwidth.

The coupling between donor and coverage antenna should be minimized to avoid instability and oscillation. This favors a directional transit antenna with physical isolation from the access antenna. More advanced L1 relays include automatic phase correction to control stability in high-gain settings.

L1 relaying can work very effectively subject to the previously noted constraints, but often requires careful integration into the network with field optimization of gain and antenna settings. This becomes more demanding in complex urban environments with multiple servers at the donor antenna and in the coverage area. Repeaters with self-configuration to control both stability and base station noise rise are now widely available. Network supervision and control of RF repeaters via inband signaling will further improve coordination and reduce the risk of their large-scale deployment.

In L2 and L3 relay techniques, the relay node decodes the received signal before retransmitting it. It is able to precisely select which signals it will retransmit. It is also possible to reschedule UEs before transmission.

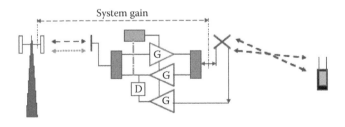

Figure 5.5   L1 relay. (D—delay; G—gain)

As a sort of DF relay, an L2 relay involves protocol layers above the PHY layer, such that advanced functions are employable at the RN to improve system performance. An L2 relay node will have at least medium access control (MAC) layer functionality, that is, decoding of received signals and recoding of transmit signals are possible in order to achieve higher link quality in the relay cell area. Scheduling and hybrid automatic repeat request (HARQ) processing are two important functions that are available in the L2 relay. The L2 relay also receives and forwards radio link control (RLC) service data units (SDUs). The performance gain of the L2 relay comes at the expense of higher complexity (cost) of the relay and will also add delay to the communication link. L2 relay processing is illustrated in Figure 5.6.

In L2 relaying technology, communication from the RN to and from the DeNB is coordinated to minimize interference to mobile traffic; leverages elevated antennas, multiple-input, multiple-output (MIMO), and higher modulation to maximize spectral density and minimize consumption of donor RF capacity.

Since L2 relay is designed for throughput enhancement when the relay node has the overlapped coverage area with the donor eNB, it assists the donor eNB in physical DL shared channel (PDSCH) and physical uplink shared channel (PUSCH) transmission, which rely on the donor eNB scheduler for both downlink (DL) and uplink (UL) scheduling. An L2 relay node does not transmit the physical DL control channel (PDCCH) and cell-specific reference signal (CRS) of the donor eNB to ensure that the cell coverage area of the donor eNB remains the same. In the absence of PDCCH at the L2 relay node, there is no cross-interference of the control channel between the relay node and the donor eNB. An L2 relay node can be considered similar to an RF repeater or L1 relay; however, performance is improved by regenerating and modifying the signal at the baseband, avoiding the problems

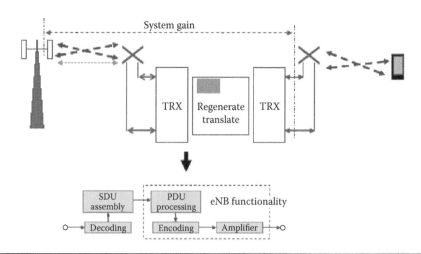

Figure 5.6   L2 relay. (PDU—Protocol data unit; TRX—transceiver)

of amplified noise, inband interference, and the instability that limit RF repeated gain, at a cost of higher complexity.

L2 relay nodes can capitalize on mechanisms such as beamforming and adaptive modulating and coding. However, L2 relays cannot provide traffic session–level quality of service (QoS) differentiation to their UE traffic, since the knowledge of data radio bearers (DRBs) and the associated QoS requirements are maintained by the radio resource control (RRC) protocol above the MAC layer. Handover enhancement and advanced interference management techniques are also not applicable to L2 relays.

L3 relaying is also a form of DF relay, which performs demodulation and decoding of RF signals received from the eNB. But L3 relay nodes go on to perform further user data processing, such as ciphering and user data concatenation, segmentation, and reassembly. Similar to the L2 relay, the L3 relay can improve throughput by eliminating intercell interference and noise and, additionally, by incorporating the same functions as an eNB, can have a small impact on the standard specifications for radio relay technology and on implementation. Its drawback, however, is the delay caused by user-data processing in addition to the delay caused by modulation/demodulation and encoding/decoding processing.

Because L3 relay nodes receive and forward IP packets (packet data convergence protocol [PDCP] SDUs), user packets at the IP layer are viewable at L3 relay nodes. Generally speaking, an L3 relay node has all the functions of an eNB, and it conventionally communicates with its donor eNB through an X2-like interface.

An L3 relay node would include functionality such as mobility management and session setup and handover, and as such acts as a full-service eNB. This adds more complexity to the implementation of such a relay node and the delay budget is further increased.

The various relay types according to the protocol layer where the signal/data is forwarded are shown in Figure 5.7.

In summary, L1, L2, and L3 relays are associated with different functions, requiring various complexities in terms of control channel, data processing, and high-layer interface. Table 5.1 shows the technical features of the three types of radio relay technologies, as well as their respective advantages and disadvantages.

5.1.1.3 Inband and Outband Relay

In relaying senarios, the backhaul link may operate both inband and outband relaying, as shown in Figure 5.8. In outband relaying mode, the backhaul link operates on a different frequency band than the access link. Comparatively, in inband relaying mode, the backhaul link operates in the same frequency band as the relay access links and the direct access links. During inband relaying, the RN might cause interference to its own receiver because the relay transmitter could be transmitting on the same frequency band as its own relay UEs. This implies backhaul link transmission and access link reception, or backhaul link reception

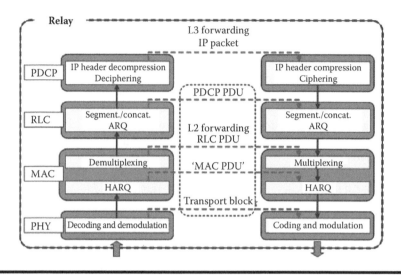

Figure 5.7   Radio relay technologies.

and access link transmission cannot happen simultaneously. Therefore, half-duplex mode should be deployed for inband relaying, unless a sufficient isolation of the outgoing and incoming signals is provided, such as by means of well-isolated antenna structures.

It is obvious that inband relay operation is more complex, since isolation in the time domain needs sophisticated configuration of the Un interface. In contrast, outband relay operation is straightforward, since it only requires adequate frequency planning. Typically, the backhaul link is assigned the lower carrier frequency so that it suffers less from distance-dependent attenuation and can be located close to the edge of the macro cell. In outband relaying mode, no additional functionality beyond Rel-8 is required (i.e., the Un interface behaves in the same way as a legacy Uu link).

5.1.1.4 Type 1 and Type 2 Relay Nodes

As mentioned previously, an RN can be either *transparent* or *nontransparent* with respect to knowledge in the UE. In transparent relay mode, the UE is not aware of whether it communicates with the network via the relay. In nontransparent relay mode, the UE is aware of whether it is communicating with the network via the relay. Accordingly, two types of RNs have been defined in 3GPP LTE-A standards, Type 1 and Type 2, which are nontransparent and transparent, respectively.

Type 1 relay nodes control their cells with their own physical cell identity, and transmit their own synchronization channels and reference signals. A Type 1 relay node behaves similarly to an eNB with wireless backhauling since it has its own cell ID, which is different from the adjacent donor cell. In other words, a Type 1 relay

Table 5.1   Comparison to Relay Technologies

Relay Category	Technical Feature
L1 Relay	1. PHY layer only
	2. Used for coverage extension or cover isolated areas
	3. Amplify-and-forward devices based on analog signal
	Desired signal cannot be separated; interference and noise is amplified as well
	Immediate forwarding is done (within the CP length), delay can be neglected; looks like multipath
	Strong RF isolation required to minimize the leakage; repeater gain is at least limited by the RF isolation
	4. "Smart" repeaters use power control or self-cancellation
	5. Alternatively, signal can be forwarded at another frequency
L2 Relay	1. Functionalities up to MAC and simplified upper layer
	2. RNs are introduced at cell edge
	3. RX and TX times require some multiplexing (TDD/FDD, coordination/cooperation among nodes required)
	4. Decoding, scheduling, and re-encoding
	5. Interference coordination needed
	6. Delay of a few subframes
	7. Clear advantage compared to L1 relay
L3 Relay	1. Full functionalities up to PDCP
	2. No new nodes defined, but new cells are created
	3. Backhaul via LTE technology; X2 protocol reused or S1
	4. Same or different spectrum could be used
	5. High spectral efficiency needed for backhaul
	6. Spatial coordination with beams possible
	7. Signaling overhead from encapsulation
	8. No need to change specification
	9. Relay as complex as Home NB
	10. Only solution for group mobility scenario

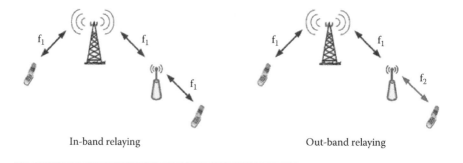

In-band relaying   Out-band relaying

Figure 5.8   Inband and outband relaying.

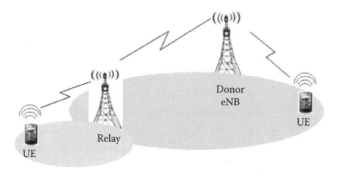

Figure 5.9   Type 1 relay.

node creates a new cell (as shown in Figure 5.9) and is thus, from a UE point of view, indistinguishable from a real eNB. As a result, an LTE-A Type 1 relay node appears as if it is a Rel-8/9 eNB to Rel-8/9 UEs. This ensures backward compatibility. However to LTE-A UEs, it should be possible for a Type 1 relay node to appear different from a Rel-8/9 eNB to allow for further performance enhancement.

A Type 1 RN is essentially a low-power eNB. By relaying both control signal and data traffic, the Type 1 relay works well for coverage extension for remote UEs.

Functionally, a Type 1 RN appears as an eNB to UEs. However, it has significant differences from a macro-cell eNB. First of all, a Type 1 RN has much lower transmit power and antenna gains, and the number of antennas might be very limited (e.g., ≤ 2) due to the size of relay sites. Also, an RN's backhaul link is wireless and its channel capacity is generally inferior to wire backhaul, especially during the two-hop process.

Since a Type 1 RN communicates with the DeNB through the backhaul link (Un interface) and with the UE through the access link (Uu interface) at the same time, sufficient isolation must be achieved between the transmitter part on the one link and the receiver part on the other. Otherwise, the RN would create such a large amount of self-interference that it might severely damage the signal reception from the attached UEs and the DeNB. The isolation in the RN can be achieved either in time, frequency, or antenna configuration. Therefore, 3GPP further distinguishes three different classes of Type 1 relays.

- **Type 1 relay:** This relay is an inband relay in which the backhaul link and access link share the same carrier frequency. Isolation is done in the time domain, implying that some of the subframes are reserved for the backhaul link and cannot be used for the access link to the relay-attached UEs.
- **Type 1a relay:** This relay is an outband relay in which separate carrier frequencies are used for the backhaul and access link. Isolation is already achieved in the frequency domain.
- **Type 1b relay:** This relay is also an inband relay; however, isolation is not done in the time domain, but via adequate antenna configuration.

Features of Type 1, 1a, and 1b relays are shown in Figure 5.10.

Type 2 relay nodes do not have their own cell identity and look just like the main cell. UEs that are in range are unable to distinguish a relay from the main eNB within the cell.

No new cell will be created for a Type 2 RN without its own cell identity; therefore, the UE will not be able to distinguish between signal transmission from the donor eNB and the relay. Consequently, a Type 2 RN is transparent to Rel-8/9 UEs (i.e., a Rel-8/9 UE is not aware of a Type 2 RN's presence). Control information can be transmitted from the eNB and user data can be transmitted from the relay (as shown in Figure 5.11). Type 2 RNs can transmit the PDSCH but do not transmit the CRS and PDCCH.

The Type 2 relay is categorized as an L2 relay. The main objective of a Type 2 relay is to increase the system capacity. It is mainly used to increase data throughput for local UEs. Type 2 relays can eliminate propagating the interference and noise to the next hop, so they can reinforce signal quality and achieve much better link performance.

Since UE is also unable to provide channel quality indicator (CQI) feedback for relay signals in a Type 2 relay, effective link adaptation is not possible. Moreover, as UE is unaware of the existence of the relay, it is unable to provide a measurement report to aid the eNB in selection of the best relay node.

Table 5.2 shows a comparison of Type 1 and Type 2 relays.

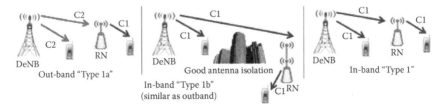

Figure 5.10   Type 1, 1a, and 1b relays.

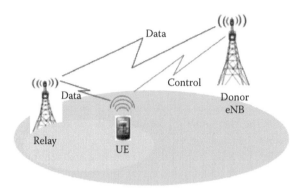

Figure 5.11   Type 2 relay.

Table 5.2   Comparison Table for Type 1 and Type 2 Relays

Item	Type 1	Type 2
PHY Cell ID	Own cell ID; create new cell (another eNB)	No cell ID; no creation of new cell
Transparency	Nontransparent relay node to UE	Transparent relay node to UE
Layer	Layer 3	Layer 2
RF parameters	Optimized parameters	N/A
Handover (HO)	Intercell HO (generic HO)	HO transparently to UE
Control Channel Generation	Generate synch. channel; Reference signal (RS), HARQ channel and scheduling information, etc.	Does not generate its own channel but decodes/forwards donor eNBs signal to UE
Cooperation	Intercell cooperation	Intracell cooperation
Usage Model	Coverage extension	Throughput enhancement and coverage extension
Overhead and Handover	Normal handover as macro	Less control signal overhead; transparent intracell handover for UE

5.1.2 Relay in LTE-A

At least Type 1 and Type 1a relay nodes will be supported in LTE-A networks. Type 1a relay nodes are characterized by the same set of features as the Type 1 relay node. The difference is that the Type 1 relay is inband and the Type 1a relay is outband. An outband Type 1a relay node is expected to have little or no impact on LTE specifications, while in order to allow inband Type 1 relaying, some functions specific to RN operations have to be defined in LTE-A.

LTE-A adopts nontransparent L3 relay nodes. To the UE, an LTE-A relay node looks like a regular eNB, that is, it has its own cell ID and its own synchronization, broadcast, and control channels. The wireless access link is compliant with the standard Uu interface in 3GPP, which ensures backward compatibility to Rel-8/9 UEs in the cell.

The DeNB is an enhanced eNB that supports RN operation. The RN is connected with the DeNB through the Un interface. This connection can only be half duplex, as illustrated in Figure 5.12, which means that the relay access link and the backhaul link cannot operate at the same time. The RN acts both as an eNB and as UE, the DeNB acts both as an eNB and as the RN's mobility management entity (MME), serving gateway (SGW), and packet data network gateway (PGW). The RN is managed by its own operation and maintenance (O&M), which is different from the DeNB's O&M. The RN supports UE handover to the DeNB, just like handover to a normal eNB from a neighboring eNB's perspective. A two-step RN startup procedure is defined to ensure that the RN connects to an eligible DeNB that needs an upgrade by the MME. The RN maps UE bearers on the Uu to Un bearers. The mapping of the Uu bearers to the Un bearers satisfies the QoS requirements of each bearer.

For the access link connecting the RN and UE, all protocol layers (MAC, RLC, PDCP, and RRC) are terminated at the RN. The user-plane protocol stack is similar to the eNB but differences exist in the control-plane protocol stack. In the context of single-cell operation, the UE should receive scheduling information and HARQ feedback directly from the RN and send its control channels (scheduling request [SR], CQI, acknowledgment [ACK] to the RN.

Currently, LTE-A only supports fixed relay, which means RN mobility between DeNBs is not supported. Also, one-hop relay is the only scenario supported in

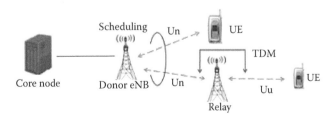

Figure 5.12   RN TDM (time-division multiplexing) operation.

LTE-A; therefore, the radio link between the eNB and UE is relayed by one RN. More than two hops will bring fewer capacity improvements in a system as increments of overhead and interference, and will be concerned about increases in end-to-end latency, which is not suitable for real-time applications. Specifically, when the channel qualities of the backhaul link and access link are comparable and use the same amount of resources, the overall rate after two hops would be only half of either link.

5.2 Relay Architecture

The deployment of relays impacts all radio access network (RAN) protocols, including RRC, S1 application protocol (S1AP), and X2 application protocol (X2AP). A normal macro eNB needs upgrading to support relay nodes, and the core network (CN) also needs to be upgraded.

Many requirements will be considered in the relay architecture design, such as providing compatibility to LTE evolved packet core (EPC) and LTE Rel-8 UEs; minimizing standardization, development, and operational complexity; providing sufficient QoS differentiation; reducing over-the-air overhead; and providing security.

5.2.1 Overall Architecture

During the study of LTE-A, many architecture alternatives for relays were investigated. The overall architecture to support RNs (shown in Figure 5.13) has finally been selected for Rel-10. In this architecture, the RN terminates the S1, X2, and Un interfaces. The DeNB provides S1 and X2 proxy functionality between the RN and other network nodes (other eNBs, MMEs, and SGWs). The S1 and X2 proxy functionality includes passing UE-dedicated S1 and X2 signaling messages and general packet radio service (GPRS) tunneling protocol (GTP) data packets between the S1 and X2 interfaces associated with the RN and the S1 and X2 interfaces associated with other network nodes. Due to the proxy functionality, the DeNB appears as an MME (for S1-MME), an eNB (for X2), and an SGW (for S1-U) to the RN.

The corresponding user-plane protocol stack used in LTE-A relay architecture is illustrated in Figure 5.14. As can be seen, there exists a GTP tunnel per UE bearer between the SGW/-GW (gateway) of the UE to the DeNB, which is switched to another GTP tunnel on the backhaul link from the DeNB to the RN (one-to-one mapping).

Furthermore, it can be observed that inside a DeNB, an RN's SGW/PGW and an RN home eNB (HeNB) GW are deployed. The embedded RN's SGW/PGW eliminates the packet delay between the DeNB and the RN's SGW/PGW, as well as the operational complexity involved in GTP tunneling. The functionalities of the embedded SGW/PGW within the DeNB include creating a session for the

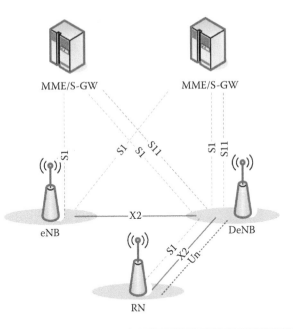

Figure 5.13 Evolved Universal Terrestrial Radio Access Network (E-UTRAN) architecture to support RN.

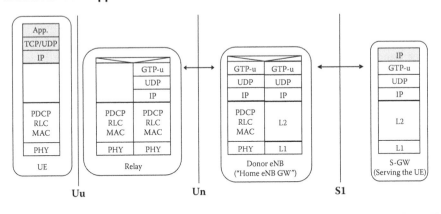

Figure 5.14 User plane protocol stack. (RLC—radio link control; TCP—transmission control protocol; UDP—user datagram protocol; GTP-u—GPRS tunneling protocol-user plane)

RN and managing evolved packet system (EPS) bearers for the RN, as well as terminating the S11 interface toward the MME serving the RN.

The embedded RN GW has the functionalities of an HeNB GW. The S1 and X2 interfaces coming from the core network terminate and get their information parsed at the RN GW within the DeNB, while S1 and X2 coming from the RN

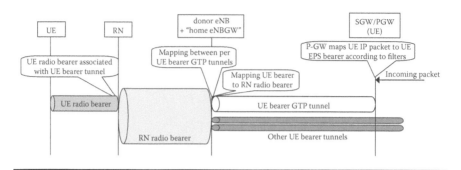

Figure 5.15 User plane packet delivery.

also terminate and get their information parsed at the DeNB. Subsequently, the DeNB will have per-UE flow visibility.

Toward the UEs in the RN cell, the RN will act as a normal eNB, which terminates the radio protocols of the Evolved Universal Terrestrial Radio Access (E-UTRA) radio interface, and the S1 and X2 interfaces. The RN also supports a subset of the UE functionality, including the physical layer, Layer 2, RRC, and non-access stratum (NAS) functionality, in order to wirelessly connect to the DeNB. The RN and DeNB also perform mapping of signaling and data packets based on existing QoS mechanisms defined for the UE and the PGW onto EPS bearers that are set up for the RN.

The packet delivery procedure in the user plane is illustrated in Figure 5.15. There is one GTP tunnel per UE bearer, spanning from the SGW/PGW of the UE to the donor eNB, which is switched to another GTP tunnel at the DeNB, going from the DeNB to the RN (one-to-one mapping).

As the first step at the PGW serving the UE, the downlink UE packet is mapped to the UE bearer and the packet is sent in the corresponding UE bearer GTP tunnel to the donor eNB. The donor eNB then classifies the incoming packets into RN radio bearers based on the QoS class identifier (QCI) of the UE bearer, and switches the UE bearer GTP tunnel from the SGW/PGW to another UE bearer GTP tunnel toward the RN (one-to-one mapping). EPS bearers of different UEs connected to the RN with similar QoSs are mapped in one radio bearer over the Un interface. Finally, the RN associates the received packet with the corresponding UE radio bearer based on the per-UE bearer GTP tunnel.

In the uplink, the RN performs the UE bearer-to-RN bearer mapping, which can be done based on the QCIs of the UE bearers.

5.2.2 S1 Protocol Stack

The S1 user-plane protocol stack for supporting RNs is shown in Figure 5.16. There is a GTP tunnel associated with each UE EPS bearer, spanning from the SGW

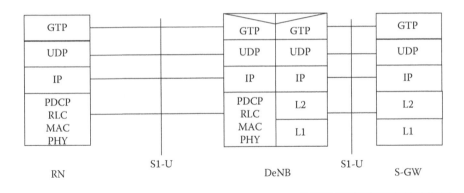

Figure 5.16   S1 user plane protocol stack for supporting RNs.

associated with the UE to the DeNB. In the DeNB, the GTP tunnel is switched to another GTP tunnel, going from the DeNB to the RN (one-to-one mapping). The S1 user-plane packets are mapped to radio bearers over the Un interface. The mapping can be based on the QCI associated with the UE EPS bearer. The Un link implements normal LTE L2 protocols such as PDCP, RLC, HARQ, and RRC, with minor additions to support Un operation.

For the S1 signaling, the S1AP messages are sent between the MME and the DeNB, and between the DeNB and the RN. The DeNB processes the S1 messages between the RN and the MME for UE-dedicated procedures. When the DeNB receives the S1AP messages, it translates the UE IDs between the two interfaces by means of modifying the S1AP UE IDs in the message but leaving other parts of the message unchanged. This operation corresponds to an S1AP proxy mechanism and would be similar to the HeNB GW function. The S1AP proxy operation would be transparent for the MME and the RN. That is, as seen from the MME, it looks as if the UE is connected to the DeNB, while from the RN's perspective it would look as if the RN is talking to the MME directly. The S1AP messages encapsulated by stream control transmission protocol (SCTP)/IP are transferred over an EPS data bearer of the RN where the PGW functionality for the RN's EPS bearers is incorporated into the DeNB (as local breakout functionality for HeNBs). The control-plane protocol stack for S1 is illustrated in Figure 5.17.

The S1 interface relations and signaling connections are shown in Figure 5.18. There is one S1 interface relationship between the RN and the DeNB and between the DeNB and the MME (serving the UE), where the S1 signaling connections are processed by the DeNB. The RN has to maintain only one S1 interface (to the DeNB), while the DeNB maintains one S1 interface to each MME in the respective MME pool. There is also an S1 interface relation and an S1 signaling connection corresponding to the RN as a UE, going from the DeNB to the MME serving the RN.

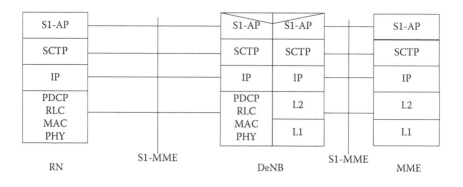

Figure 5.17 S1 control plane protocol stack for supporting RNs.

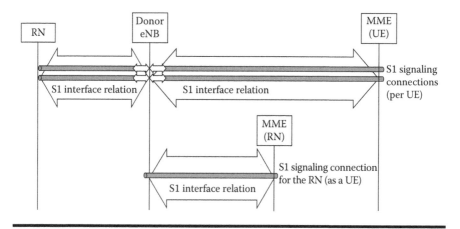

Figure 5.18 S1 interface relationships.

5.2.3 X2 Protocol Stack

The X2 user-plane protocol stack for supporting RNs is shown in Figure 5.19. The DeNB also acts as an X2 proxy. There is a GTP forwarding tunnel associated with each UE EPS bearer subject to forwarding, spanning from the other eNB to the DeNB, which is switched to another GTP tunnel in the DeNB, going from the DeNB to the RN (one-to-one mapping).

The X2 control-plane protocol stack for supporting RNs is shown in Figure 5.20. For X2 signaling, the DeNB processes the X2 messages between the RN and other eNBs for UE-dedicated procedures. The processing of X2AP messages includes modifying the X2AP UE IDs and GTP tunnel endpoint IDs (TEIDs) in the X2AP messages. For X2 control-plane signaling there is one X2 interface relationship between the RN and the DeNB. When UE under the RN performs handover (HO), an HO request is received from the RN by the DeNB, which reads the target

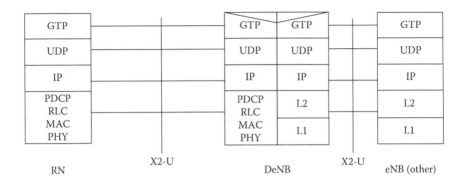

Figure 5.19 X2 user plane protocol stack for supporting RNs.

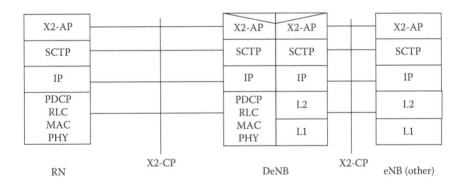

Figure 5.20 X2 control plane protocol stack for supporting RNs.

cell ID from the message and forwards it to the appropriate target eNB. Forwarding tunnels between the RN and the target eNB also are established via the donor eNB.

5.2.4 Radio Protocol Stack

The RN connects to the DeNB via the Un interface using the same radio protocols and procedures as a UE connecting to an eNB. The radio control-plane protocol stack is shown in Figure 5.21.

In the control plane, the RRC layer of the Un interface has functionality to activate specific subframe configurations for transmissions between an RN and a DeNB for RNs that are not able to transmit/receive from their DeNB in all subframes. The DeNB is aware of which RNs require specific subframe configurations. The RRC layer of the Un interface also has the functionality to send updated system information in a dedicated message to RNs that are not able to receive signaling from its DeNB in all subframes.

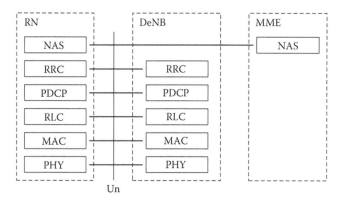

Figure 5.21 Radio control plane protocol stack for supporting RNs.

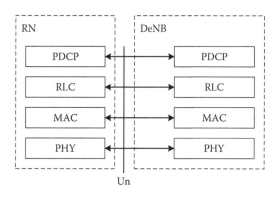

Figure 5.22 Radio user plane protocol stack for supporting RNs.

For the user plane, the DeNB acts as a proxy between the RN and the SGW (i.e., the GTP tunnel on S1 is moved to another GTP tunnel on the Un). The PDCP layer of the Un interface has functionality to provide integrity protection for the user plane. The integrity protection is configured per DRB. The radio user-plane protocol stack is shown in Figure 5.22.

5.3 Relay Radio Protocol

The design of the backhaul link and access link in LTE-A relaying could be different from the conventional radio link due to several factors. The main reason is that an RN may use a larger number of antennas than a UE, would have more strong processing power than a UE, and could thus allow more advanced transmitters and receivers. Also, due to implementation limitations, it is difficult for an RN to transmit and receive on the same frequency band at the same time.

Based on the previous observations, some new designs could be considered to enhance both the uplink and downlink throughputs of the relay backhaul link between the RN and the DeNB and at the same time to reduce the interference of the backhaul link to the access link. Hereafter some radio design aspects of LTE-A relaying are discussed, which include RN subframe configuration, the control channel for relay, and channel multiplexing in the uplink.

5.3.1 Backhaul Subframe Configuration

For the Type 1 inband relaying adopted in LTE-A, the eNB-to-RN backhaul link operates in the same frequency spectrum as the RN-to-UE access link. Due to the RN's transmitter causing interference to its own receiver, simultaneous eNB-to-RN and RN-to-UE transmissions on the same frequency resource may not be feasible unless sufficient isolation of the outgoing and incoming signals is provided by means of specific, well-separated, and well-insolated antenna structures. Similarly, at the RN it may not be possible to receive UE transmissions simultaneously with the RN transmitting to the eNB.

5.3.1.1 TDM between Un and Uu

The scheme in LTE-A to handle the interference problem is to realize time-division multiplexing (TDM) between the backhaul link and access link within the same spectrum, as shown in Figure 5.23. The scheme involves operating the RN such that the RN is not transmitting to UEs when it is supposed to receive data from the donor eNB (that is, to create gaps in the RN-to-UE transmission. During the gaps, the UEs are not supposed to expect any downlink transmission from the RN. On the other hand, RN-to-eNB uplink transmissions can be facilitated by not allowing

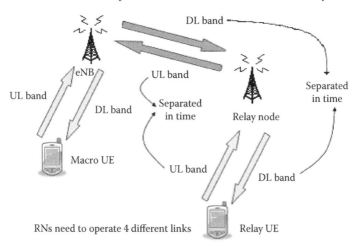

Figure 5.23   TDM between backhaul and access links.

any UE-to-RN transmissions in some subframes, which can be implemented easily in the RN uplink scheduler without any change in specification.

However, something must be done to naturally create the gaps in the RN-to-UE transmission; otherwise Rel-8/9 UEs, without any LTE-A relaying knowledge, will lose their connections to the RN after failing to detect cell-specific reference signals, and the physical control format indicator channel (PCFICH) and physical broadcast channel PBCH in some downlink subframes would be used as the gaps. Therefore, to maintain backward compatibility, the design of the Un interface on the backhaul link must be based on the assumption that the access link can operate with legacy UEs with Rel-8 functionality only.

To make RN-to-UE downlink transmission gaps without losing backward compatibility, a mechanism already provided in the LTE Rel-8 specification, a multicast-broadcast single-frequency network (MBSFN) type of subframe to support the multimedia broadcast multicast service (MBMS) protocol, is reused in LTE-A for downlink transmission on the access link to avoid conflict with backhaul link transmission (Figure 5.24). This means that when it is the time for the eNB to transmit to the RN, the UEs supported by the RN should, according to configuration via RRC signaling, assume an MBSFN type of subframe as the current subframe and do nothing other than receive the first one or two OFDM symbols as the control channel. In the MBMS protocol, each MBSFN subframe is divided into a non-MBSFN region and an MBSFN region. The non-MBSFN region, which can be configured to have one or two OFDM symbols, may be used by the RN to send control channels to Rel-8/9 UEs in the RN cell, while the MBSFN region will be ignored by Rel-8/9 UEs in the RN cell. This allows the donor eNB to use the MBSFN region of an MBSFN subframe for backhaul downlink transmission.

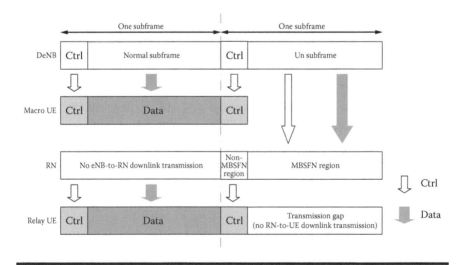

Figure 5.24 TDM between Un and Uu link by MBSFN.

As shown in Figure 5.24, in a normal downlink subframe from the eNB's perspective, the eNB may do downlink transmission to macro UEs, but no downlink transmission from the eNB to RNs is allowed. This can be implemented by the eNB downlink scheduler. In eNB semistatically defined Un subframes, which are configured as an MBSFN in the RN and are informed to relay UEs, the eNB may do downlink transmission in the RN's MBSFN region, during which no RN-to-relay UE transmission is allowed. More specifically, in the non-MBSFN region of an MBSFN subframe, the RN may transmit control information to relay UEs, but the eNB should not transmit any control information to the RN simultaneously.

As a result, the RN cannot listen to the PDCCH from the DeNB in MBSFN frames. In order to signal the DL resource allocation and UL grant to the RN, a new control channel R-PDCCH has been defined in LTE-A.

In summary, in LTE-A, subframes during which eNB-to-RN transmission may take place are configured by the RRC layer. Downlink subframes configured for eNB-to-RN transmission, which are also called Un downlink subframes, will be configured as MBSFN subframes by the RN to relay UEs. MBSFN subframes may be configured with a 10-ms period or 40-ms period in the MBMS protocol already defined in LTE Rel-8.

The Un subframe configuration schemes for frequency-division duplexing (FDD) and time-division duplexing (TDD) frame structure are different.

5.3.1.2 FDD Subframe Configuration

For the FDD frame structure, the Un subframe configuration is expressed as a combination of 8 different patterns, and each pattern can be represented by an 8-bit bitmap: {00000001}, {00000010}, {00000100}, {00001000}, {00010000}, {00100000}, {01000000}, or {10000000}. The "1" in each pattern indicates the position of the first Un downlink subframe in the first frame of the 40-ms configuration period. Each combination of patterns can be identified by the decimal equivalent of the binary number representing the 8-bit bitmap. For instance, the decimal number 85 represents the configuration with a combination of 4 patterns {01010101}. The beginning of the Un subframe configuration is aligned to the beginning of 40-ms window (SFN mod 4 = 0) of the eNB cell. It is also aligned with the 40-ms period of the MBSFN configuration.

It should be noted that some downlink subframes cannot be configured as an MBSFN, which has already been declared in the MBMS protocol. The specific restriction also depends on the frame structure type.

For the FDD frame structure, subframes 0 and 5 are used for synchronization channel (SCH) and broadcast channel (BCH) in Rel-8, and subframes 4 and 9 are used for paging in Rel-8. The downlink transmission on the access link in these 4 subframes within each radio frame should be guaranteed, so they cannot be configured as an MBSFN by the RN.

The relationships between each pattern and the configured downlink subframes within the 40-ms period are listed in Table 5.3.

Table 5.3 Patterns for Un Subframe Configuration

Patterns for Un Subframe Configuration	DL Subframes within 40-ms Period
0 (10000000)	8,16,32
1 (01000000)	1,17,33
2 (00100000)	2,18,26
3 (00010000)	3,11,27
4 (00001000)	12,28,36
5 (00000100)	13,21,37
6 (00000010)	6,22,38
7 (00000001)	7,23,31

Un subframe configuration is performed through the RRC message "RNReconfiguration" sent from the DeNB to the RN. RNReconfiguration contains an 8-bit width "subframeConfigPatternFDD" whose meaning is the combination of 8 patterns as mentioned previously.

After getting the Un subframe configuration in RNReconfiguration, the RN should inform its UEs that the corresponding UN downlink subframes are configured as an MBSFN through the RRC message "MBSFNAreaConfiguration," in which the configuration of MBSFN for the 40-ms period is represented as a 24-bit width bitmap indicating the MBSFN subframe allocation in four consecutive radio frames. The "1" denotes that the corresponding subframe is allocated for MBSFN. For FDD, the bitmap is interpreted as follows: starting from the first radio frame and from the first or leftmost bit in the bitmap, the allocation applies to subframes 1, 2, 3, 6, 7, and 8 in the sequence of the four radio frames.

Take the decimal number 85 as an example for "subframeConfigPatternFDD." It represents the combination of 4 patterns: {01000000} + {00010000} + {00000100} + {00000001} = {01010101}. As shown in Figure 5.25, this 8-bit subframe configuration pattern is applied cyclically to all the subframes within the 40-ms period. All the subframes corresponding to "1" in the pattern within the 40-ms period, excluding the subframes that cannot be configured as MBSFN, are configured as Un downlink subframes. Therefore, conforming to Table 5.3, the Un subframes configured by pattern 85 are {1, 3, 7, 11, 13, 17, 21, 23, 27, 31, 33, 37}.

Accordingly, the 24-bit bitmap in "MBSFNAreaConfiguration" to configure MBFSN to relay UEs should be set to {101010101010101010101010} by the RN.

In theory, for FDD, UL backhaul transmissions can take place in any subframe, provided the RN can configure "blank" UL subframes by not allocating UL grants or canceling any UE-to-RN UL transmission during these subframes. It is clearly

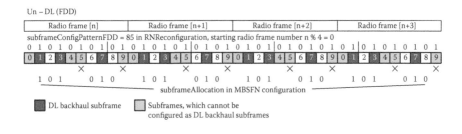

Figure 5.25 FDD Un subframe configuration.

desirable that UL backhaul subframes are preconfigured, for example, by semistatic higher-layer signaling, so that the RN can avoid scheduling its relay UEs in such subframes.

In FDD, the set of uplink backhaul subframes is derived implicitly from the downlink backhaul subframes and the HARQ timing. Specifically, if subframe k is configured as the Un downlink subframe, then subframe $k + 4$ is implicitly configured as the Un uplink subframe. The eNB can dynamically schedule the relay uplink on this set of semistatically configured uplink backhaul subframes.

5.3.1.3 TDD Subframe Configuration

For TDD frame structure, due to the same restriction as in FDD, subframes 0, 1, 5, and 6 within each radio frame cannot be configured as MBSFN. Furthermore, because TDD configuration 0 has only subframes 0 and 5 as downlink subframes, it cannot be used for the RN cell because it does not support any MBSFN subframes. In addition, in TDD configuration 5, there is only one single uplink subframe, subframe 2, which implies that TDM between the backhaul link and the access link cannot be implemented. Eventually, only TDD configurations 1, 2, 3, 4, and 6 can be used for constructing the relaying network. The supported TDD configurations for backhaul link transmission are summarized in Table 5.4 (from 3GPP 36.216). For each subframe in a radio frame, "D" indicates that the subframe is configured for Un downlink transmissions, and "U" indicates that the subframe is configured for Un uplink transmissions.

"SubframeConfigPatternTDD" in RRC message "RNReconfiguration" is represented as an integer with valid values ranging from 0 to 18.

The MBSFN subframe configuration for TDD in "MBSFNAreaConfiguration" is still represented as a 24-bit bitmap indicating MBSFN subframe allocation in four consecutive radio frames, but the interpretion of the bitmap is different from that for FDD. Considering that uplink subframes will not be allocated to be MBSFN and that in all TDD configurations subframe 2 is always configured as an uplink subframe, the 24-bit bitmap for TDD frame structure is interpreted as follows: starting from the first radio frame and from the first or leftmost bit in the

Table 5.4 Supported Configurations for Backhaul Transmission (TDD)

Subframe Config Pattern TDD	TDD Configuration	Subframe Number n									
		0	1	2	3	4	5	6	7	8	9
0	1					D				U	
1					U						D
2						D				U	D
3					U	D					D
4					U	D				U	D
5	2			U						D	
6					D				U		
7				U		D				D	
8					D				U		D
9				U	D	D				D	
10					D				U	D	D
11	3				U				D		D
12					U				D	D	D
13	4				U						D
14					U				D		D
15					U					D	D
16					U				D	D	D
17					U	D			D	D	D
18	6					U					D

bitmap, the allocation applies to subframes 3, 4, 7, 8, and 9 in the sequence of the four radio frames. The last four bits are not used.

Taking subframeConfigPatternTDD = 0 as an example, this indicates that subframe 4 in each radio frame is configured for Un downlink transmission, as shown in Figure 5.26.

Accordingly, the 24-bit bitmap in "MBSFNAreaConfiguration" to configure MBFSN to relay UEs should be set to {01000010000100001000xxxx} by the RN.

For TDD, both asymmetric and symmetric DL/UL Un subframe allocations are supported. The set of Un UL subframes is explicitly configured.

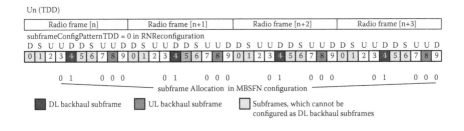

Figure 5.26   TDD Un subframe configuration.

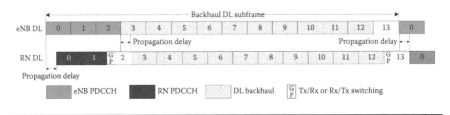

Figure 5.27   Regular RN DL Timing.

5.3.2 Relay Timing

For RNs with Type 1 half-duplex relay mode, TX/RX and RX/TX switching time will be reserved in a backhaul subframe. Meanwhile, as specified in LTE-A, the RN should ensure that the access link downlink subframe boundary is aligned with the backhaul link downlink subframe boundary, except for possible adjustment to allow for RN transmit/receive switching. The synchronization requirements between the eNB and the RN may impact the number of available OFDM symbols for backhaul transmission in a subframe. Therefore, design of the RN DL/UL frame timing will consider maximizing the number of available OFDM symbols for backhaul transmissions.

For backhaul transmissions, the RN acts a UE to the donor eNB. In regular RN DL frame timing, the RN DL frame lags behind the eNB DL frame by the corresponding propagation delay between the eNB and the RN, as shown in Figure 5.27.

When the RN has to reserve both TX/RX and RX/TX switching time, two OFDM symbols, 2 and 13, cannot be used for Un downlink transmission. Due to this ineffectiveness, a delayed RN DL frame timing was introduced in LTE-A, such that only one OFDM symbol is needed for both TX/RX and RX/TX switching, as shown in Figure 5.28.

In Figure 5.28, the RN cell DL timing is delayed based on reception of DeNB signal, so that at the end of the subframe there is still room for the RN to do RX/TX switching before transmission of symbol 0 in the next subframe. The amount

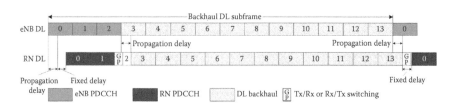

Figure 5.28 RN DL timing with a fixed delay.

Figure 5.29 RN DL timing for TDD.

of delay should be at least equal to the RN switching time. Symbol 2 is still not available due to TX/RX switching, but it does not matter since that symbol is used for PDCCH in the DeNB, the total number of OFDM symbols available for backhaul transmission is eleven, and is not impacted by the DeNB cell PDCCH symbol number.

In the fixed delay scenario, the amount of fixed delay will be sufficiently large for RX/TX switching. Furthermore, the total TX/RX and RX/TX switching time shall not exceed the duration of an OFDM symbol.

With fixed delay RN DL frame timing, more OFDM symbols are available for backhaul usage, which is clearly beneficial. However, it may only be suitable for FDD operation. In TDD networks, the network-wise alignment of subframe boundaries of all eNBs and RNs may be necessary for proper TDD operation, which means the fixed delay cannot be applied to RN DL framing timing. The TDD RN DL timing is illustrated in Figure 5.29.

LTE-A specifications provide flexibility to configure backhaul downlink timing depending on the backhaul link timing relationship between the eNB and the RN. There are three different configurations of frame structure for the first slot and two different configurations for the second slot, shown in Tables 5.5 and 5.6, respectively.

Depending on the different control region sizes in the donor cell and relay cells, the DeNB can start PDSCH transmissions to an RN on the second, third, or fourth OFDM symbol in the Un downlink subframe. The start point is semistatically configured to the RN by the RNReconfiguration message.

For FDD, the Un PDSCH transmission ends at the last OFDM symbol of the Un downlink subframe. While for TDD, Un PDSCH transmission ends at the second to last OFDM symbol of the Un downlink subframe.

Table 5.5 Un DL Frame Configuration (First Slot)

Configuration	DL Start Symbol	End Symbol Index	Useful Symbols
0	1	6	6
1	2	6	5
2	3	6	4

Table 5.6 Un DL Frame Configuration (Second Slot)

Configuration	Start Symbol Index	End Symbol Index	Useful Symbols
0	0	6	7
1	0	5	7

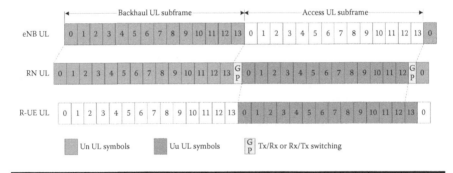

Figure 5.30 RN UL timing.

R-PDCCH transmissions intended for a relay always start at symbol 3, as a fixed starting position may simplify the specification structure.

For RN UL backhaul transmissions, both RX/TX and TX/RX switching time may be needed in a UL backhaul subframe. So far, the backhaul/access link UL timing scheme, as shown in Figure 5.30, has the agreement to be supported in LTE-A without additional specification requirements.

As shown in Figure 5.30, the Uu UL and the Un UL subframe boundary is staggered by a fixed gap and the RN TX/RX switching time is achieved by forbidding transmission on symbol 13 of the Un UL subframe. As a result, the Uu uplink transmission from relay UE to RN is delayed. This method is beneficial because it can utilize all the single-carrier frequency-division multiple access (SC-FDMA) symbols in a backhaul subframe, which is helpful in resolving the backhaul shortage problem.

To configure the relay UE not to transmit the last SC-FDMA symbol in front of a backhaul subframe on the Uu link, the RN can configure the access link subframe as a cell-specific SRS subframe, but does not configure the relay UE with any UE-specific SRS parameters.

5.3.3 Backhaul Physical DL Control Channel

Because PDCCH in the Un downlink subframe cannot be monitored by RNs, LTE-A introduced R-PDCCH, which is used by the donor eNB to dynamically or semipersistently assign to RNs the resources for DL as well as UL data transmission. R-PDCCH should be supported both in the donor eNB and in the relay node. As a simple description, R-PDSCH is used hereafter to denote the PDSCH sending from the eNB to the RN.

5.3.3.1 R-PDCCH Resource Assignment

The R-PDCCH is placed at a fixed location in the 2-dimensional OFDM resource grid of the MBSFN subframe. The fixed location is semistatically configured via RRC signaling. In the RRC message RNReconfiguration, an R-PDCCH region can be allocated by 4 different resource assignment formats: type0, type1, type2Localized, and type2Distributed, which are totally the same as the resource assignment to PDSCH of a macro UE.

The DL grant to an RN is always transmitted in the first slot of a Un subframe within the R-PDCCH region. If a DL grant is transmitted in the first physical resource block (PRB) of a given PRB pair within the R-PDCCH region, then a UL grant may be transmitted in the second PRB of the PRB pair, as illustrated in Figure 5.31.

Since the R-PDCCHs for DL grants are located within the first slot, low decoding latency is allowed, and can reduce the required buffer size for R-PDSCH because R-PDSCH (shown as PDSCH in Figure 5.31) decoding can start from the beginning of the second slot. Putting the UL grants in the second slot is reasonable since the actual UL transmission will occur several milliseconds later.

For DL backhaul, the actual resources used for R-PDCCH transmissions can vary dynamically between subframes. It can happen that not all the semistatically assigned PRBs are used for R-PDCCH in certain subframes. The PRBs used for R-PDCCH transmissions depend on the number of scheduled RNs, preferred PRBs for PDSCH transmissions for macro UEs, and so on. Whether to configure contiguous PRBs or scattered PRBs or scattered groups of contiguous PRBs can be decided by the donor eNB. Also, in view of frequency and interference diversity, the TDM/FDM (frequency-division multiplexing) hybrid structure used by R-PDCCH can achieve higher diversity gain as more PRBs are used to transmit an R-PDCCH.

The format on the R-PDCCH is identical to the conventional Rel-8 PDCCH for macro UEs, which includes resource block (RB) allocation, modulation and coding scheme (MCS) assignment, HARQ attributes, and MIMO-related attributes. Each

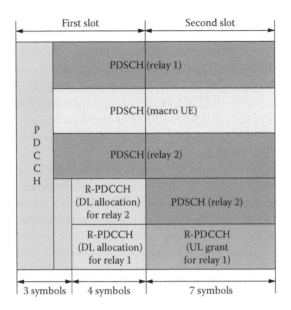

Figure 5.31   R-PDCCH multiplexing.

RN should perform blind decoding, as each Rel-8 UE in the macrocell does, in order to identify on which PRBs among the preconfigured resources its R-PDCCH is sent.

As the Rel-8 downlink control information (DCI) formats are reused for RN DL and UL scheduling without changing the DCI payload size, it is reasonable to reuse Rel-8 control channel element (CCE) size (number of resource elements [REs] in the CCE) for the R-PDCCH. Based on the DL grant R-PDCCH region ranging from the fourth symbol to the last symbol in the first slot, 44 REs can be used per PRB when CRS is used for R-PDCCH demodulation. In order to have similar CCE size as for the Rel-8 PDCCH, which is composed of 9 resource element groups (REGs) or 36 REs, it is reasonable to define 1 relay CCE (R-CCE) per PRB.

5.3.3.2 R-PDCCH Mapping

An R-PDCCH can be transmitted on one or several PRBs. The same aggregation levels (one, two, four, or eight) for Rel-8 PDCCHs are also applied to R-PDCCH for RNs. However, the mapping of the R-PDCCH to time–frequency resources is different. LTE-A defined two different mapping schemes: R-PDCCH without cross-interleaving and R-PDCCH with cross-interleaving.

Without cross-interleaving, one R-PDCCH is mapped to one set of virtual resource blocks according to the aggregation level. No other R-PDCCHs are transmitted using the same set of resource blocks. If the resource blocks are located sufficiently apart in the frequency domain, frequency diversity can be obtained, at least for the higher aggregation levels.

In LTE Rel-8, PDCCHs for UEs are extensively interleaved across the bandwidth to combat the interference as well as the deep fading in the frequency domain. The basic granularity in Rel-8 PDCCH interleaving is the REG. Cross-interleaved R-PDCCH reuses the Rel-8 PDCCH processing functionality to the extent possible. R-PDCCHs of several RNs can be cross-interleaved in a manner similar to Rel-8, but the cross-interleaving is not done over the entire bandwidth but only over the allocated R-PDCCH PRBs.

The cross-interleaved R-PDCCH should only be used in combination with cell-specific reference signals, while the non-cross-interleaved R-PDCCH can use dedicated reference signals as well for demodulation. Therefore, non-cross-interleaved mapping is useful for beamforming of the backhaul transmissions. The cross-interleaved R-PDCCH mapping method may obtain frequency diversity also for the lowest aggregation level. However, the non-cross-interleaved R-PDCCH is able to get benefits from applying frequency-selective scheduling to the R-PDCCH.

5.3.3.3 R-PDCCH Search Space

In LTE Rel-8, the UE needs to perform blind decodings for PDCCH detection in both the common and the UE-specific search spaces. There are no common search spaces for R-PDCCH detection as there is no need for RNs to receive broadcast information.

The number of R-PDCCH candidates monitored by an RN is the same as for a macro UE, that is, 6, 6, 2, and 2 for aggregation levels 1, 2, 4, and 8, respectively. For an R-PDCCH without cross-interleaving, the starting CCE position of the RN-specific search space is the first PRB of the configured RN-specific R-PDCCH region. For a R-PDCCH with cross-interleaving, the Rel-8 PDCCH search space design is reused, in which the set of CCEs corresponding to DL/UL grant candidate m of the grant search space is given by

$$L \cdot \left\{ (Y_k + m) \bmod \left[N_{\text{CCE},j}^{\text{R-PDCCH}} / L \right] \right\} + i$$

where $N_{\text{CCE},j}^{\text{R-PDCCH}}$ is the total number of CCEs, $j = 0$ (DL) or 1 (UL), derived based on RN-specific semistatic R-PDCCH configuration.

5.3.3.4 RN PDSCH Resource Allocation

Similar to the downlink access link, the R-PDSCH intended for an RN shall be processed and mapped to resource elements in the same way as the Rel-8 PDSCH from the eNB to macro UE. All three resource allocation types 0, 1, and 2 are supported for the R-PDSCH. However, different from a PDCCH located at the dedicated control region, which is separated from the data region used by the PDSCH, the R-PDCCH locates within the PRBs in the conventional data region. Therefore, the PRBs, or part of PRBs that can be used for R-PDSCH transmission according to R-PDCCH indication, must be further clarified. An RN will, upon detection of an R-PDCCH intended for it in a Un DL subframe, decode the corresponding PDSCH in the same subframe according to the following assumptions.

If the RN receives a resource allocation that overlaps a PRB pair in which a DL grant is detected in the first slot, the RN assumes that there is PDSCH data transmission for it in the second slot of that PRB pair.

If the UL grant is allocated in the second slot of the PRB pair, the PRB pair cannot be used for PDSCH transmission to avoid collision between the PDSCH and the UL grant.

When a demodulation reference signal (DMRS) is used for R-PDCCH demodulation, the DL grant and UL grant in a PRB pair will be for the same RN. That is, no REs in such a PRB pair can be used for a different RN.

When CRS is used for R-PDCCH demodulation, the DL grant and UL grant in a PRB pair can be for the same or different RNs.

When UL grants are transmitted only in the second slot, there should be no data transmission in the first slot.

5.3.4 Backhaul Reference Signals

Backhaul downlink transmissions use the same reference signals as defined for LTE-A macrocells, including CRS, DMRS, and channel state information reference signal (CSI-RS). Different reference signal types can be used for decoding the R-PDCCH and R-PDSCH. If the R-PDCCH is received using DMRS, generally in a beamforming scenario, then DMRS should be used for the R-PDSCH decoding as well. However, if CRS is used for the R-PDCCH with cross-interleaving, it is still reasonable to use DMRS for R-PDSCH.

An extra restriction exists for using DMRS in Un downlink transmission with TDD, in which the last OFDM symbol of a Un subframe is muted for RX/TX switching. So the reference signal sequence of antenna ports 7, 8, 9, and 10 used for lower layers will only be mapped to resource elements in the first slot of a PRB pair allocated for Un downlink transmission. Antenna ports 11 to 14 used for lower layers will not be used for Un downlink transmission.

5.3.5 Relay HARQ Process

5.3.5.1 Backhaul Downlink HARQ

In LTE Rel-8 and Rel-9, asynchronous HARQ is applied in the downlink. Likewise, asynchronous HARQ can be applied to the Un downlink as well. For backhaul downlink HARQ in FDD, if the R-PDSCH is transmitted in downlink backhaul subframe n, it is implicitly determined that subframe n+4 is an uplink backhaul subframe, where the RN can transmit the UL ACK/NACK. Therefore, the timing sequence is such that uplink backhaul subframe is 4 ms after downlink backhaul subframe, which is aligned with LTE Rel-8, and it does not affect the HARQ process on the access link.

With asynchronous HARQ, HARQ retransmission can be scheduled in any Un DL subframe, considering the available PRB resources for R-PDCCH and R-PDSDCH transmissions for the RN. Asynchronous HARQ also provides good flexibility in configuring Un subframes.

An example of FDD backhaul downlink HARQ procedures taking "subframe-ConfigPatternFDD" = 85 is shown in Figure 5.32. Taking HARQ process P1 for example, after R-PDSCH is transmitted in subframe 1 of radio frame n, the corresponding feedback will be certainly received in subframe 5 of radio frame n, since it works implicitly as a backhaul UL subframe. However, subframe 9 of radio frame n is not allowed to be configured as a Un DL subframe in FDD. The next available Un DL subframe, subframe 1 of radio frame n+1, is scheduled for HARQ process P1 to implement asynchronous HARQ processing. The same process also applies to HARQ processes P2 and P3.

We can see that the 8-ms round-trip time (RTT) of DL HARQ in Rel-8 may not be followed in relay backhaul downlink HARQ processing. It may be that DL HARQ retransmissions cannot occur 8 ms after initial transmission due to the lack of a Un subframe, but that is not a problem for asynchronous HARQ operation in DL. The actual backhaul DL HARQ RTT depends on which subframes are configured as Un DL subframes in a 40-ms period.

For TDD, considering that each of the supported backhaul DL/UL subframe configurations is a subset of one TDD configuration, DL HARQ procedures are similar to Rel-8 with regular 10-ms RTT.

The downlink HARQ feedback procedure in Un is almost the same as in Rel-8. The major exception is that the PUCCH resource index for ACK/NACK transmission is semistatically configured via RRC signaling, and will not change as the subframe number changes, rather than calculated based on the RN's cell radio network temporary identifier (C-RNTI) in different subframes.

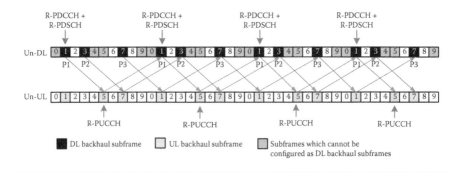

Figure 5.32   Backhaul downlink HARQ.

5.3.5.2 Backhaul Uplink HARQ

In LTE Rel-8 and Rel-9, synchronous HARQ is applied in the uplink for both FDD and TDD. If the Un uplink can also adopt synchronous HARQ, for a given HARQ initial transmission, the timing of corresponding Un UL HARQ retransmissions and Un DL ACK/NACK transmissions is predetermined, and as a result, Un uplink HARQ processing may be simplified. However, due to the restriction on MBSFN subframe locations and due to the limited number of DL and UL subframes in TDD systems, the synchronous HARQ time sequence seems difficult to meet in the Un link. Fortunately, the synchronous HARQ process in the Un uplink is still possible for FDD and TDD, as discussed in the following text.

The number of UL HARQ processes is statically defined in Rel-8. For FDD, eight UL HARQ processes are repeatedly used in each UL subframe. For TDD, the number of UL HARQ processes is predefined for each TDD configuration. In the PDCCH containing the UL grant, no HARQ process ID is specified. Because the 8-ms RTT may not always be met in the Un uplink, some additional work must be done to map HARQ process IDs and Un uplink subframes.

For FDD, thanks to the implicit Un uplink subframe allocation, if a UL grant is transmitted in subframe k, the corresponding UL data transmission must be able to happen in subframe k+4. However, the UL HARQ retransmission may not happen in subframe k+8 because it is not configured as the Un uplink subframe.

As discussed in Section 5.3.1.2, the Un subframe allocation for FDD is represented as a configuration pattern that is repeated within the 8-ms period, so there are 256 different Un HARQ timelines, each of which corresponds to one of 256 Un subframe patterns. Within the 256 Un subframe configurations, a number of configurations form a group in which one configuration is nothing but a circular-shifted version of another configuration. For instance, configuration 170 is just a 1-bit shift variation of configuration 85. In essence, all the configurations in the same group result in an equivalent Un resource allocation, and the number of UL HARQ processes and the Un HARQ timeline also become the same.

Generally, the number of Un uplink HARQ processes for each Un subframe configuration can be calculated by

$$N_{HARQ} = \max_{i=0...N} \sum_{j=i}^{i+8} \begin{cases} 1 & Un\,subframe \\ 0 & Uu\,subframe \end{cases}$$

where N is the Un subframe configuration period of 40 ms.

Moreover, it is beneficial to change the UL HARQ process number for some Un subframe configurations in a group if it leads to a constant RTT. In summary, Table 5.7 (from 3GPP 36.216) lists the number of Un UL HARQ processes for each Un subframe configuration.

We can see that the maximum number of Un UL HARQ processes is 6. HARQ processes are sequentially assigned to available subframes. Figure 5.33 shows a Un

Table 5.7 Number of Un UL HARQ Processes for FDD

Configuration	Number of HARQ Processes
1,2,4,8,16,32,64,128	1
3,5,6,9,10,12,17,18,20,24,33,34,36,40,48,65,66,68,72,80,96,129, 130,132,136,144,160,192	2
7,11,13,14,19,21,22,25,26,28,35,37,38,41,42,44,49,50,52,56,67, 69,70,73,74,76,81,82,84,85,88,97,98,100,104,112,131,133,134, 137,138,140,145,146,148,152,161,162,164,168,170,176,193,194, 196,200,208,224	3
15,23,27,29,30,39,43,45,46,51,53,54,57,58,60,71,75,77,78,83,86, 87,89,90,91,92,93,99,101,102,105,106,107,108,109,113,114,116, 117,120,135,139,141,142,147,149,150,153,154,156,163,165,166, 169,171,172,173,174,177,178,180,181,182,184,186,195,197,198, 201,202,204,209,210,212,213,214,216,218,225,226,228,232,234,240	4
31,47,55,59,61,62,79,94,95,103,110,111,115,118,119,121,122,123, 124,125,143,151,155,157,158,167,175,179,183,185,187,188,189, 190,199,203,205,206,211,215,217,219,220,221,222,227,229,230, 233,235,236,237,238,241,242,244,245,246,248,250	5
63,126,127,159,191,207,223,231,239,243,247,249,251,252,253, 254,255	6

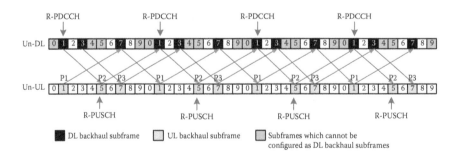

Figure 5.33 Backhaul uplink HARQ.

UL HARQ sequence example for Un subframe configuration 85, which has three Un UL HARQ processes.

On the Un link in FDD, UL HARQ retransmissions are synchronous with regard to the HARQ process. UL retransmissions are transmitted in the subframe that corresponds to the same UL HARQ process as the initial transmission. The HARQ RTT is not fixed but depends on the RN subframe configuration. Since both the donor eNB and the RN will use all the Un UL HARQ processes sequentially, there is no misunderstanding between the donor eNB and the RN with respect to which Un UL HARQ process to use. Therefore, the UL HARQ process ID is not required to be indicated by the R-PDCCH, and there is no fixed relation between subframe number and the Un UL HARQ process ID.

For TDD frame structure, since there is only one or two Un UL subframes that may be configured, the number of Un UL HARQ processes can be easily determined to be one or two, respectively. A Un UL HARQ retransmission will occur in a subframe with the same subframe number as the original transmission whenever applicable.

Finally, as the RN cannot receive eNB downlink control information in the non-MBSFN region of each MBSFN, it will not expect HARQ feedback on the physical HARQ indicator channel (PHICH) in Un downlink subframes. The RN will deliver an ACK to its higher layers for each transport block transmitted on the R-PUSCH. Consequently, every Un UL HARQ retransmission has to be triggered by an R-PDCCH.

5.3.5.3 Access Downlink HARQ

Considering that there is no difference between relay UEs and macro UEs from the UE's perspective, downlink HARQ processing on the access link will be compatible with LTE-Rel-8 as much as possible.

For FDD, as defined in Rel-8, if the RN transmits the PDSCH in Uu downlink subframe k, its corresponding feedback will be sent by relay UE in Uu uplink subframe k+4. Since the 4-ms interval is exactly the same as the interval between a Un downlink subframe and its corresponding feedback Un uplink subframe, it is guaranteed that whenever the RN transmits the PDSCH in a subframe k that is not used for backhaul downlink transmission, subframe k+4 must be available for Uu uplink transmission. Therefore, Uu downlink HARQ processing is easy to follow in the same way as in Rel-8. Also, although it is possible that subframe k+8 is not available for HARQ retransmission, it can be handled by the asynchronous nature of downlink HARQ procedure.

For TDD, when an uplink subframe is configured as the Un uplink subframe, it is not available anymore for a relay UE to transmit Uu downlink HARQ feedback. That will lead to losing the feedback for some Uu downlink transmissions, as illustrated in Figure 5.34. In the example, subframeConfigPatternTDD is set to 6, which means TDD configuration 2 is used; meanwhile, subframe 3 is configured as the Un downlink subframe and subframe 7 is configured as the Un uplink

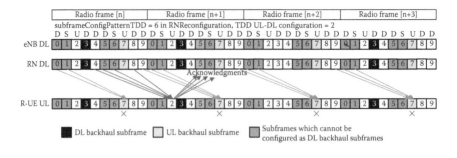

Figure 5.34 ACK/NACK not available on uplink access link.

subframe. It can be seen that because no uplink transmission on the access link is allowed in subframe 7, downlink transmissions in three subframes of each radio frame on the access link cannot receive any downlink HARQ acknowledgment.

Although the RN downlink scheduler can restrict downlink transmissions in such subframe(s), it is less resource effective. Another way to handle this scenario is to assume all feedback for such downlink transmissions are ACK, and perform RLC retransmissions if necessary.

5.3.5.4 Access Uplink HARQ

On the Uu link, due to desired backward compatibility, the uplink HARQ timing should be identical to that in Rel-8. The synchronous uplink HARQ process of a relay UE can be scheduled by physical hybrid ARQ indicator channel (PHICH) for nonadaptive retransmission or by PDCCH for adaptive retransmission. The PHICH and PDCCH can be transmitted by the RN even in the MBSFN subframes.

However, for FDD frame structure, although the HARQ acknowledgment can be received, the subframe where the retransmission should take place may be used by the backhaul link and not be available for the access link. Under this condition the corresponding uplink HARQ process retransmission needs to be suspended; otherwise the synchronous nature of the uplink HARQ process will be destroyed. Suspension of an uplink HARQ process can be achieved by transmitting a positive acknowledgment on the PHICH, irrespective of the PUSCH decoding result. A retransmission can be scheduled by explicit PDCCH for the same HARQ process in a later subframe available for access link uplink transmission.

For TDD frame structure, the 10-ms RTT makes it certain that if the initial transmission of an uplink HARQ process occurs in a uplink subframe (not a Un uplink subframe) on the access link, its retransmission is always able to happen after 10 ms in the same subframe number of the next radio frame, so no extra specification work is required.

5.3.6 Relay Procedures

Because the RN behaves like an eNB from the relay UE's perspective, and like the UE from the DeNB's perspective, some relay-specific procedures exist in relay operations.

5.3.6.1 RN Startup Procedure

Since relaying was introduced after LTE Rel-8, some MMEs in already established LTE networks may not support relay functionality. To help the DeNB find an MME with relay functionality during the RN attach procedure, the MME provides an "RN support indication" to the DeNB at S1 setup.

The RN startup procedure is based on the normal UE attach procedure and consists of two phases.

In the first phase, the RN attaches to the Evolved Universal Terrestrial Radio Access Network (E-UTRAN) evolved packet core (EPC) as a UE at power-up and retrieves initial configuration parameters (e.g., a list of DeNB cells from the RN operation and maintenance [O&M]). After this operation is complete, the relay node detaches from the network as a UE and triggers phase II, as shown in Figure 5.35.

In the second phase, as shown in Figure 5.36, the RN connects to a DeNB selected from the list acquired during phase I to start relay operations. For this purpose, the normal RN attach procedure is applied. The procedure is the same as the normal UE attach procedure with the exception that the SGW/PGW functionality is performed by the DeNB and that during the RRC connection setup the RN signals an RN indicator to the DeNB. The "RRC Connection Setup Complete" message sent from the RN to the DeNB includes "rn-SubframeConfigReq" information element (IE), which indicates that a connection is established for the RN

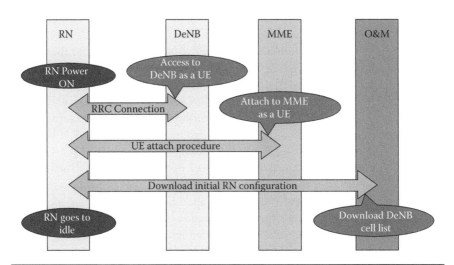

Figure 5.35 Phase I: Attach for RN preconfiguration.

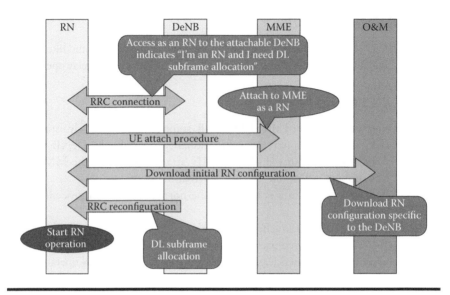

Figure 5.36 Phase II: Attach for RN operation.

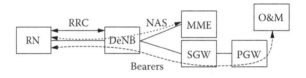

Figure 5.37 RN operation as DeNB.

and whether the RN subframe configuration is required. The MME indicates to the DeNB that the RN is authorized to attach as a relay node. After the DeNB initiates the setup of the bearer for S1/X2, the RN initiates the setup of S1 and X2 associations with the DeNB.

After all Un configuration is in place, the RN will start operating its eNB functionality broadcasting over Uu. In general, there are three steps for RN setup: RRC setup, NAS setup, and bearer setup, as illustrated in Figure 5.37. The RN will also add and configure the DRB for the S1/X2 message exchange. The QCI/QoS for the S1/X2 bearer may be assigned by the network operator via O&M. The RN will initiate SCTP setup after the DRB is set up.

The DeNB is aware of which RNs require specific subframe configurations and initiates the RRC signaling for such configurations. After an RN becomes aware of its own need for such a specific subframe configuration, at the last step during phase II, the RN indicates this need to the DeNB, which may initiate the RN reconfiguration procedure for such configuration if needed. The RN applies the configuration immediately upon reception.

The RNReconfiguration message includes the RN subframe configuration and/ or system information updates from the DeNB. The RN reconfiguration procedure is applicable only for half-duplex RN in the RRC_CONNECTED state. Through the RNReconfiguration message, the DeNB configures subframes during which Un UL and DL transmissions may take place. The RN will configure Un DL subframes to be MBSFN subframes on the Uu. The Un subframes can be reconfigured based on the traffic load.

After the RN has started operating like an eNB, UEs may connect to the RN over Uu. At this time, the RN will run S1-MME (to an MME), S1-U (to a SGW), and X2 for signaling and bearer traffic related to UEs connecting to the RN over Uu. The DeNB itself acts as an S1 proxy and an X2 proxy for this signaling and bearer traffic. The DeNB hosts an MME proxy function that terminates the SCTP from the RN (corresponding to S1 signaling traffic for UE connected to the RN), and hosts an SGW proxy function that terminates GTP-u from the RN (corresponding to S1-U bearer traffic for UE connected to the RN).

5.3.6.2 System Information Update

An RN configured with an RN subframe configuration does not need to apply the system information acquisition and change monitoring procedures. Upon change of any system information (SI) relevant to an RN, the DeNB transmits the SI blocks containing the relevant SI to the RN configured with an RN subframe configuration via dedicated signaling using the RNReconfiguration message. For RNs configured with an RN subframe configuration, the SI contained in this dedicated signaling replaces any corresponding stored SI and takes precedence over any corresponding SI acquired through the system information acquisition procedure. The dedicated SI remains valid until it is overridden.

On the relay Uu link, SI can change only at the modification period boundaries. Notification of SI changes in modification period n+1 has to be provided to the UE in period n. Therefore, subframe configuration on the Un interface and the Uu interface can temporarily be misaligned, as illustrated in Figure 5.38, where a new subframe configuration can be applied by the RN on Un earlier than on Uu, since Un subframe reconfiguration is activated immediately on Un by the RN.

5.3.6.3 Radio Link Failure on Un

It is reasonable to assume that the backhaul link is very stable, but radio link failure (RLF) cannot be avoided completely. The RLF detection procedure for the RN is the same as for normal UE.

When Un link failure occurs, the RN will fall back to UE mode (not go into the RN's RN mode). The RN will release the RN subframe configuration and release all the connections from UEs attached to the RN, while trying to recover the backhaul link to the eNB by using the physical random access channel (PRACH),

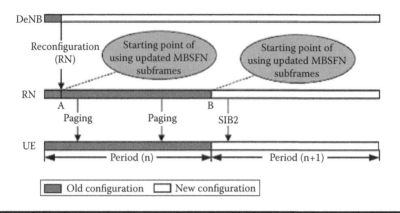

Figure 5.38   System information update in RN.

and the initial access procedure as a normal UE. Before the Un link is completely restored, the RN will not accept any UE connection.

The RRC connection reconfiguration procedures and Un subframe reconfiguration procedure after successful RRC connection reestablishment should be the same as procedures used during initial RN startup. If reestablishment fails, the RN goes to IDLE and tries to recover.

5.3.6.4 QCI Mapping on Un

In LTE-A, a total of eight different bearers will be supported on the backhaul link between the RN and the DeNB (Un interface) to carry different types of traffic, including S1/X2AP control signaling, O&M control signaling, user data traffic at different QoS levels (i.e., QCIs).

Due to the limited number of DRBs on the Un, it must be possible, especially for user data traffic, to map multiple Uu bearers of different QCIs to a single Un bearer (QCI mapping). QCI mapping can be controlled by the operator (O&M).

Generally, Uu bearers with the same QCI are multiplexed on a single RN bearer with the appropriate QoS level. Bearer differentiation at the demultiplexer is achieved via the differentiated services code point (DSCP) field in the GTP header. QCI-to-DSCP mapping is provided to the RN via O&M.

The eNB part of the RN assigns a DSCP value to uplink packets, based on the QCI of the Uu bearer that delivered the packet. Subsequently, the UE part of the RN applies traffic flow template (TFT) filtering to uplink packets and map them to a Un bearer.

The RN and DeNB should have a common understanding of QCI mapping and knowledge of the Uu to which the QCI are mapped. A Un QCI may assist the DeNB with UL scheduling. O&M configures a QCI-to-DSCP mapping table at the relay node, which is used to control the mapping in the uplink of multiple Uu bearers of

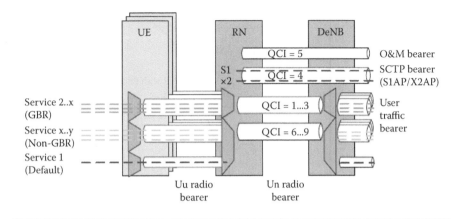

Figure 5.39 Bearer mapping on the backhaul link. (GBR—guaranteed bit rate)

different QCIs to a single Un bearer. Figure 5.39 shows an example of the mapping with three different classes of user data traffic mapped to two classes of Un traffic.

5.4 Relay Performance Analysis

As we have discussed, relay nodes can be used to extend coverage or to enhance the cell throughput of the macro eNB. Type 1 relay nodes are inband with their own cell ID, and the deployment of Type 1 relay nodes is similar to that of the macro and micro overlaid system with the exception of wireless backhauling. The inband wireless backhauling of relay deployments set additional constraints of TDM TX/RX at the relay node. Type 2 relays do not have their own cell ID and can therefore be used to enhance throughput within the coverage area of the macro eNB.

Type 1 relay nodes also could be deployed as moving nodes, such as on trains and airplanes, which would extend service to LTE-A users in high-speed environments and remote areas without wireline backhaul connections. However, the deployment of Type 1 relay nodes presents challenges in system design in the areas of interference management, downlink control channels, and resource management of backhauling in network operations. In this section we focus on performance analysis of the Type 1 relay.

Generally, in LTE-A networks, relaying is a key to offering potential capacity and coverage enhancement. With proper interference management, the network capacity could be substantially improved with a high relay density. On the other hand, adding RN(s) to the network will decrease capacity from the DeNB Uu interface. It is important to quantify this reduction. The additional load on the Un is created by performance management data from the RN, additional NAS and RRC signaling, and the increase in cell capacity due to the additional RN(s).

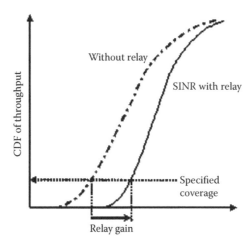

Figure 5.40 Calculation of relay gain. (CDF—cumulative density function)

A preferred method to calculate relay gain is illustrated in Figure 5.40. First we increase the cell radius until the SINR of the multihop network at the specified coverage is the required value. The relay gain is the difference between the cumulative density curves of SINR with and without relays at the specified coverage.

The radio performance, physical parameters, and basic characteristics of a relay node may differ depending on the deployment scenario for outdoor or indoor coverage improvement. The radio performance of the relay node is significantly affected by the deployment scenario, coexistence with other systems, and the other RNs within the same donor cell.

5.4.1 Fixed Relay Gain

Assume that the access link (eNB–UE and RN–UE) and backhaul link (eNB–RN) are operated in the same band, an intersite distance (ISD) 500 m case (3GPP Case 1) is simulated. For the RN deployment scenario, the evaluation is made for a fixed RN deployment where RNs are deployed close to the macrocell edge. Figure 5.41 shows the average user throughputs of a fixed deployment scenario in which we can see there are cell throughput gains in relay systems compared to a system without relays. When the number of backhaul subframes is more than 2, in the case with one RN there is about an 8% gain, and in the two- and four-RN cases 14–17% gains are obtained. The aggregated cell throughputs slightly increase when the number of backhaul subframes is increased.

5.4.2 Backhaul Link Performance

It is important to have good inband backhaul link quality in order to realize meaningful performance improvement with relays and to avoid poor backhaul links

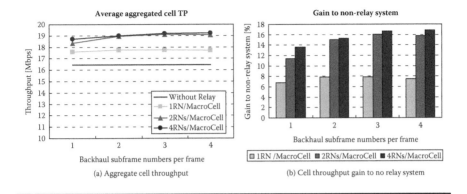

(a) Aggregate cell throughput

(b) Cell throughput gain to no relay system

Figure 5.41 Cell throughput in fixed RN deployment scenario. (a) Aggregate cell throughput. (b) Cell throughput gain to no relay system.

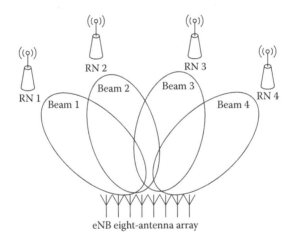

Figure 5.42 SDMA between the eNB and multiple RNs.

becoming bottlenecks. It is worth mentioning that even though resource reuse in an access link greatly increases the total throughput of the access link, the total throughput is the minimum of the backhaul and access link. So it is necessary to improve backhaul link and cell-edge performance through different techniques, such as interference coordination and mitigation techniques.

Backhaul link spectrum efficiency should be at least 5 bps/Hz for good relay performance. Spatial division multiple access (SDMA) is a promising technique for improving the spectrum efficiency of the backhaul link in relay systems. Multiple beams are allocated to the different RNs, which also increases the interference level in the backhaul link, but can provide multiplexing gain. Figure 5.42 is an illustration of SDMA between an eNB and multiple RNs.

5.4.3 Interference of Relay

Type I relays can be used to extend coverage and improve capacity with inband backhaul, but these benefits come with the cost of additional interference. The interference can be categorized as interference between RNs and UEs served by the eNB, interference from the eNB to UEs served by RNs, and interference between the RN and UEs served by other RNs.

Interference may significantly impact the performance of a relay network. It will lead to performance degradation, especially regarding MIMO, and shrink the RN coverage due to degradation of link quality in the access channel.

For a Type I inband relay, the data interference may be reduced by coordinating the schedulers at the eNB and RN. For control signal, the interference can be mitigated through reasonable cell planning or employing interference coordination technologies such as collaborative power allocation. For the Type 2 relay, it can work more flexibly through scheduling at the eNB (e.g., cooperation or non-cooperation mode); it is possible to mitigate interference by properly selecting operation mode in different scenarios.

The RN can be placed deep in the coverage of the eNB to support coverage holes, hot spots, or temporary needs. Most of the UEs within the RN's coverage area have better visibility toward the RN than the eNB and consequently are aided by the presence of the RN. But some UEs, under the control of the RN or the eNB, have good visibility to both the eNB and the RN and experience high interference from either the RN or the eNB, as illustrated in Figure 5.43.

Interference may also exist between RNs due to the concurrent backhaul and access link transmissions, as illustrated in Figure 5.44. In the uplink, the aggressor

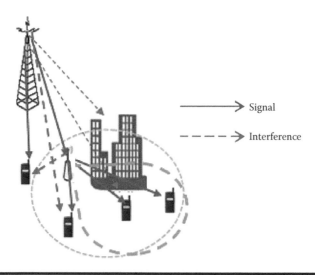

Figure 5.43   Interference from the eNB to UE served by RNs.

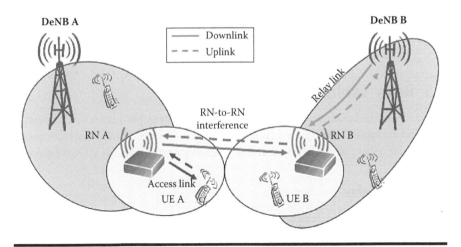

Figure 5.44 **Backhaul: access (RN to RN) interference.**

RN is transmitting on a backhaul link, whereas a victim UE in a nearby RN cell is transmitting on its access link. Meanwhile in the downlink, the aggressor RN is transmitting on the access link to its UE, whereas a victim RN is receiving a backhaul link transmission from its DeNB. This kind of interference can be much higher than the received signal; therefore, the transmission on the backhaul and access links needs to be coordinated. In addition to network geometry-related parameters, such as path loss, transmit power, and inter-RN distance, downlink RN-to-RN interference depends mainly on two factors: subframe allocation and subframe timing. The latter can be different for FDD and TDD systems. For example, TDD requires more stringent timing between neighboring cells. For FDD, the network can be operated in either asynchronous or quasi-synchronous mode. In asynchronous FDD, timing offset between neighboring eNBs is arbitrary (i.e., uniformly distributed over time). For quasi-synchronous FDD, the synchronization tolerance can be one OFDM symbol. The subframe timing mentioned here is the combination of DL timing schemes and the propagation delay difference of interfering RNs.

There is also interference on physical signals and channels. The data channel from the eNB will interfere with the RS from the RN and vice versa. This effect will be magnified in UEs that can clearly hear from the eNB. SCH, physical broadcast channel (PBCH), and PDCCH also interfere with each other.

Interference management can play an important role in ensuring that gains from the deployment of relays are maximized. Usually RN-to-RN interference has a small impact on backhaul link capacity when the RN-to-RN propagation environment is non-line-of-sight (NLOS). Current solutions orthogonalize interference by static (or semistatic) partitioning of resources between the eNB and the RNs. Scheduling of transmissions should guarantee that eNB–UE and RN–UE links that do not interfere can use the same resources, and that eNB–UE and RN–UE links that interfere will use orthogonal resources.

5.5 Relay Evolution

To maximize the benefits of relaying, much effort has been put into relay technology evolution, including carrier aggregation (CA) relay, mobile relay, multihop relay, cooperative relay, relay pairing schemes, and relay enhancement with MIMO technology.

5.5.1 CA Relay

Up-to-date studies on relays focus mainly on single-carrier inband scenarios, and specification work and performance evaluation in LTE-A is focused on inband operation. In general, the bottleneck of relay performance is the backhaul capacity, and carrier aggregation is an approach to improving the backhaul link capacity. For example, an operator has more than 20 Mhz available in one frequency band for backhaul or spectrum in multiple bands, so carrier aggregation over the Un interface should be investigated with both inband and outband relay configurations.

The configurations in Figure 5.45 have been the main interests for CA-over-Un studies.

5.5.2 Mobile Relay

High-speed public transportation is being deployed worldwide at an increasing pace. Therefore, providing multiple services with good quality to users on high-speed vehicles is important but more challenging than typical mobile wireless environments due to reduced handover success rate and degraded throughput due to high Doppler effects. Mobile relay (a relay mounted on a vehicle wirelessly connected to the macrocells) is a potential technique to solve these problems.

Moving relays can experience frequent handovers between different DeNBs. Radio conditions on the backhaul link to a moving RN can change significantly

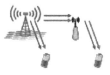

In-band relay: in-band on both CCs

Out-band relay: out-band on one CC

Mixed relay: in-band on one CC and out-band on the second CC

Figure 5.45   CA over Un.

as the vehicle moves across the DeNB coverage area. The backhaul link is volatile, whereas access links are stable and of a good quality. Moving relays mounted on a high-speed train have gained attention recently, and the following issues need to be considered:

- The Doppler frequency shift
- 20–30 dB penetration loss due to the train carriages
- The low handover success rate

Mobile relay (see Figure 5.46) can improve the handover success rate by performing a group mobility procedure instead of individual mobility procedures for every UE and can also improve spectrum efficiency by exploiting more advanced antenna arrays and signal processing algorithms than normal UEs. In addition, separate antennas for communication on the backhaul and access links can be used to effectively eliminate the penetration loss through the vehicle.

5.5.3 Multihop Relay

Increased system capacity can be achieved through the use of multihop links, as opposed to a single-hop link. The intention of multihop links was to reduce the transmitter-to-receiver distance, to achieve higher data rates as compared to long single-hop links. Furthermore, multihop relays can improve performance in dead spots and shadowing, and support spatial reuse, which can lead to increased overall system capacity.

The multihop relaying architecture for LTE-A networks is composed of one or more relays between an eNB and a UE. A single eNB serves one or more multihop

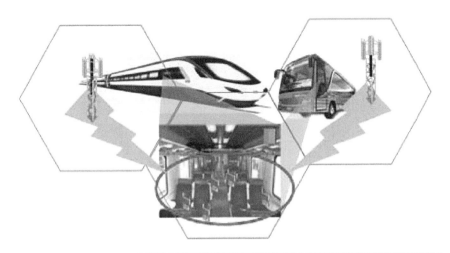

Figure 5.46 Moving relay to group mobility.

chains in its cell, which forms a tree structure with the eNB as the parent. Multihop relays are expected to have an impact on relay architecture and protocols.

5.5.4 Cooperative Relay

From a system-level point of view, relaying can be performed in either a conventional or cooperative/collaborative fashion. In conventional relaying, the UEs receive data from the serving eNB or the RN. In collaborative relaying, on the other hand, the eNB can multicast the data to cooperative RNs, which in turn forward them to the destination UE, where they are combined (possibly with the data directly received from the eNB). Both amplify and forward (AF) and decode and forward (DF) RNs can be used in conventional and cooperative systems. The conventional relaying scheme is preferred for its simplicity.

When two-hop relay transmission is adopted, an adaptive cooperative relay scheme can be used according to current user density in a certain cell. The cooperative operation may be realized with parallel transmission between the eNB and an RN, or by adopting parallel transmission between two RNs (as illustrated in Figure 5.47). By fully utilizing cooperative communication, it can effectively increase user capacity as well as improve system performance.

5.5.5 Relay Pairing

In a network with multiple RNs and multiple UEs in each cell, having collaboration between RNs and eNBs in data transmission is important to improve the coverage and throughput of multihop relay networks. Therefore, it is very important to develop effective pairing schemes to select appropriate RNs and UEs to collaborate in relay transmissions.

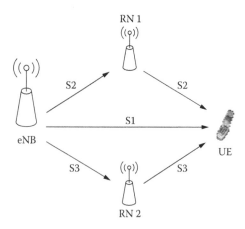

Figure 5.47 Cooperative relay.

Some pairing schemes have been developed. With the simplest pairing scheme, an RN will randomly select any UEs in its service coverage without considering such things as UE location, gains, or achievable data rates in its decision making.

Another so-called opportunistic pairing scheme is suitable for DF and AF relaying under an aggregate power constraint. The scheme reveals that relays are useful even when they do not actively transmit, provided that they adhere to the opportunistic cooperation rule and give priority to the "best" available relay, which is chosen naturally to maximize the instantaneous channel strength.

The other two pairing schemes are the centralized pairing scheme and the distributed pairing scheme, which imply that the pairing procedure can be executed in either a centralized or distributed manner. In a centralized pairing scheme, an eNB will serve as a control node to collect the required periodic report from the RNs including the CSI for RN–UE and RN–eNeB links. The eNB will, based on the link quality information, construct a pairing and broadcast it to all RNs and UEs under its coverage and can maximize the number of served UEs. However, RNs feeding back CSI for all the links to the eNB causes huge overhead in the network and leads to a latency problem.

However, in a distributed pairing scheme, each RN selects an appropriate UE by using local channel information and a contention-based MAC mechanism. The feedback overhead is reduced since there is no need for the RN to report the CSI of the RN–UE link. Nevertheless, optimization for resource allocation cannot be achieved due to the nonexistence of a central node to efficiently control and coordinate resource usage among all the links in the network.

Generally speaking, the centralized scheme requires more signaling overhead, but can achieve better performance gains than the distributed scheme.

5.5.6 Relay Enhancement

Downlink data transmission for relay links could have closed loop spatial multiplexing for stationary or slow-moving relay nodes, open loop spatial multiplexing for relay nodes on high-mobility objects such as a train, or fixed or adaptive beamforming with single or multilayer transmission, which could be semistatically configured by higher-layer signaling. For relay nodes that are stationary or moving relatively slow, optimization could be considered for data transmission on a relay link with the aspects such as codebook-based precoding optimizations, CQI and PMI feedback granularity in both time and frequency, adoption of a larger codebook, or non-codebook-based precoding. Some advanced techniques could also be considered for relay link transmission such as high-order MIMO with more than 4x4 MIMO, CoMP, and multilayer beamforming.

Chapter 6

Self-Organizing Network

Today, in order to build and maintain highly reliable high-performance networks, we have to face the challenge of the mobile radio networks and complex operational tasks. With the introduction of Long Term Evolution (LTE) technology, operators have the opportunity to optimize their network management operations and reduce their operational expenses. The self-organizing network (SON) is the technical solution to achieve management of complexity in networks and thus provide operating expense (OPEX) reduction and revenue protection. The SON is the combination of several functions in different areas (configuration, optimization, healing).

In order to minimize preprovisioning during the deployment of LTE networks, it is necessary to automate the configuration of network parameters. Three approaches are being considered based on the location of the functions: centralized, decentralized, and hybrid (shown in Table 6.1 and Figure 6.1). For centralized implementation, the evolved Node-B (eNB) assigns parameters to each of its cells, which are allocated by the operation and maintenance (O&M) system. For decentralized (or distributed) implementation, it will be possible for the eNB to assign parameters from the O&M system and then automatically configure each of its supporting cells. For hybrid implementation, parts of the optimization algorithms are executed in the O&M system, while others are executed in the eNB. This brings together the best of both centralized and decentralized architectures and is flexible in supporting different kinds of optimization cases.

Encompassing aspects of network optimization include automated base station deployments, physical cell ID (PCI) assignments, self-optimization via antenna tilting, traffic load balancing, dynamic reconfiguration of network planning, and automatic repair functions. The main SON functionalities consist of self-configuration, self-organization, and self-optimization mechanisms that are collectively referred

Table 6.1 SON Frame Structure Category

Centralized	Distributed	Hybrid
• The SON system is centrally executed at the network management level; all data has to be forwarded on a central level • Multiple cells involved • All data flow is into and out of the network management level; longer update • Simpler multivendor solution • Stable and easy to implement and upgrade • Addresses multiple cells	• The SON system is distributed in each node; however, cells communicate with each other • Scalable; fast and flexible updates • Short-term statistics • Lower backhaul impact • Low latency, high availability • Addresses only one cell	• The SON is executed partly at the network operations level and partly at the cell level • Advantages and disadvantages of previous solutions • Low latency, high availability • Requires additional interface definition • More complex handing

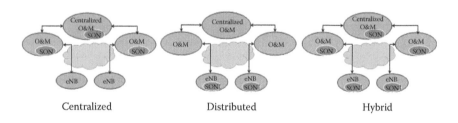

Figure 6.1 SON frame structure.

to as *SON capabilities* and are expected to improve operational efficiency and provide enhanced network performance. A detailed description of SON functionalities is provided in Figure 6.2.

- **Self-configuration:** Self-configuration involves automated network integration of new eNBs by automatic connection and automatic configuration, core connectivity (S1), and automatic neighbor site configuration (X2).
- **Self-optimization:** Self-optimization involves automated tuning of the network with the help of the user equipment (UE) and eNB measurements on the local eNB level and/or network management level.
- **Self-healing:** Self-healing involves automatic detection and localization and removal of failures.

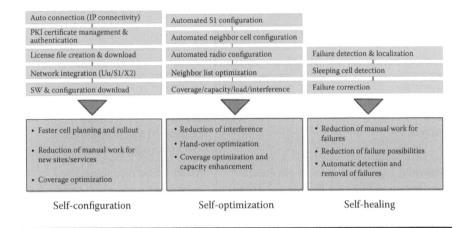

Figure 6.2   Main SON functionalities.

LTE provided the catalyst for the development of the SON, work on which was initiated by the Next-Generation Mobile Networks (NGMN) Alliance. The standardization of the SON is still in process within the Third-Generation Partnership Project (3GPP). A number of SON features are supported in 3GPP Rel-8 and will expand in scope with subsequent releases. LTE Rel-10 will also contain enhancements to the technologies introduced in Rel-8 and Rel-9. The requirements for the SON are set to be more fully defined in Rel-9 and Rel-10 and are listed in Table 6.2.

6.1 Self-Configuration

Self-configuration is a broad concept that involves several distinct functions that are covered through specific SON features, such as automatic software management, self-testing, and automatic neighbor relation configuration. During power-up, the eNB discovers all installed hardware and runs appropriate hardware tests to validate that the hardware is functioning correctly. Alarms will be triggered indicating the nature of each failure associated with any failed test. The eNB will support plug-and-play inventory management of all equipped hardware. On installation and power-up, the eNB is able to identify equipped boards and inventory, and configuration information will be updated accordingly without requiring manual reconfiguration or installation actions. On removing or inserting new boards (e.g., baseband), the eNB inventory and configuration data will also be automatically updated without manual intervention. All changes to hardware configuration will be reported automatically to the operation and maintenance center (OMC) on initial connection and then on an ongoing basis driven by either hardware changes at the eNB or on request from the OMC.

The self-configuration algorithm should take care of all soft configuration aspects of the eNB once it is commissioned and powered up for the first time.

Table 6.2 SON Standardization in 3GPP Rel-8, Rel-9, and Rel-10

Rel-8		Rel-9 and Rel-10
Self-Configuration	*Self-Optimization*	*Self-Optimization*
Dynamic configuration of S1 (eNB–core NW) interface	Basic mobility load balancing	Coverage/capacity optimization: Optimization of system parameters to maximize (adjust to the desired balance between) system coverage and capacity
Dynamic configuration of X2 (inter-eNB) interface	Load-information exchange between eNBs over X2 interface for interference mitigation	Mobility load balancing: Optimization of cell reselection/handover parameters to distribute traffic load across the network
Framework for PCI (Physical cell ID) selection	Standardized eNB measurements for multivendor SON	Mobility robustness optimization: Optimization of cell reselection/handover parameters to minimize radio link failures due to mobility
Automatic neighbor cell discovery		Common channel configuration optimization: Optimization of common channel configuration, such as random access channel configuration based on eNB measurements
Self-configuration of eNBs; automatic software management;		Minimization of drive tests: Logging and reporting of various measurement data (e.g., location information, radio link failure events, and throughput) by the UE and collection of data in a server to minimize the number of drive tests run by operators

It should detect the transport link and establish a connection with the core network elements, download and upgrade the corresponding software version, set up the initial configuration parameters including neighbor relations, perform a self-test, and finally set itself to operational mode. In order to achieve these goals, the eNB should be able to communicate with several different entities, as depicted in Figure 6.3.

To be able to successfully achieve all functions, the following prerequisites should be met prior to the installation of the new node:

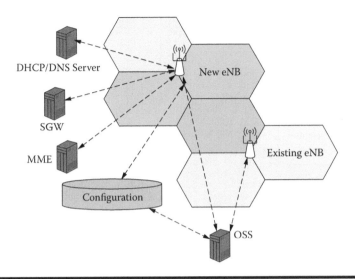

Figure 6.3　Self-configuration of eNB.

1. A network planning exercise for the cell should have been completed resulting in a set of radio frequency (RF) parameters, including location, cell identities, antenna configuration (height, azimuth, and type), transmit power, maximum configured capacity, and initial neighbor configuration. This information should be made available in the configuration server.

2. The transport parameters for the eNB should be planned in advance, including bandwidth, virtual local area network (VLAN) partition, IP addresses, and so on. The IP address range and serving gateway address that correspond to the node should be made available in the configuration server.

3. An updated software package download should be made available from the operations support system (OSS).

6.1.1　Automatic Neighbor Relations

The primary purpose of automatic neighbor relations is to automatically learn the required information about neighbor cells and eNBs without requiring any interaction or planning on the part of the network operator. The basic approach is that in LTE, unlike in previous technologies, the UE does not have a list of neighbors, but is always looking for other cells that are candidates for handover. Whenever it finds one, it reports the physical cell ID of the candidate to the eNB. If the eNB is aware of the reported cell, then normal handover decisions will proceed. Otherwise, a radio resource control (RRC) procedure (system information acquisition) is initiated by the UE to obtain the global cell ID for the reported cell.

The neighbor relation table (NRT) provides each cell with a list of suitable neighbors for handovers (HOs) for the mobile terminals it serves. Currently, these NRTs

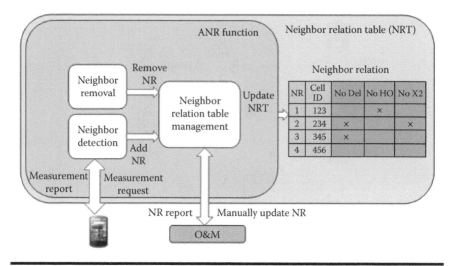

Figure 6.4 Interaction of eNB and O&M for automatic neighbor relation.

have to be set up manually, supported only by network planning tools and costly drive tests. In LTE, the system will generate NRTs for the cells it serves by triggering selected UE for neighbor cell measurements and will interpret the reported results. This function is called automatic neighbor relation (ANR) detection. ANR is based on measurements from real UEs and manages intra-LTE neighbor cell relations; the feature can add and delete neighbor relations (NRs) and is shown in Figure 6.4.

The automatic neighbor discovery mechanism of ANR allows an eNB to learn information on its neighbors. Neighbor cell detection optimization can be controlled by neighbor cell signal strength threshold. The discovery mechanism can utilize the assistance of the UE as well as the exchange of information over network interfaces.

The eNB will also set up an X2 connection to the target eNB so that subsequent handovers can be performed directly. To do this, the eNB needs to learn the X2 address of the target eNB, which it does by sending an eNB interaction message to the mobility management entity (MME) to return its IP address for X2 interfaces.

Autointegration and ANR are the earliest recognized features that all vendors currently have. Specific demands, multivendor multitechnology scenarios, and femtocells will be the long-term challenges.

The ANR function resides in the eNB and manages the conceptual neighbor relation table (NRT*). For each cell, the eNB has and keeps an NRT. For each

---

* This table gathers all the NR of a given cell, or all the NR of the cells of a given eNB. It is updated dynamically (relations can be added or removed) thanks to ANR functionality. Neighbor relations include the NR white list and black list. The NR white list contains the NRs that cannot be removed from the NRT. This list is built by the operator and then comes from the O&M. The NR black list contains the NRs that cannot be used for mobility (HO) purpose. This list is also built by the operator and then comes from the O&M.

NR, the NRT contains the target cell identifier (TCI), which identifies the target cell. For the Evolved Universal Terrestrial Radio Access Network (E-UTRAN), the TCI corresponds to the E-UTRAN cell global identifier (ECGI) and the physical cell identifier (PCI) of the target cell. The PCI is used by the physical layer to identify and separate data coming from different transmitters. The ECGI is a system-level parameter that identifies the cell. Furthermore, each NR has three attributes: the NoRemove, the NoHO, and the NoX2, as described below.

■ NoRemove = White list of neighbor cells, ANR will not remove this relationship
■ NoHO = Black list of neighbor cells, ANR will not add this relationship
■ NoX2 = NoX2 link; handover is undertaken via the S1 link

ANR will also support and allow flexible preplanning and an automated configuration and update of neighbor cell information, which will create key operational benefits such as reduced maintenance and improved performance with no need for neighbor relation planning for new sites, reduction in dropped calls, and so on.

6.1.1.1 Physical Cell Identifier

Automatic configuration of the physical cell ID is one of the first features targeted in implementation of self-organizing and self-optimizing functionality. This is important for reducing the amount of preplanning and provisioning that the operator has to do for an LTE network, and also for reducing the replanning needed as additional eNBs are deployed. The PCI is used in the LTE network as a way for mobile devices to distinguish among different cells. There are 504 available PCI numbers but an LTE network may contain up to 30,000 cells. The PCIs are grouped into 168 unique physical-layer cell-identity groups, each group containing three unique identities. So the PCI can be expressed by: $PCI = 3j + k$, where group number $j = 0 \dots 167$; $k = 0 \dots 2$ (shown in Figure 6.5). Thus, the same PCI must be used by several cells. However, a mobile device cannot distinguish between two cells if they both have the same PCI. The physical layer cell ID collision happens when eNBs with the same cell ID exist within the detection range of the UE. When the PCI collision happens, the UE receives the superpositioned primary synchronization

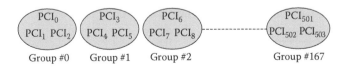

Figure 6.5 Physical layer cell-identity groups.

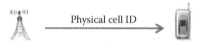

Physical cell ID

Figure 6.6   PCI acquisition.

signals (PSS) and secondary synchronization signals (SSS) of synchronization channels from multiple eNBs. The UE decodes the superpositioned PSS/SSS from multiple eNBs as if they were a multipath channel from a single eNB. The UE will further listen to the physical broadcast channel (PBCH) to decode the required system information for the cell. The UE has a high probability of failing the PBCH decoding and being unable to gain further access to the system. The PCI is widely used in the mobility procedure. The PCI allocation/optimization is a use case and service of the SON.

The LTE cell identifies itself on the radio interface in two ways. The LTE cell identifies itself first with the PCI (see Figure 6.6), which can be retrieved from PSS and SSS decoding. This ID has a limited range of 504 distinct values and is therefore not unique. As the PCI is the primary anchor point for a UE camping in a cell, the value must be unique within the coverage area of the cell itself as well as all neighbor cells that could be received by a UE. Radio network planning must guarantee the proper reuse of the same PCI without any conflicts.

The LTE cell also identifies with the global cell ID (GID), which is broadcast as part of the system information block. The system information block is part of system information block 1, which sent out by a longer repetition cycle of 80 ms and repeated every 20 ms. This ID should be unique in the network with respect to the public land mobile network (PLMN), the eNB, and cell identity.

The LTE SON system should automatically assign physical cell IDs for each of its supporting cells, ensuring that each ID is unique when compared with neighbor* cells and neighbors' neighbor cells. In the meanwhile, PCI assignment rules should avoid allocating the same PCI within the minimum reuse distance and signal strength threshold. During allocation of the physical cell ID it must be ensured that the assigned physical cell ID is collision and confusion free, that is, that it is not identical to immediate neighbor cells and also neighbors' neighbor cells (see Figure 6.7).

The PCI selection mechanism of the SON allows an eNB to select its PCI. The selection can be based on either a centralized or distributed PCI assignment algorithm. The eNB could have the UE cell search capability to detect the cell ID in the neighborhood. Once the given numbers of physical layer cell ID are detected,

---

* What is a neighbor? A neighbor is a topology radio neighbor cell or mobility neighbor cell. The mobility type of neighbor cell could be used as a target cell with a mobility purpose such as mobile relay nodes; collisions can occur with the existing macrocell ID when the mobile relay node passes by. The mobility neighbor cell cannot be the topology neighbor.

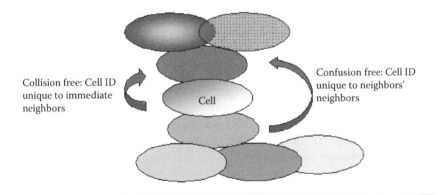

Figure 6.7   PCI assignment rules.

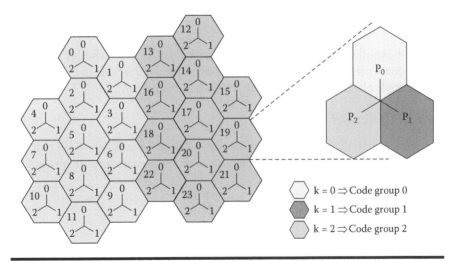

Figure 6.8   The example of PCI planning.

the eNB would select the physical layer cell ID from the remainder of the pool randomly or systematically. The eNB could also send the request to the neighbor eNB through X2 or the MME through S1 for PCI assignment. The eNB or MME could assign the PCI based on the knowledge of the existing PCI in the neighborhood.

In live network PCI planning, we can assign a color group to each sector and a code group per site. Typically 10–15 three-sector sites are in a cluster and use a subset of the code groups in each cluster. If there are around 70 code groups available, PCIs may be repeated every fifth or sixth cluster, which is shown in Figure 6.8. Structured planning like this can eliminate the risk of having conflicting k or frequency shift in the same site, in adjacent cells, or pointing at each other; the risk of having conflicting SSS sequences in adjacent cells is also reduced, although this may appear at cluster borders.

Relations between neighboring cells need to be planned carefully. Incorrect configurations can cause handover failures and dropped calls. The normal handover procedure is described as follows (refer to Figure 6.9):

1. The UE checks the serving cell NRT and launches measurement report to cell 3; when cell 3 is detected as a good cell signal, the UE sends a report to the serving cell.
2. The serving eNB checks the NRT and finds the neighbor cell list for cell 3 (ECGI = 103). If there is an X2 interface, the eNB launches a Handover Request message including the ECGI in the target cell ID to the target eNB for handover resource preparation. If there is no X2 but an S1 interface, the eNB launches a Handover Required message including the ECGI in the target ID to the MME to request preparation of the resources at the target.
3. If step 2 is successful, the serving eNB then launches an RRC Connection Reconfiguration message toward the UE with the carrier frequency, PCI, and other information of the target cell 3.

For the conflict cell, such a reconfiguration will not be successful because there is no resource preparation. Thus, there will be handover failure due to PCI conflict. One example of the PCI conflict issue is shown in Figure 6.9.

6.1.1.2 UE-Based ANR

This feature is 3GPP compliant and supports multivendor radio networks that use the UE's capability to identify neighbor cells with appropriate measurements. The

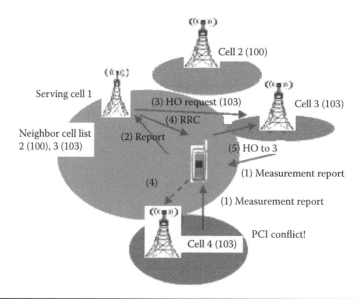

Figure 6.9   One example of PCI conflict issue.

3GPP variant of automated detection and configuration of unknown LTE cells supports self-configuration of neighbor cell information without operator planning work and requires proper measurement support from the UEs.

The self-configuration of relations avoids manual planning and maintenance as well as subsequent planning errors. The eNB uses UE measurements of neighbors to detect new neighbors as per the ANR functionality defined in 3GPP for intra-LTE and inter-RAT (radio access technology) neighbor cells. On identification of potential neighbor cells, the eNB will update its NRTs. Each NRT will be compared to preferred and nonpreferred neighbors. If preferences are determined by the operator, X2 Setup messages that include all identified neighbor relations will be sent for intra-LTE scenarios to potential neighbor eNBs. In general, intra-LTE and inter-RAT ANR procedures are completed in three steps:

1. The radio part with neighbor cell discovery
2. Neighbor site's X2 transport configuration discovery
3. The X2 connection setup with neighbor cell configuration update

6.1.1.2.1 Step 1: The Radio Part with Neighbor Cell Discovery

When a UE changes from the idle state into the connected state, it receives the measurement configuration and an instruction to report all detected/strongest cells above a given threshold. Therefore, it may report strong cells with PCIs that are not known to the eNB. Should this happen, the eNB sends a measurement request to the UE to discover and report the cell GID for the unknown PCI. The step of neighbor cell discovery for intra-LTE and inter-RAT ANR can be found in Figures 6.10 and 6.11.

The intra-LTE intrafrequency NR autodetection and optimization procedure is described as follows and is shown in Figure 6.12.

1. The UE reports the neighbor measurement to cell A; PCI = 5 is best.
2. Cell A concludes that PCI = 5 is not known.
3. Cell A orders the UE to read the CGI of cell B.
4. The UE reads the CGI broadcasted by cell B, and reports cell B's CGI to cell A. The neighbor cell is added in cell A.

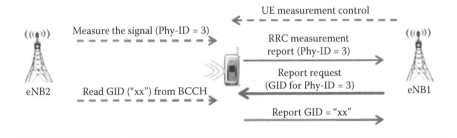

Figure 6.10 Neighbor cell discovery for intra-LTE ANR. (Phy—physical)

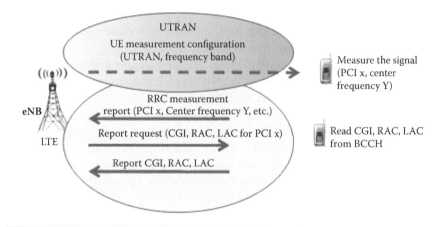

Figure 6.11 Neighbor cell discovery for inter-RAT (radio access technology) ANR (automatic neighbor relation). (RAC—routing area code; CGI—CSG [closed subscriber group] identifier; LAC—location area code; BCCH—broadcast control channel)

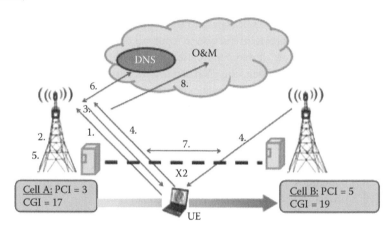

Figure 6.12 Intra-LTE NR autodetection and optimization procedure. (PCI = Layer 2 cell identity; CGI = Layer 3 cell identity)

5. Cell A checks if an X2 connection to cell B is allowed.
6. Cell A gets the IP address for cell B (CGI) from the domain name system (DNS).
7. The X2 connection is established.
8. Cell A updates the OSS parameters and observation data. O&M is notified for the NRT update.

The inter-RAT and interfrequency NR autodetection and optimization procedure is described as follows:

1. Each cell contains an interfrequency search list. This list contains all frequencies that will be searched.
2. The serving eNB instructs the UE to look for neighbor cells in the target RATs/frequencies.
3. The UE reports the PCI of the detected cells in the target RATs/frequencies in the following ways:
   - By the carrier frequency and the primary scrambling code (PSC) in the case of a UTRAN frequency-division duplexing (FDD) cell
   - By the carrier frequency and the cell parameter ID in the case of a UTRAN time-division duplexing (TDD) cell
   - By the band indicator + the base transceiver station identity code (BSIC) + the broadcast control channel (BCCH) Absolute Radio Frequency Channel Number (ARFCN) in the case of a Global System for Mobile Communication (GSM)/Enhanced Data Rates for GSM Evolution (EDGE) (GERAN) cell [SM/EDGE radio access network]
   - By the pseudo-random noise (PN) offset in the case of a code division multiple access (CDMA) 2000 cell
4. The serving eNB instructs the UE, using the newly discovered PCI to read the following information:
   - The CGI and the routing area code (RAC) of the detected neighbor cell in the case of GERAN detected cells
   - The CGI, the location area code (LAC), and the RAC in the case of UTRAN detected cells
   - The CGI in the case of CDMA2000 detected cells
   - For the interfrequency case, the ECGI, the tracking area code (TAC), and all available public land mobile network (PLMN) ID(s) of the interfrequency are detected from the broadcast channel of the detected inter-RAT/interfrequency neighbor cell.
5. The serving eNB adds the target cell to the neighbor cell list and the neighbor relation between the serving cell and target cell. O&M is notified for the NRT update.

If the UE is able to deliver the GID, the next step is to derive the IP connectivity information needed to address the neighbor eNB.

6.1.1.2.2 Step 2: Neighbor Site's X2 Transport Configuration Discovery

The basics of the neighbor site's X2 transport configuration discovery is described in this section and shown in Figure 6.13.

The derived GID cannot be used to directly address neighbor sites at the IP transport level because it is not structured as a fully qualified domain name. Therefore, the GID needs to be resolved to the transport network layer (TNL) configuration,

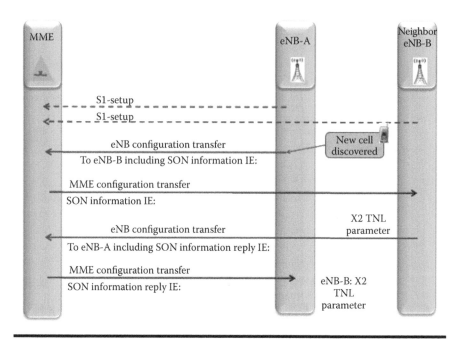

Figure 6.13 IP address resolver flow. (IE—information element)

which is needed to establish the IP–IP security (IPsec) connection with the stream control transmission protocol (SCTP) toward the new-found neighbor site.

The 3GPP specification defines an information transfer procedure* to exchange the TNL configuration and IP addresses between two neighbor sites with the help of the MME. With the combination of the eNB configuration transfer and the MME configuration transfer procedures, the two neighbors exchange their transport network layer configurations, with the MME acting as a relay (the MME is transparent). As each eNB that is connected to an MME has previously registered to the MME with the S1 setup procedure, the MME is able to find the right SCTP route to the neighbor eNB.

6.1.1.2.3 Step 3: The X2 Connection Setup with Neighbor Cell Configuration Update

The X2 connection setup with neighbor cell configuration updates is described in this section and shown in Figure 6.14.

When the TNL configuration is received, the X2 connectivity is set up and the eNB exchanges the list of served cells with its new neighbor.[†] The exchanged cell information covers all of a site's served cells and is stored in the configuration.

---

* Refer to 3GPP standard TS 43.413.
† Refer to 3GPP standard TS 36.423.

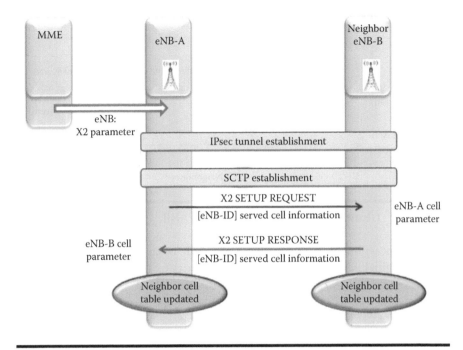

Figure 6.14 X2 setup and configuration exchange.

The UE may start reporting unknown neighboring cells when it leaves the coverage area of the serving cell (event-based reporting; event A3 or A4 for example, during intra-LTE, intrafrequency ANR, or periodic reporting with intra-LTE, interfrequency, or inter-RAT UTRAN/GERAN ANR). However, this approach raises the risk that the first handover to this newly detected cell may not be successful because the UE that reported the unknown cell is already at the serving cell's edge and may be handed over immediately following detection of the cell. If this happens, the UE may be handed over to the newly detected neighbor cell even though the X2 interface setup is incomplete. In this case the handover is highly likely to fail. To solve this, the vendor implements ANR mechanisms that can detect any unknown neighboring cell as early as possible and not when a UE is already subject to handover.

6.1.1.3 UE Measurement Requirements

A feature of UE-based ANR will use the UE's capability to identify neighbor cells with appropriate measurements. Actually, there is little difference as compared with non-SON implementation.

This feature of UE-based ANR will impact the battery power consumption of the UE that conducts the measurement and sends out reports. To reduce the battery life impact, we must minimize the amount of measurement that is required and minimize the measurement reports to be sent to the network. Most likely the

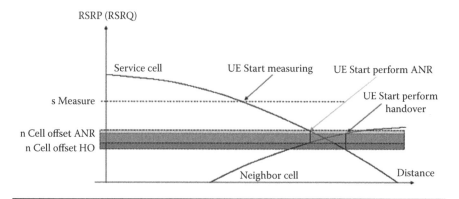

Figure 6.15   UE measurement for handover and ANR.

measurement will need to be done more frequently when the eNB is newly added or during initial network deployment. The UE actions in ANR are described as follows:

- The UE reports the PCI of the detected cell.
- If unknown to the serving cell, the UE is requested to read the GCI that is broadcast by the unknown cell and report it to the serving cell.
- The X2 between eNBs can then be set up and data exchanged.
- The serving cell may add the unknown cell to the neighbor list.

The UE measurement procedure for handover and ANR PCI reporting, and handover and ANR-specific thresholds can be found in Figure 6.15.

To be considered for CGI measurement and reporting, the neighbor cell will meet the following conditions:

$$- RSRP_{Neighbor} > RSRP_{Serving} + Offset$$
$$- RSRP_{Neighbor} > cellAdd\ RSRP\ Threshold$$
$$- RSRQ_{Neighbor} > cellAdd\ RSRQ\ Threshold$$

Where RSRP is reference signal received power and RSRQ is reference signal received quality.

6.1.2 S1 MME and X2 Dynamic Configuration

This feature is meant to ease the work in early technology trials for establishing neighbor cell relations, and the mechanism allows an eNB to establish an S1 interface toward the MME and/or an X2 interface toward another eNB. During startup of the eNB, the X2 connections are established to all planned neighbor base stations. All additional needed cell configuration information is exchanged between the requesting eNB and all responding neighbor eNBs.

6.1.3 Autonomous Component Carrier Selection

Autonomous component carrier selection is another SON feature. Once a new LTE-A eNB is switched on, it starts by selecting one of the component carriers as its primary component carrier. UEs cannot connect to the eNB before the primary carrier has been selected, and no signals are transmitted from it. The information that is available for selection of the primary component carrier is therefore mainly local eNB measurements and potential information from surrounding active eNBs. The initial selection of the primary component carrier is based on the following information: The new eNB measures the average received interference power in the uplink on each component carrier and receives information from the immediate surrounding eNBs, which expresses which component carriers they have selected as their primary and secondary carriers (could also include information concerning the transmit power used for primary and secondary carriers). Then the new eNB measures the average path loss toward the immediately surrounding eNBs. This can be achieved by having the new eNB measure the reference signal received power (RSRP) from the surrounding eNBs, while assuming that the reference signal transmit power of the cells is known by the new eNB. Given this information, the new eNB is capable of autonomously selecting its primary component carrier. It will basically try to avoid selecting the same primary component carrier as the surrounding eNBs. If this is not possible (i.e., if there are more neighboring eNBs than there are component carriers), then it will make a selection that causes minimum interference coupling between the cells by the inter-eNB path loss measurement, which is shown in Figure 6.16.

Once the eNB has selected its primary component carrier, it can start to carry traffic. The quality of the primary component carrier used by the UE is hereafter monitored by the eNB, and if quality problems are detected, it may trigger a reselection, where another component carrier is selected as the primary. As the offered traffic increases

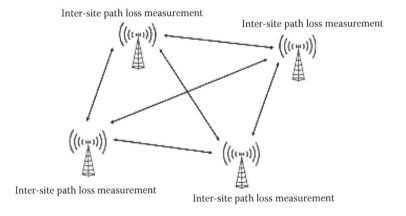

Figure 6.16   The eNBs measure the intersite path loss.

for the cell, the eNB starts to allocate additional secondary component carriers, if this can be allowed without seriously degrading the performance of surrounding cells.

6.2 Self-Optimization

Constant change is the characteristic of a mobile communication network. The network architecture, wireless environment, distribution, and behavior of users are always changing, which requires continuous network adjustment to adapt to the changes. The purpose of network optimization is to identify what is impacting the network quality by data acquisition and analysis, to reach the best network running state and get the best benefits of network resources through technical means or network parameter adjustments, and to grasp the trend of network growth to provide the foundation for network expansion. Traditional network optimization makes optimal decisions by manual data collection and analysis, and cannot react quickly and accurately because of the technical complexity. The self-optimization program provided by the SON makes it possible for a network to track and perceive its state changes, make accurate judgments based on the large amounts of data parallel processed by expert system, and adjust to the optimal state in smaller granularity and shorter time.

The self-optimization process is defined as the process in which UE and eNB measurements and performance measurements are used to autotune the network for coverage and capacity optimization. This process works in the operational state, which starts when the RF interface is switched on.

In LTE-A, effective, seamless edge-to-edge capacity optimization will become essential. The LTE network must be optimized to maximize coverage and capacity, and to provide sufficient quality of service (e.g., coverage hole detection, capacity bottleneck identification.

6.2.1 General Description

When deploying a new air interface technology, operators are typically concerned with maximizing coverage during the initial rollout. As time progresses and capacity demands on the network increase, operators will focus on enhancing capacity. Maximizing coverage and capacity simultaneously is traditionally done with planning tools before rollout. For LTE, the infrastructure will optimize the network by employing a number of different techniques. These features include the following:

- **SON automatic neighbor relation:** This feature is used to add and optimize neighbor relationships continuously as eNBs are added to the network.
- **SON intercell interference coordination:** This feature is used to improve cell edge performance in dense networks.
- **Mobility robust optimization:** This feature is used to automatically adjust the mobility parameters and solve the mobility problems, for example, to

identify unsuitable cells, to identify problematic settings of cell selection/reselection parameters, and to minimize handovers when a UE is transitioned from idle to active mode.

- **SON load balancing:** This feature is used to ensure that traffic is distributed among eNBs and congestion is avoided. SON load balancing also occurs across LTE carriers and across other radio access technologies (2G/3G).
- **Energy saving:** This feature is used to cut operation expenses by switching off cells, adapting the transmission power, and adapting the multiantenna schemes, including the following:
  - Switch off cells when they are not required to provide capacity or coverage (especially needed when a larger number of home eNBs are deployed in the network).
  - Reduce energy consumption by the eNB during normal operating circumstances.
  - Adapt the transmission power while ensuring no influence on the coverage, handover, and load balancing.
  - Adapt the multiantenna schemes; some of the antennas can be switched off to save power. Adapt among single-input multiple-output (SIMO), multiple-input multiple-output MIMO, and beamforming schemes to achieve the maximum capacity with the minimum transmission power.

6.2.2 Self-Optimization of Neighbors

In a legacy radio network, neighbor relation planning and optimization are major O&M tasks that occupy about 30% of the overall O&M workload. The legacy approach is based on centralized analysis and planning, and is not precise or done in real time. Some operators may employ a third-party optimization service, but that still relies on human interaction, which is slow, error prone, and costly. The ANR is a solution to tackle legacy issues and has relevant mature standards. The ANR can dynamically update the neighbor relation list on the basis of UE measurement reports. The ANR is in the self-establishment process for initial neighbor relation detection and configuration, and in the self-optimization process for continuous neighbor relation optimization (see Figure 6.17). Some operators promote special requirements that include control of neighbor relation addition/removal, inter-radio access technology (IRAT)/interfrequency ANR.

6.2.3 Mobility Robustness

Mobility robustness (MRO) is a part of the 3GPP SON use cases.* In the 3GPP recommendation to eliminate unnecessary handover (HO) and provide appropriate handover timing, MRO automatically adjusts the thresholds related to cell

---

* Refer to 3GPP TR 36.902 V9.0.0.

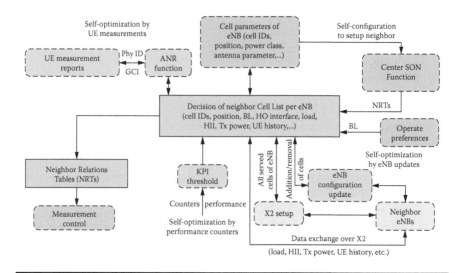

Figure 6.17 Self-configuration and self-optimization of neighbor relations. (BL—black list; HII—high interference indicator)

reselection and handover. The adjustment is triggered by the related key performance indicator (KPI) degradations and is processed while identifying the root causes of the degradations, such as a handover that is too early or too late or the ping-pong effect. The goals of the mobility robustness optimization function are as follows:

■ Reduce the number of handover-related radio link failures.
■ Detect handovers that are too early or too late.
■ Detect handovers to the wrong cells and detect inefficient use of network resources due to unnecessary handovers.
■ Avoid drive tests needed for the detection of handover problems.

The goal of this feature is to improve system performance by optimizing the intra-LTE (intrafrequency) radio network handover configuration to reduce the events noted previously. Reducing such events limits the number of call drops and radio link failures, as well as the amount of exchange signaling.

During the initial network rollout, the operator defines different sets (usually default configuration) of mobility-related parameters for different LTE base station deployments, such as rural, urban, and hotspot LTE base stations. This initial set of mobility parameters (default parameterization) may not always be the optimum configuration for different base transceiver station (BTS) deployments, raising the probability that mobility procedures will not work at all or will work only with a reduced performance. In this case, mobility robustness will detect the underperforming mobility procedures and calculate improved parameter values to fine-tune the mobility configuration and improve the overall mobility performance.

Detection of poor mobility performance is based on the long-term evaluation of certain performance management (PM) counters and KPIs, as follows:

1. **Input for mobility robustness feature:** Performance counters/KPIs
   - Intra/inter_handover_fail
   - Intra/inter_handover_success
   - Intra/inter_handover_drops per radio link failure (RLF) cause per neighbor cell
   - Additional PM counters/KPIs
2. **Output of mobility robustness feature:** Mobility robustness may propose new values for optimizing the following parameters:
   - Hysteresis (events A1, A2, A3, A4, A5, B1, B2)
   - Threshold (events A1, A2, A4, A5, B1, B2)
   - Offset (events A3, A4, A5, B1, B2)
   - Time-to-trigger (A1, A2, A3, A4, A5, B1, B2)
   - Filter coefficient (A1, A2, A3, A4, A5, B1, B2)
   - Time-to-trigger speed scale factors (A1, A2, A3, A4, A5, B1, B2)

The operator can specify policies to control how mobility robustness behaves. For example, the operator can specify the trigger and exit thresholds for activating mobility robustness and set priorities for running parallel SON use cases.

6.2.3.1 Automatic Handover Parameter Optimization

Handover is one of the key mechanisms for mobility and load management. To reduce the need for the time-consuming task of manually setting handover parameters, ongoing optimization of handover parameters after initial deployment will be started. The handover algorithm is based on event-driven or periodic UE measurements and considers handover thresholds, hysteresis values, cell individual offset (CIO), and time to trigger. Ping-pong effects can be minimized by coordinating handover margins and/or cell reselection parameter changes between the cells that aid ANR by supporting the identification and subsequent avoidance of unsuitable neighbors.

The mobility robustness optimization function may reside in the O&M and the synchronization of LTE and universal mobile telecommunications system (UMTS) handover parameters can be handled via the O&M. The main task of this function is to detect problems concerning mobility and adjust the corresponding parameters to improve performance. Figure 6.18 illustrates how the mobility robustness optimization function works.

- ▪ The eNB maintains counters for handover issues like early, late, and rapid handovers. It also maintains counters for access failures and drops that happen due to failed handovers.

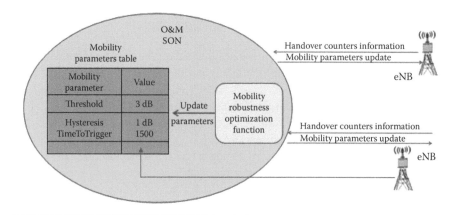

Figure 6.18 Mobility robustness optimization function.

- This information is sent to a centralized mobility robustness optimization function entity in the O&M.
- This function, based on the counter information and other available internal information, automatically adjusts the mobility parameters (thresholds, hysteresis, time-to-trigger, etc.) to improve the mobility performance in the network.

6.2.4 Mobility Load-Balancing Optimization

Mobility load balancing (shown in Figure 6.19) is a subfunction within the radio resource management function under eNB control. The objective of mobility load balancing (MLB) is to intelligently spread user traffic across the system's radio resources as necessary to provide quality end-user experience and performance, while simultaneously optimizing system capacity. This feature can provide real-time optimization of cell overload by intelligently handing over cell-edge users to a cell with spare capacity. This optimization adjusts the thresholds related to the cell reselection/handover parameters to cope with the unbalanced traffic load and lead some UEs on the edge of a congested cell to reselect or handover to the less-congested adjacent cells. Load balancing should be done by using a minimum number of cell reselection or handovers without causing a mobility problem. It should also minimize the total investment in capacity by taking into account the different kinds of load, such as radio load, transport network load, and hardware processing load.

Prior to attempting to balance traffic across various radio access networks (RANs), the LTE network will attempt to balance traffic across LTE eNBs in idle mode and connected mode. The basic strategy is to move traffic from overloaded sectors to underloaded sectors. Idle mode and connected mode traffic will be offloaded to nearby LTE cells from overloaded LTE cells by adjusting idle mode

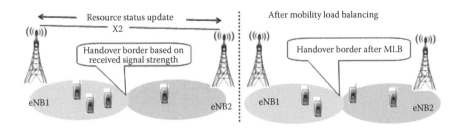

Figure 6.19   Mobility load balancing.

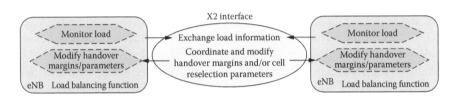

Figure 6.20   Load-balancing function.

parameters and handover parameters. Intercell interference coordination (ICIC) techniques will be employed to minimize interference created when UE is moved from the nearest eNB to one that is farther away (from a path loss perspective) in concert with this approach.

The algorithm for load balancing can be based on radio load, transport network load, or both. If both radio load and transport network load are considered, the one that will have higher priority must be determined. The algorithm will also consider differentiation among quality of service class identifiers (QCIs).

The mobility load-balancing optimization function resides in the eNB. Figure 6.20 illustrates how this function works.

1. Each eNB monitors the load in the controlled cell.
2. The load information is exchanged over the X2 interface with surrounding eNBs.
3. The load-balancing function in each eNB decides to distribute the UE camping and/or delay, or advance handing over of the UEs between cells, to balance the traffic load between cells. To do this, the algorithm needs to compare the current loads among the cells, the type of ongoing services and their quality of service (QoS) requirements, the cell configuration, and so on. Based on all these details, the handover margins and/or cell reselection parameters between the adjacent eNBs are modified in a coordinated manner. This results in distributing the load between adjacent cells.

4. When making a decision to handover a UE to another cell/carrier frequency, the following factors are considered: UE measurement reports of its serving and neighbor cell signal strength, the UE's current signal-to-interference ratio, the UE's serving cell and neighbor cell loading conditions, the UE's QoS/application profile, and the UE's mobility level.

From a heterogeneous network point of view, balancing the load between these heterogeneous cells is expected to improve the overall performance. A bias in cell selection and handover decision is obtained by boosting the reference signal power along with reducing the eNB transmit power.

The other load-balancing feature in load management is carried out between LTE and other radio access technologies such as Universal Terrestrial Radio Access (UTRA) and GSM when traffic has been balanced within LTE and the utilization is still high. Depending on the actual radio network topology, the UE population, and the typical services in use, load balancing is a tool that could be used to achieve more efficient use of network resources for the benefit of overall network capacity and end-user satisfaction.

6.2.5 RACH Optimization

The random access channel (RACH) is an uplink unsynchronized channel that is used for initial access or uplink synchronization. The random access process allows multiple UEs to simultaneously gain access to a cell by using different random access preamble sequence codes. RACH optimization is designed to minimize access delays for all UEs in the system, minimize uplink interference due to the RACH and physical uplink shared channel (PUSCH), and minimize interference among RACH attempts. The RACH optimization of automatic root sequence reallocation is designed to relate the root sequence index to the PCI. By self-optimization of random access, the access delay can be controlled, the coverage can be improved, and the call setup success rate and handover success rate will be increased. The self-optimization can further result in better utilization of uplink resources among the physical random access channel (PRACH) and the PUSCH, reduced intercell interference caused by the RACH, and lower preamble false detections due to the same preamble being used in two neighboring cells. The random access channel needs to be configured per cell due to different cell sizes by adjusting the RACH parameters, and its goal is to reduce access delay and intercell interference.

The required data for RACH optimization is provided in Table 6.3.

The target of RACH optimization is to adapt the number of RACH access slots to the RACH traffic load in order to reduce call setup and handover delays and to achieve high success rate of call setup and handover. The RACH load measurements are used as the basis to optimize RACH configurations. The required control parameters for RACH optimization is provided in Table 6.4.

Table 6.3   Statistic Information for RACH Optimization

Stats Information	*Collection Unit*	*Interval (Minimum)*	*Note (Minimum)*
Number of transmissions of random access preamble and kinds of random access procedures at the completion of the random access procedure	UE	—	Per completion of random access procedure
Average interference power per subframe	Subframe	5 minutes	Several minutes
Estimated transmit power of random access preamble; the statistics of UEs' report are utilized to estimate the random access success probability and compare with a given target threshold, and based on this comparison result, the power of the initial RA preamble is adjusted.	UE	—	Per completion of random access procedure
Neighbor list	Cell	—	Each neighbor list update
Number of times that the CAZAC code for noncontention was not allocated at the execution of the noncontention random access procedure when handover is from an adjacent eNB to this eNB	Cell	5 minutes	5 minutes
Number of attempts of the noncontention random access procedure when handover is from an adjacent eNB to this eNB (does not care whether it is successful or a failure)	Cell	5 minutes	5 minutes

The dynamic RACH power control parameter adjusts the initial random access preamble power based on the reports from the UEs, which completes the random access procedure by using the RRC UE information transfer procedure.

The RACH optimization function resides in the eNB. Figure 6.21 illustrates how this function works.

Table 6.4 Control Parameters for RACH Optimization

Settable Parameters	Notification Interval
Back-off parameter	Dozens of minutes
Random access preamble transmission power parameter	Same as above
Power increase range	Same as above

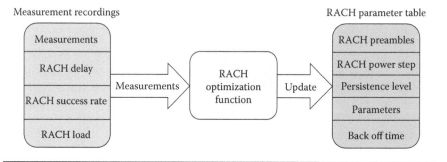

Figure 6.21 RACH optimization example.

1. The eNB records measurements such as the RACH delay, RACH success rate, and RACH load (number of received RACH preambles) per preamble range.
2. These measurements are sent to the RACH optimization function, which may set some thresholds for these recordings.
3. If any of these recordings crosses the threshold, the function performs a corresponding analysis and adjusts the RACH configuration parameters.

6.2.6 Minimum Drive Test

The drive test (DT) collects network coverage and throughput measurements, which will be used in network planning, optimization, and dimensioning. The DT is normally triggered by a new base station, cell deployment, or construction of a hot spot such as a big building, user complaints, or regular maintenance by operator.

Operators always invest a considerable amount of money in drive test equipment and the execution of drive tests during delivering rollout, replacement, modernization, and optimization projects where huge costs are incurred due the resources required to perform the drive test.

The minimum drive test (MDT) feature is designed to replace the current expensive drive testing and is intended for use during network deployment, optimization, and operation. By commanding UEs in certain areas to collect the measurements, rigorous drive tests and the related OPEX can be reduced. In addition, such an approach can greatly reduce response time and provide quick optimization to ensure a quality end-user experience. Take China, for example, where the MDT will help even more in complex geographical areas or in high-population areas where vehicle is hard to approach.

There are two levels of measurements collected by the MDT for optimization: radio environment measurement and eNB measurement. Radio environment measurements such as RSRP, carrier to interference plus noise ratio (CINR), downlink (DL) physical throughput, DL application throughput, and uplink (UL) application throughput are to be used for coverage optimization and network dimensioning. The eNB measurements such as call drop rate, handover failure rate, and connection delay are indirectly used for coverage, mobility, capacity, and common channel status evaluation and optimization. Unlike the self-optimization functions defined in the context of the SON, where network nodes can trigger automatic operational changes, minimizing the drive test for network performance analysis is mainly achieved by supporting the collection of measurements at the network management level.

The MDT is designed to automate collection of trace information from the eNB in a preconfigured use case manner, including UE measurements that would be postprocessed and analyzed quickly. MDT features are as follows:

- Use current subscribers in the system to gather RF information. RF information will be postprocessed to determine whether RF optimization is needed and if so, the eNB will be instructed to optimize its RF.
- Eliminate designated drive test.
- Logged MDT (periodic measurements in idle mode) configured through the RRC procedure in idle mode
- Immediate MDT (event period-based measurements in active mode) configured through the measurement report procedure
- Radio link failure data collection (collect RF data, UE position, and other information at the time of radio link failure)
- Use trace functionality to start MDT for individual UE or O&M to start MDT for a region.
- Data retrieved for logged MDT when the UE reconnects
- Data processed by the O&M server, which generates RF change instructions

The MDT can provide operating expenses (OPEX) and capital expenditures (CAPEX) savings since not as many resources are needed to evaluate the performance of the service and network, thereby eliminating the need to perform an expensive drive test. In a live network, we should still consider the MDT key success factors, such as how to select MDT subscribers (how to select subscribers that give comprehensive RF picture of the system), the kinds of MDT parameter settings to be collected that include comprehensive RF information, and which server RF adjustment algorithm will be chosen.

6.3 Self-Healing

6.3.1 eNB Self-Healing

According to statistics, one wireless network may report up to 10,000 alarms every day. It is difficult for operation and maintenance personnel to cope with this number of alarms, and thus a large number of faults reflected by the alarms cannot be addressed. In order to improve the efficiency and quality of the network, it is critical to reduce the burden on operation and maintenance personnel. The network itself should be able to sense, identify, locate, and associate alarms while starting the self-healing mechanism to eliminate the corresponding failure and return the network to normal working condition.

One of the targets of a SON is to maintain network performance and quality with minimum operator supervision and maintenance costs. Self-healing is a SON functionality that detects problems and resolves these problems without user impact. This significantly reduces the maintenance costs by reducing unplanned site visits. The self-healing functionality performs automatic built-in tests and system functionality tests. For each detected test failure, appropriate alarms will be generated by the faulty network entity. The self-healing functionality monitors the alarms and resolves them automatically. It performs analysis using the correlated information (test results, measurements, etc.) and according to the results, takes appropriate recovery action to solve the problem automatically.

In 3GPP, self-healing has been focused on the recovery of components such as fallback to the previous software version, reducing the output power in case of temperature failure, or switching to backup units in case of board unit faults, through analyzing the alarms and faults reported. The self-healing function aims at automatic detection and localization of most of the failures and applies self-healing mechanisms to solve several failure classes.

BTS provides facilities for ensuring and restoring the availability of the system and its services.

- **Hardware and software supervision:** The functionality of the system's software and hardware is permanently supervised to ensure that no part of the system drops out of service without being detected. Some error types are actively supervised by a central software component; other types are detected by one of the affected distributed components during normal operation. For the latter kinds of errors, an internal alarm reporting system alerts the central component.
- **Recovery:** If a hardware component fails or a software error is detected, the system takes action to restore functionality. The recovery action is chosen to minimize the impact on other system functions. If the system is not able to restore service on its own, an alarm is raised giving a clear indication of the problem to enable the operator to undertake appropriate repair action.

■ **Diagnostics:** Online and offline tests are available to enable the operator and/or service personnel to pinpoint problems if the symptoms given in the alarm report are inadequate.

6.3.2 Cell Outage Detection

A problematic cell-to-cell relationship can occur now and then in typical microcell urban deployments. Network performance is supervised by performance counters that supervise the problematic cell. These counter values are monitored for each individual cell and standard traffic conditions for the cell are extracted (traffic ongoing in that cell, at that specific time of the week). Actual counter values for a cell are compared with its standard traffic scheme. If there is no traffic when there should be, the cell is considered as possibly having a problem. If there is no change over a set period of time, a quality of service alarm is raised with an indication that this cell may be no longer be performing correctly.

The network performance will be supervised by performance counters. Counters indicating the performance of a cell will be used to supervise the functioning of a cell. An alarm will be raised with an indication that this cell may not be performing correctly anymore.

This feature will raise an indication that there may be a problem, but the operator still needs to check on the situation in that cell and take whatever actions are needed. This feature can be switched on or off networkwide.

The following failure scenarios may result in a sleeping cell and will be covered by cell outage detection:

1. Physical channel failures including RACH failure, synchronization channel (SCH) and reference signal failure, broadcast information transmission failure, paging failure, user plane transmission failure, etc.
2. Cell/LTE base station failures including power outage, tower-mounted antenna (TMA) failure, TX antenna failure (for example, wrong position or tilt or direction, caused by a storm), MIMO outage or degradation (for example, one MIMO out of service), etc.
3. Transport channel failures including S1 or X2 connection failures that result in breaks in user data transmission, although the radio connection remains intact.

6.4 SON Outlook

The SON is a promising approach to significantly reduce network installation and operating costs, and can improve the user experience. SON is extremely important and is being used today. Many automation improvements may be required for future SONs, which are described as follows:

- LTE layer management including femto, pico for mobility, load, speed, and interference optimization
- Service-optimized IRAT layer management including Wi-Fi
- Multitechnology and multivendor radio hardware and transport management
- LTE relay configuration and stability
- Making SON better, faster, and more dynamic and efficient than manual optimization
- SON for the LTE transport network

The following SON use cases and scenarios will be considered in the future:

- Mobility robustness optimization (MRO) will need further enhancements. The extended solutions for MRO for inter-RAT scenarios should be considered, especially UMTS, LTE, and intra-UMTS scenarios. Detection and correction of mobility problems are needed that do not lead to connection failure or cause unnecessary signaling at the network side and battery consumption at the terminal in macro and heterogeneous network (HetNet) deployments.
- The SON will support selection of the proper RAT based on QoS-related information exchanged between RANs of different RATs.
- Extensions of existing ANR mechanisms will enable exchange of neighbor cell relations of cells belonging to different RATs, especially between GSM/UTRAN and LTE, and allow ANR for closed subscriber group cells and other improvements for ANR in UMTS.
- Further coordination between MRO and other traffic control mechanisms (MLB or traffic steering) should provide the robustness of the overall SON solution. The SON will provide the information necessary to make MRO aware of the reasons for mobility decisions that may be subject to MRO procedures. Also, when applying corrective actions, MRO should be aware of possible mobility load-balancing or traffic steering policies.

The features of the SON for the LTE transport network are foreseen as the following:

- Dynamic bandwidth allocation; bandwidth allocation shall automatically follow the changes in traffic load
- Automatic tuning of QoS parameters, e.g., differentiated services model parameters, scheduler parameters, buffer management parameters per QoS class
- Optimal configuration/reconfiguration of eNB connectivity with the serving gateway (SGW) for the multi-S1 scenario (i.e., one eNB connected with multiple GWs); how to configure in case of one SGW fails
- Optimal load balancing, e.g., to apply the multiprotocol label switching (MPLS) traffic engineering method
- Transport admission and congestion control optimizations.

Chapter 7

Heterogeneous Networks

At present, various wireless communication systems have formed many relatively independent networks, including a variety of standard cellular mobile networks, wireless local area networks (LANs), wireless digital home networks, and vehicle wireless communication systems. These network patterns are longitudinally independent; each network has its own specific resource composition and, based on the cyber source, provides a specific function and application. Such a network pattern has gradually exposed its inherent problems: a variety of complex protocols, complex wireless coexistence, and very high wireless network management and maintenance costs. The system design and performance evaluation of the Third-Generation Partnership Project (3GPP) Long Term Evolution (LTE) radio access networks have so far been generally based on homogeneous cell layouts. Figure 7.1 illustrates a basic homogeneous deployment where each hotspot is covered by many high-powered nodes (i.e., a macrocell) with equal transmission power.

Given a continued exponential trend in traffic, the one layer of the macro network will hit a capacity limit sometime over the next decade. Expanding the macro layer network to a much denser multilayer network will be an option to further improve the network capacity in a cost-efficient way. With growing demand for data services, it is becoming increasingly difficult to meet the required data capacity through traditional cell-splitting techniques, which require deployment of more wide area evolved Node-Bs (eNBs). Instead, cell-splitting gains and therefore significant capacity improvements can be achieved in a cost-effective manner by deploying network nodes within the local area range, for example, low-power pico eNBs, Home eNBs (HeNBs) and closed subscriber group (CSG) cells, and relay nodes. We refer to a network deployment incorporating one or more of these local area range node categories as a *heterogeneous network deployment*. Those node categories may be not subject to the same level of deployment planning as wide area eNBs (and sometimes no planning at all, as in the

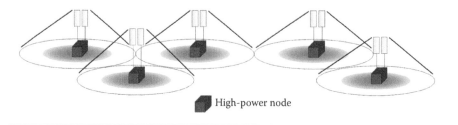

High-power node

Figure 7.1 Homogeneous deployment.

case of HeNBs). Moreover, each individual node typically has a small footprint due to its low transmit power, resulting in highly variable traffic loading. The introduction of new low-power nodes benefits the system average throughput due to the cell-split gain, and most people believe that significant capacity improvements can be achieved in a cost-effective manner by deploying LTE heterogeneous networks (HetNet).

A heterogeneous network is a classical hierarchical cell technique that represents cellular deployments with a mixture of cells of differently sized and overlapping coverage areas; for example, a number of micro- and picocells overlaid by a macrocell, where wide area coverage is provided by a *macro* layer while a lower *pico* layer provides capacity by introducing a number of rather isolated cells within each macrocell. The HetNet as a deployment principle has been known for decades as hierarchical cell structures in the Global System for Mobile Communication (GSM) system. HetNets have been of particular interest for complementing macrocell layouts to handle nonuniform traffic distributions, for example, considering primarily macrocells in certain areas with underlaid smaller cells for high capacity needs at traffic hotspots. Such deployments are considered to be useful to offload capacity from the macrocells, in which an operator can offer high-speed Internet access in certain areas. Compared with homogeneous networks, the distribution of cells in a heterogeneous network may or may not be uniform.

While heterogeneous network deployments can significantly improve system capacity, they also pose some challenges from the system performance perspective. In particular, interference may vary significantly and be more severe as compared with wide-area-only deployments. It is therefore essential to ensure robust network operations. Interference conditions are expected to change from location to location (due to the possibly lower level of network planning of these deployments) and from time to time (due to the variable traffic load at each node).

In heterogeneous networks, low-power nodes are distributed throughout a macrocell network. Low-power nodes can be micro eNBs, pico eNBs, HeNBs (for femtocells), and relays and distributed antenna systems (DASs), the last of which may employ remote radio head (RRH) cells. These types of cells operate in low-geometry environments and produce high interference conditions. The placement of low-power nodes and user equipment (UE) distribution has a significant influence on the performance of HetNet scenarios and on the type of needed interference coordination mechanisms. The various types of heterogeneous deployments are summarized in Figure 7.2 and the definitions of the various low-power nodes are summarized in Table 7.1. In addition to collaborative multipoint (CoMP), intercell interference coordination (ICIC) techniques can play a critical role in obtaining good performance within heterogeneous deployments.

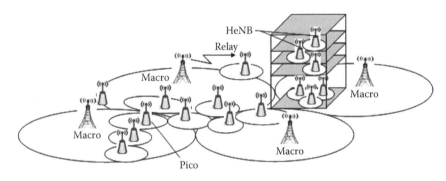

Figure 7.2 Heterogeneous network.

Table 7.1 Characteristics of Low-Power Nodes

Type of Nodes	LPA Power	Number of TX/RX Antennas	Backhaul Characteristics	Comments
Micro eNB	30-dBm, 10-MHz carrier	2/2 or 4/4	X2	Open to all UEs; placed outdoors
RRH node	30-dBm, 10-MHz carrier	2/2 or 4/4	Several μs latency to macro	Open to all UEs; placed indoors or outdoors
Pico eNB	30-dBm, 10-MHz carrier	2/2 or 4/4	X2	Open to all UEs; placed indoors or outdoors
Home eNB	20-dBm, 10-Hz carrier	2/2 or 4/4	Home broadband	Closed subscriber group, placed indoors
Relay	30-dBm, 10-MHz carrier	2/2 or 4/4	Wireless outband or inband	Open to all UEs; placed outdoors

LTE Advanced (LTE-A) technology provides tools for efficient macrocell–picocell resource sharing and interference coordination. It might now be time for hierarchical cell structure (HCS) and HetNet deployments due to new traffic patterns and higher data rates, and the challenge to possibly mix open and closed subscriber groups in the same spectrum. Heterogeneous networks are deployments that use LTE and LTE-A technology components, which are shown in Figure 7.3.

As we previously mentioned, the importance of the self-organizing network (SON) increases significantly within heterogeneous networks. The solid foundation within the current fourth-generation (4G) product will be highly leveraged in the heterogeneous solution including all elements of autonomics, self-discovery, self-configuration, dynamic interference management, and so on.

In the cell of a macro base station, the data rate decreases exponentially from the center to the cell edge. By introducing multilayer nodes, one is able to shift capacity and performance where needed. Assuming a uniformly distributed capacity demand, the data rate should be equal over the entire cell.

The definition of low-power nodes in heterogeneous networks is described as follows:

■ **Macro:** Conventional base stations that use dedicated backhaul and are open to public access.
■ **Relay:** Base stations wirelessly connected to the radio access network via a donor cell.
■ **Pico:** Low-power base stations that use dedicated backhaul connections and are open to public access.
■ **Femto:** Consumer-deployable base stations that utilize the consumer's broadband connection as backhaul. Femto base stations may have restricted access.

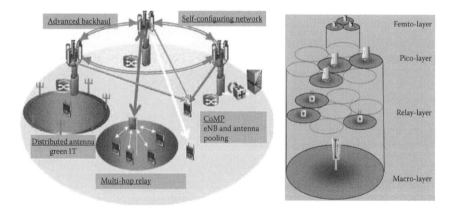

Figure 7.3   LTE-A implementation.

HetNet will be a main deployment solution to tackle high throughput in the future, and only pico remote radio units (RRUs) and fiber-optics need to be deployed. In this scenario, the baseband will be shared by indoor and outdoor cells; interference can be mitigated with centralized processing and handover can be managed with joint processing. Pico RRUs can also provide multilayer multiple-input, multiple-output (MIMO) to provide high throughput.

A picocell is used for an open subscriber group (OSG)*, and any UE is eligible to access the pico eNB. There is no difference in downlink if a UE is associated with an eNB from which it receives the highest downlink (DL) signal. But from the uplink (UL) perspective, a UE is associated with a nonoptimal cell. The network can assign a cell selection bias to a picocell that has the potential to improve the throughput performance of the UE, but this leads to harsh interference scenarios for UE in the expanded area of the pico cell.

The OSG cell is a normal cell that can be accessed by all the users. The CSG cell is a cell that is a part of the public land mobile network (PLMN); it broadcasts a CSG indication that is set to "True" and a specific CSG identity. A CSG cell is accessible by the members of the closed subscriber group for that CSG identity.

7.1 Features of the Heterogeneous Network

In heterogeneous networks, the throughput gain of heterogeneous deployment is the effect of cell splitting. As more and more users are attached to picocells, the loading of those cells will increase while the macro eNB loading will decrease. Meanwhile, the interference problem may become serious due to the introduction of lower-power nodes (overlaid by a macrocell), which leads to low geometries, especially in co-channel deployment scenarios. The low geometries seen in heterogeneous deployments necessitate the use of interference coordination for both control and data channels to enable robust operation. In addition, because unbalanced transmitting power is provided between different cell layers, the control channels suffer more serious interference than in the case of homogeneous networks.

The heterogeneous network includes the following features:

■ Strong interference scenarios.
■ For large variability in loading across small cells, intercell fairness becomes more important in dense heterogeneous networks and needs to equalize performance of users with similar quality of service (QoS) requirements.

---

* Open subscriber group (OSG): All users have access to the cell. Support for hybrid access modes. Closed access mode: Only UEs belonging to the CSG are entitled to access the cell. Hybrid access mode: All UEs are allowed to access the cell, but UEs belonging to the CSG are entitled to access with priority.

◾ *Intelligent UE association* requires more protection against interference, resulting in a better spectral efficiency and network capacity. In LTE Rel-8, a UE associates with an eNB with the best downlink signal; sometimes UE association with an eNB with a weaker signal-to-interference plus noise ratio (SINR) is required under certain conditions.

◾ In frequency reuse 1/1 mode, the downlink SINR value at the cell edge is about −10 dB, and advanced interference management techniques are required to provide robust performance, improve intercell fairness, and enable gains in spectral efficiency.

7.1.1 Future Network Deployments Based on Heterogeneous Networks

The heterogeneous LTE network contains network nodes, such as macro eNBs for initial coverage only, and picocells with different characteristics such as transmission power and radio frequency (RF) coverage area. The eNBs with different transmission powers are used to support large and small RF coverage areas. The macro eNB with a large RF coverage area is deployed in a planned way for blanket coverage of urban, suburban, or rural areas. Local nodes with small RF coverage areas aim to complement the macro eNB for coverage extension or throughput enhancement and a richer user experience. The RF coverage areas of the heterogeneous network nodes could be overlapped or disjointed as shown in Figure 7.4. The overlapped RF coverage design of large cells and small cells aims to enhance the system performance in throughput, accessibility, privacy, and service. The design of disjointed RF coverage areas in heterogeneous networks, or relays, intends to extend the RF coverage area to smaller local regions with a heavy data load and fill coverage holes with no incremental backhaul expense. A heterogeneous network will need a flexible and low-cost network deployment using a mix of macro, pico, relay, RRH, and Home eNBs, and the deployment of microcells along with macrocells could be done through nonhomogeneous cell splitting. Heterogeneous

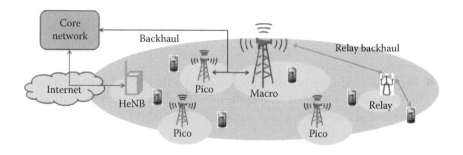

Figure 7.4 Generic heterogeneous network and RF coverage area of network nodes.

network applications include nonhomogeneous cell deployment by operators and hybrid network deployment of public and private or semiprivate networks.

According to several heterogeneous scenarios—macro + femto, macro + outdoor pico, macro + indoor pico, macro + outdoor relay, macro + indoor relay—the transmit power of the different types of cells are 30 dBm for hot zone cells, 20 dBm for femtocells, and 30 dBm or above for relay nodes. Either a 2Tx/2Rx or a 4Tx/4Rx antenna configuration could be used for any of these cell types. Another important characteristic of these different cell types is the type of backhaul connection: generally, hot zone cells have a backhaul X2 connection, whereas the existence of an X2 connection for femtocells depends on the deployment. Relay nodes of course have an over-the-air backhaul connection. The characteristics of low-power nodes can be found in Table 7.1.

We believe that in the future we cannot meet the traffic demand with only one macro layer; but with heterogeneous networks, we could reach capacity far beyond the capacity limit of macro networks at an affordable cost. Different operators may have different deployment strategies depending on their own spectrum and backhaul resource situations. We have to solve numerous challenges to realize heterogeneous networks. Key challenges for realization of heterogeneous networks are listed here:

- Wireless backhaul (inband, outband) is a key enabler
- Interference avoidance and cancellation between network layers and interference mitigation among femtocells
- Seamless mobility between different technologies, spectrum bands, and network layers
- Single, simple (for users) authentication scheme
- Low-cost, plug-and-play (SON) base stations
- The need for efficient resource allocation balancing between macro- and femtocells to adapt to load dynamics for better spectral efficiency

7.1.2 Home eNB

The co-existence of HeNBs within a macro eNB coverage area is one scenario of heterogeneous network deployment. The HeNB is a private device operating in an operator's licensed spectrum along with macro eNBs in the heterogeneous network deployment and with relatively little control by the operator.

HeNBs as defined in Rel-8 have multiple access control mechanisms:

1. Closed access (residential deployment): Access is only allowed for the subscribed user. The HeNB is defined as a closed subscriber group cell and access control is located in the gateway (GW).
2. Open access (enterprise deployment): All users are allowed access to the HeNB and receive the offered services.

Access control is based on the HeNB cell ID, which is called the CSG* identifier (CGI). Rel-8 defines basic CSG provisioning and access control. The UE will need to support automatic CSG selection in idle mode as well as manual CSG selection. Autonomous search is performed to find CSG member cells, whereas the algorithms are left completely to UE implementation. The problem of increased UE power consumption by excessive cell search can be addressed by using the physical cell ID (PCI) to identify a CSG cell. A number of PCIs are reserved to identify CSG cells within the same carrier frequency in case of a mixed deployment (other HeNBs using the same frequency). A nonstandardized autonomous cell search, left to UE implementation, was chosen to minimize impact on non-CSG UEs and the regular base station. Mobility of closed access HeNBs supports only outbound handover from the HeNB to a macrocell by the normal handover procedures, but not inbound mobility from the macrocell to the HeNB.

HeNBs can broadcast one and only one CSG. HeNBs can be in open, closed, or hybrid mode, allowing all subscribers, only CSG subscribers, or a controlled combination of all subscribers to register. Multiple HeNBs can belong to the same CSG. CSG lists per subscriber are kept in the home subscriber server (HSS). These are downloaded to the mobility management entity (MME), which does CSG checking when the subscriber registers on a CSG cell. CSG lists are also kept in the subscriber identity module (SIM) so that the UE knows to which CSGs it belongs. HeNBs are not listed in the macro neighbor cell list. Macro eNBs broadcast the PCIs of HeNBs that may be within their coverage areas. UEs can then search for HeNBs that they may be able to access with these PCIs. PCIs may not be unique for the group of HeNBs (i.e., multiple HeNBs may share the same PCI). During handover into the HeNB, the UE must send in both the PCI and the CGI to uniquely identify the HeNB. Since there are no X2 interfaces to HeNBs or the HeNB GW, the source cell sends the handover request to the MME, which forwards it to the HeNB/HeNB GW. LTE HeNB architecture is depicted in Figure 7.5.

LTE Rel-9 provides further functionality to support more efficient HeNB operation and to provide a better user experience. The key functionalities added to the radio access network for HeNBs in Rel-9 include hybrid cells; inbound mobility support; access control; HeNB operation, administration, and maintenance; as well as RF requirements for time-division duplexing (TDD) and frequency-division duplexing (FDD) HeNBs. Significant work has led to a more consistent framework to efficiently and reliably deploy LTE HeNBs and provide a satisfying user experience.

Hybrid access is a new access concept introduced with Rel-9 in addition to closed access and open access. Basically, the cell provides open access to all users but still acts like a CSG cell. Subscribed users can be prioritized compared to unsubscribed users and can be charged differently. According to the specification, a *hybrid cell* is defined as a cell that has the CSG indicator set to false and yet broadcasts a CGI.

---

* Closed Subscriber Group (CSG): Only a subset of users have access to the femtocell.

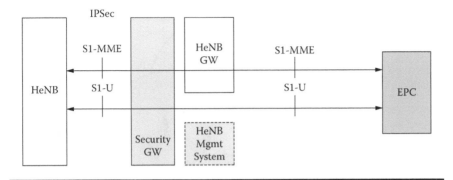

Figure 7.5   LTE HeNB architecture. (EPC—evolved packet core)

Table 7.2   HeNB Concept in 3GPP RAN

HeNB	*Concept Progress*
Rel-8	Architecture (functional split of core network [CN], HeNB GW)
	Closed subscriber group (CSG) and idle mode mobility
	Hand-out active mode mobility
Rel-9	Mobility (handover between HeNBs, open and hybrid access modes)
	Introduction of operator CSG list HeNB RF requirements
Rel-10	HeNB mobility enhancements, X2 type direct interface for HeNBs Enhanced ICIC for HeNBs, TDM
	eICICI concept using muting patterns (almost blank subframes [ABS] or multimedia broadcast/multicast service single frequency network [MBSFN] subframes)
	SON for use cases with HeNB

In LTE-A, the HeNB provides further functionality for mobility enhancements, an enhanced ICIC (eICIC) feature, and SON use cases with HeNB. Table 7.2 gives a simple summary of HeNB evolution.

Although only a few registered UEs are allowed to access the CSG, there are also harsh interference scenarios for macro UEs that are located within the HeNB coverage. In particular, control channel performance for macro UEs under HeNB/pico coverage may be severely degraded by harsh interference from the HeNB/pico, which is shown in Figures 7.6 and 7.7.

The issues of access control, handover, resource allocation, and interference management that occur in the deployment of microcells when overlaid on macrocells also applies to the deployment of HeNBs. Because there is little or no coordination between the HeNB and the macro eNB, HeNBs will introduce a high level of uncertainty in heterogeneous network deployment due to being privately owned by the end users.

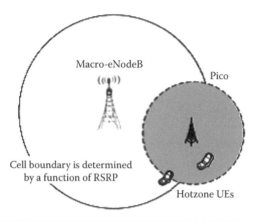

Figure 7.6   Interference from pico cell. (RSRP—reference signal received power)

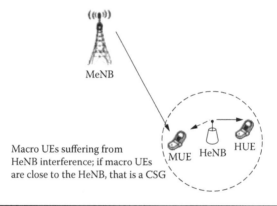

Figure 7.7   Interference from HeNB. (MeNB—macro eNB; MUE—macro UE)

7.1.3 QoS of Heterogeneous Network

Because the characteristics of radio bearers differ according to the radio propagation environments, the radio bearers in macro- and picocells also differ from each other. These bearer differences should be adequately exploited to deliver different QoS data streams. For example, a high data rate transmission requires high transmit and receive power, so the propagation distance and cell radius should be short, like a radio bearer in a picocell. Conversely, it is natural for low data rates to be transmitted and received in macrocells. As a result, different radio bearer channels can be exploited for macro- and picocells by assigning different values of the QoS-related parameters such as QoS class identifier (QCI), allocation retention priority (ARP), and so on.

For example, assume a UE simultaneously communicates through a mobile video phone and browses the web. In this case, the QoS required for the video phone and the web browsing data is different; thus, different QoS parameter values must be assigned for each radio bearer. In the same manner, the H.264 (ITU-T H.264: Advanced video coding for generic audiovisual services) data stream can be split into different QoS data streams and two types of data stream layers are assumed in this contribution. One is the base layer that consists of a low-rate but highly reliable data stream; the other is an enhanced layer that consists of a high-rate but less reliable data stream. These layers are assigned for different radio bearers with different QoS parameter values. That is, the base layer is transmitted from the radio bearer of macrocells and the enhanced layer is transmitted from that of picocells. The example of UE receiving multiple radio bearer data streams with CoMP or joint processing (JP) in macro- and picocell environments is shown in Figure 7.8. By matching the characteristics of each layer with those from each radio bearer, the utilization efficiency can be maximized for the radio spectrum.

Another example is a streaming video system that consists of a video server located in a wired network and a video user terminal (UE) in an LTE network. The system can be seen as a service system in a unified wired and wireless network. In the same manner as described above, the total QoS can be easily controlled by mapping the multiple stream layers of the H.264 with different QoSs for each layer. At first, in the IP network, each scalable video coding (SVC) layer stream can be matched with each QoS by mapping the priority flow table in the Internet Protocol version 6 (IPv6) message format. Then, in the LTE network, the corresponding QCI and ARP can be defined and mapped for each layer stream.

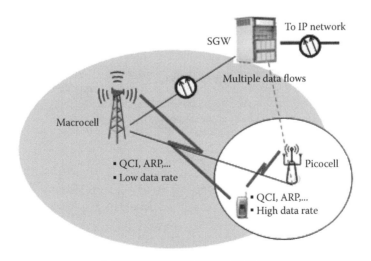

Figure 7.8 Example of UE receiving multiple radio bearer data streams with CoMP/JP in macro- and picocell environments. (SGW—serving gateway)

7.2 Heterogeneous Network Combined with Cloud RAN

In 2020 and beyond, communication link throughput may approach the Shannon limit (the theoretical maximum information transfer rate of a communications channel, for a particular noise level), even in the challenging time-varying wireless channel. In order to increase throughput and promote green communications, it is necessary to put access points close to users. The most promising potential methods to improve the system capacity are network MIMO and CoMP. At the same time, wave-division multiplexing (WDM) passive optical network (PON) will advance the next-generation optical broadband access, and 400-Gbps optical transmission of each wavelength is coming. The ability of multicore processors becomes increasingly powerful, and cloud computing based on IT platforms will be very popular.

Some wireless network operators, such as China Mobile believe that centralized processing, cooperative radio, and cloud infrastructure radio access networks (C-RANs) are the answer to all these requirements. C-RAN is composed of three parts (see Figure 7.9): the distributed radio network equipped with remote radio heads (RRHs) and antennas, a high-bandwidth and low-latency optical transport network that connects RRHs and the BBU pool, and a centralized BBU pool composed of high-performance general-purpose processors and real-time virtualization technology.

After the processing units have been put in a centralized pool, it is essential to design virtualization technologies to distribute the processing units into virtual base station entities. The major challenges of virtualization include the need for a high-efficiency and flexible virtualization environment to guarantee real-time scheduling and strictly controlled processing delay and jitter, and an interconnection topology among physical processing resources in the baseband pool. This includes the interconnection among the chips on a processing board, among the boards in a physical rack and among multiple racks.

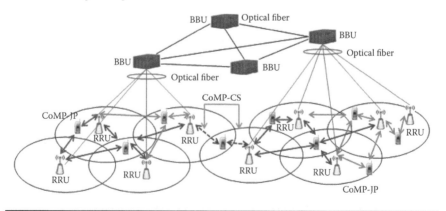

Figure 7.9   Network topology of BBU pool + RRU. (RRU—remote radio unit; BBU—baseband unit)

The benefits of a BBU pool + RRU-based C-RAN system in LTE-A are discussed in the following text.

The architecture of BBU pool + RRU can reduce costs for equipment rooms. Because it is small, the BBU can be installed in residential and business buildings. In this way, network construction can be achieved in an economical, flexible, and fast manner.

Since the C-RAN system has a very high density of RRUs, small cells with lower transmit power can be efficiently deployed. Therefore, the energy used for signal transmitting will be reduced, which is especially helpful to UE battery life and results in much lower electromagnetic radiation.

The BBU as the central processing part should have data and channel state information (CSI) information, so it can perform joint transmission, scheduling, and detection. Hence, inter-cell interference (ICI), which reduces the cell-edge throughput, will be mitigated or even beneficial as a useful signal instead of acting as interference.

BBU + RRU–based C-RAN is also suitable for nonuniformly distributed traffic by simply adjusting the density of the RRUs.

Although the serving RRU changes with the movement of the UE, there is no handoff within a BBU pool. Because the coverage of one BBU is larger than the traditional eNB, the problems of frequent handoffs will be overcome. However, the latency and overhead constraints on X2 interface do not exist in this BBU + RRU–based CoMP system.

Another operator has a similar concept using the China Mobile Communication Corporation (CMCC) C-RAN, which is a cloud computing center. In this concept, radio units and digital units are separate but connected by optical fiber. The digital unit is located at the cloud computing center. The goal of implementing a cloud computing center is to reduce capital expenditures (CAPEX) and operating expenses (OPEX), increase network capacity, improve management efficiency, and improve scalability and flexibility.

7.2.1 Cloud RAN Key Technology

Mobile cloud computing is the usage of cloud computing in combination with mobile equipment. Cloud computing exists when tasks and data are kept on the Internet rather than on individual nodes, which provides on-demand access. Mobile cloud computing is a combination of a mobile network and cloud computing, thereby providing optimal services for mobile users.

The key technologies of cloud RAN (see Figure 7.10) include centralized processing, cooperative radio, transport solutions, switch fabric, and the *real-time cloud*.

7.2.1.1 Centralized Processing

Centralized processing (see Figure 7.11) means that baseband resources are centralized in a baseband location if operators have abundant optical fibers. There is only

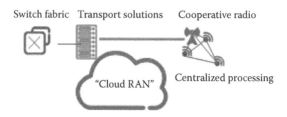

Figure 7.10   Key technology of Cloud RAN.

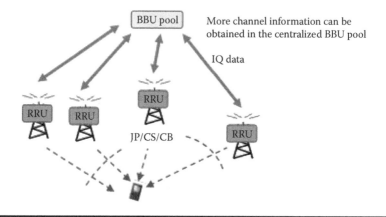

Figure 7.11   Centralized processing. (CB—coordinated beamforming)

outdoor RRU on site. Besides CMCC, many important operators such as Korea Telecom, China Unicom, China Telecom, NTT DoCoMo, and Orange think there are concrete benefits from a baseband pool and that centralized baseband processing makes it easier to implement cooperative processing, although transport bandwidth requirements will increase dramatically. Cloud baseband can significantly improve HetNet capacity through interference mitigation. The interconnection topology among physical processing resources in the baseband pool includes the interconnection among the chips on a processing board, among the boards in a physical rack, and among multiple racks.

7.2.1.2 Cooperative Radio

One key target for the cloud RAN system is to significantly increase average spectrum efficiency and cell-edge user throughput efficiency. Joint processing is the key to achieving higher system spectrum efficiency. To mitigate interference of the cellular system and increase capacity, advanced multicell joint radio resource management (RRM) and cooperative multipoint transmission schemes that can make use of special channel information and realize cooperation among multiple antennas at different physical sites should be developed. To support the CoMP joint processing algorithms, both end user data and UL/DL channel information needs to be shared among virtual eNBs. The interface between virtual eNBs to carry this information should support high bandwidth and low latency to ensure real-time cooperative processing. The information exchanged in this interface includes one or more of the following types: end-user data package, UE channel feedback information, and virtual eNB scheduling information. Therefore, the design of this interface must meet the real-time joint processing requirement with low backhaul transportation delay and overhead.

Key points that should be investigated in the next steps include the following:

- Investigation of CoMP performance gains by considering how many BBUs/ cells should be gathered together
- CoMP performance study in terms of delay requirements to evaluate whether it aligns with that of transmissions over optical in a BBU pool situation
- Feasibility of CoMP implementation in a BBU pool by considering the specific features of CoMP and requirements of transmissions
- Antenna calibration among RRUs

7.2.1.3 Transport Solutions

With the introduction of the baseband pool, the BBU is located at the edge of aggregation and access networks. The access network has to support transport between the BBU and RRU. The optical fiber between BBUs and RRUs has to carry a large amount of baseband sampling data in real time. For example, in the GSM system, a bandwidth (BW) of 1 Gbps is required for 40 carriers (each carrier has 200 kHz BW), and in time-division synchronous code division multiple access (TD-SCDMA), such bandwidth can accommodate only four carriers (each carrier has 1.6 MHz BW). The bandwidth for 20-MHz LTE systems with 8Tx/8Rx antennas is up to 9.8304 Gbps. In the evolution of LTE-A, this bandwidth requirement will sharply expand to 49.152 Gbps. As a result, it is important to investigate the feasibility of cloud RAN by taking into consideration whether an efficient data compression scheme exists to solve the transmission issues and satisfy the requirements. Current understanding of the round-trip delay of transmission is under 5 μs. An alternative way to deploy cloud RAN could be implemented by sharing the same fiber with WDM that currently only uses one wavelength in gigabit PON (GPON). The issues that need to be investigated for GPON scheme include: how to configure and allocate the wavelength, what is the impact on GPON of adding WDM components in the optical path, and so on. From the common public radio interface (CPRI)* requirements, frequency

---

* CPRI (common public radio interface) is an industry cooperation aimed at defining a publicly available specification for the key internal interface of radio base stations between the radio equipment control (REC) and the radio equipment that uses a fixed mapping of in-phase/ quadrature (I/Q) streams.

accuracy should be less than 0.002 ppm, and time delay variation should be less than 8 ns.

7.2.1.4 Switch Fabric

A high-bandwidth interconnection is needed to exchange the in-phase/quadrature (I/Q) data (or traffic data in another form) among BBUs (switch fabric), and its capacity should be able to increase according to increases in BBU pool scale. The control information needs to be exchanged for cooperative processing, and base-band data need to be exchanged for load sharing. Usually more than 1,000 10-Gbps fiber ports will be needed for a 300-cell BBU pool.

7.2.2 Cloud RAN Architecture

Cloud RAN architecture is depicted in Figure 7.12. In the diagram, the nodes of small cell clusters and relay stations were introduced in the access network. Each small cell cluster may cover a portion of a downtown area, business district, or sports stadium. WDM PON and CPRI over an optical transport network (OTN) are adopted. Cloud RAN architecture will allow the wide-area resource pooling that is inspired by the cloud computing concept. So, expanding network capacity is easier and power consumption is decreased so that devices such as eNB antennas can be close together, which reduces transmission power.

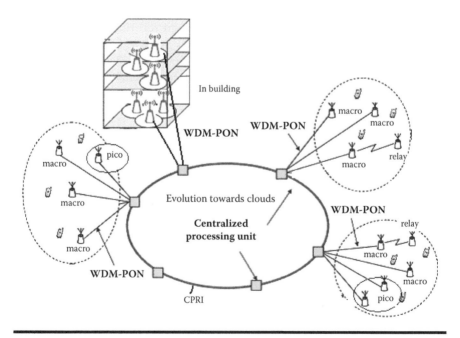

Figure 7.12   Cloud RAN architecture.

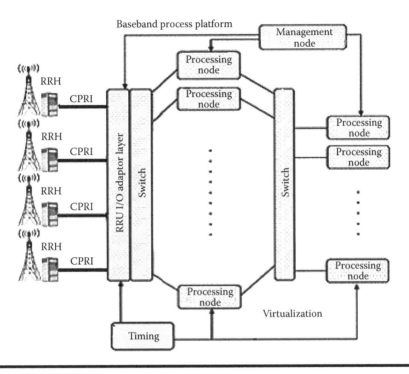

Figure 7.13 Wireless network cloud platform.

WDM PON makes use of wavelengths (frequency multiplexing) instead of dedicated fibers and would be a candidate to enable next-generation optical broadband access for 400-Gbps-per-wavelength optical transmission.

As baseband processing moves into a few central points, processors communicate via CPRI on a high-capacity optical backbone, and via a single fiber to each distant radio. The key technologies of the cloud computing platform (shown in Figure 7.13) include a software-defined radio-based multiprotocol baseband processing platform, virtualization, compute and storage resource scheduling, load balancing, and a CPRI signaling route. The requirements of the cloud computing platform of a cloud RAN includes less than 1-ms "hard" real time to finish DL or UL Layer 1 processing, RRU synchronization for MIMO, delay variation of less than around 8 ns, and Layer 1 processing that can be divided into multiple synchronization tasks on different processors within the hard real-time requirements of virtualization.

With a wireless network cloud, all signal processing, including modulation and encoding of the signals to and from the physical antennas, is carried out using software radio technology. With multicore and multithreading techniques, it is possible to use general-purpose data centers to carry out the signal processing entirely in software. The raw signals are relayed to and from multiple antennas by remote radio heads, via optical fibers from as far away as 40 kilometers.

7.3 Radio Strategy

7.3.1 Key Radio Technology Bottlenecks and Software-Defined Radio

In LTE-A, the large operating bandwidths and complex RAN architecture impose additional requirements on the radio frequency components such as processing capability, antennas, and power amplifiers. These components should be able to operate over a wide range of operating frequencies. Furthermore, high-speed processing units (digital signal processors [DSPs] or field-programmable gate arrays [FPGAs]) are needed to perform computationally demanding signal processing tasks with relatively low delay. Carrier aggregation (CA) has a major impact on eNB RF requirements, which we discussed in Chapter 2. CA has more impact on base station (BS) TX requirements than RX requirements and creates higher requirements for power amplifiers, analog-to-digital converters (ADCs), digital-to-analog converters (DACs), and so on to support working bandwidth up to 40–50 MHz.

Most of the existing ADC bandwidth does not exceed 100 MHz and has been a bottleneck for software radio. ADC and DAC determine the feasibility of software radio. ADC and DAC require a very high sampling rate, wide signal bandwidth, high effective number of quantification bits to support a wide dynamic operating frequency range, and a large spurious-free dynamic range. ADC and DAC must allow recovery of small amplitude signals with very little distortion in strong interference environments and must have low power and low price.

A higher-speed central processing unit (CPU) and more memory for digital signal processing are needed. This may be implemented in an application-specific integrated circuit (ASIC), an FPGA, a DSP, or an algorithm in a software radio. There are three kinds of logical processors: ASIC, FPGA, and DSP. ASIC has the fastest processing speed and the worst flexibility. DSP has the best flexibility and the lowest processing speed. The processing speed of FPGA is lower than ASIC and its flexibility is better than ASIC. The selection of the device is determined by the actual design requirements. Currently, the speed of the DSP can meet the requirements of software radio, but the higher the speed, the greater the power consumption and cost. Figure 7.14 shows an LTE receiver implemented in DSP with FPGA assist, which provides flexibility to implement evolving standards.

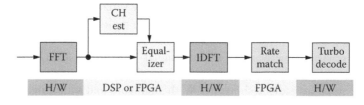

Figure 7.14 An example of LTE receiver design. (FFT—fast Fourier transform; H/W—hardware; IDFT—inverse discrete Fourier transform; CH est—channel estimation)

There are several points that impact eNB implementations in LTE-A.

- Up to 100 MHz will be supported by the eNB. The peak-to-average power ratio (PAPR) will be fixed around 8.4 dB, which is the same as white noise.
- For intermediate frequency/radio frequency (IF/RF) transmitters, multiple power amplifiers for each carrier due to ADC sampling speed for digital pre-distortion (DPD) could be reduced to one power amplifier (PA) if ADC is fast enough for DPD. The eNB needs a low insertion loss combiner to share one antenna and may require synchronization between RRUs if multiple RF chains exist.
- For IF/RF receivers, multiple RF/analogy filters (filter bank) are required to split the signal to a narrowband signal matched to the ADC capabilities and advanced digital processing needs to remove impairments from the filter bank. The number of RX RF chains could be reduced to one if a cost-effective ADC were fast enough.
- Effective algorithm and architecture is desired for performance improvement of MIMO, and power consumption reduction is required for both UE and the eNB along with cost reduction. Due to more complicated digital processing and high data rate services, DSP may require multiple CPUs.

The RF components of the eNB are shown in Figure 7.15.

There are several points that impact UE implementations in LTE-A.

- As with the eNB, up to 100 MHz will be supported by the UE. However, PAPR will be increased if more carriers are aggregated.
- For IF/RF transmitters, UE prefers one transmit chain, and a lower insertion loss combiner is required if there are multiple transmit chains. It will be a challenge to implement based on current ADC capabilities of UE. LTE and

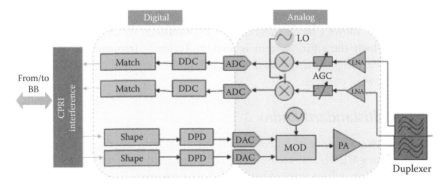

Figure 7.15 RF components. (LO—local oscillator; BB—baseband; AGC—automatic gain control; DDC—digital down converter; ADC—analog to digital converter; DAC—digital to analog converter)

LTE-A UE needs PA linearization techniques that have more power consumption; DPD may be necessary if more carriers are aggregated.

■ For receivers (IF/RF and baseband), the UE is the same as eNB.

Based on the previous discussion, the industry generally believes that software radios are emerging as the approach for multiband, multimode personal communications systems to solve the problems of LTE-A.

The standardized architecture of software-defined radio (SDR) can support both current and future applications. SDR is a radio that includes a transmitter in which the operating parameters of frequency range, modulation type, or maximum output power, or the circumstances under which the transmitter operates in accordance with commission rules, can be altered by making a change in software without making any changes to hardware components that affect the radio frequency emissions. The SDR realizes radio by software. It can be a multiband and multistandard system that supports different frequency bands with significant levels of air interference. SDR supports a multichannel system, which provides more independent transmission and receiving channels at the same time.

In order to reduce costs, it is necessary for SDR to put the ADC as close to the antenna as possible and substitute a general-purpose microprocessor for ASIC for digital signal processing. The ideal antenna should have the following features: wide frequency band, small size, high efficiency, and low price. The ideal preamplifier should have the following features: wide frequency band, low noise, improved linear performance, and low price. The ideal ADC should have the following features: high working frequency, high sampling speed, and a high resolution level, as well as low power consumption and price.

Figure 7.16 shows a block diagram of the ideal software radio. The RF carrier signal received by the antenna is analog to digital (A/D) converted immediately after it is amplified by the low-noise amplifier. After the conversion, the signal is processed digitally by software. All functions, modes, and applications can be configured and reconfigured by software.

Figure 7.17 shows a block diagram of a conventional wireless communication system in which the entire system is analog. Transmit frequencies, modulation type, and other RF parameters are determined by hardware and cannot be changed without hardware changes.

7.3.2 Multistandard Radio

When we deploy a greenfield* LTE site, LTE migration from an existing GSM site and from an existing CDMA site are likely scenarios that need to be supported. The same baseband solution will be utilized for all of these deployment types. Radio solutions

---

* In wireless engineering, a greenfield is a project which lacks any constraints imposed by prior networks. Any new networks designed from scratch to enable new radio access network technologies (i.e., 3G, 4G,) are referred to as greenfield projects.

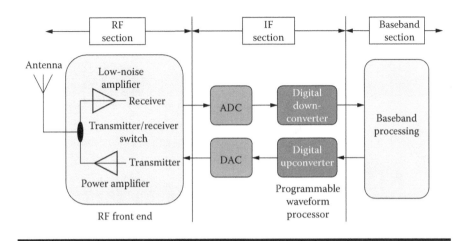

Figure 7.16   Ideal software radio transceiver.

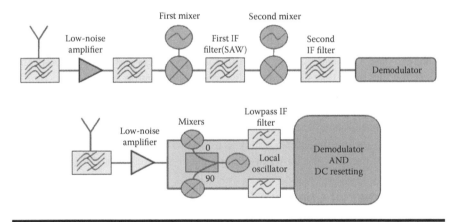

Figure 7.17   Conventional transceiver. (SAW—surface acoustic wave, is a filter that is an electromechanical device commonly used in radio frequency applications)

are required to support LTE, multicarrier GSM with the ability to be software upgraded to LTE, and CDMA 1x, evolution data optimized (EVDO), and LTE simultaneously while interfacing with different technology baseband solutions. A software-defined radio system can easily produce a radio that can receive and transmit widely different radio protocols (sometimes referred to as *waveforms*) based solely on the software used.

In LTE Rel-8 and Rel-9, it is recognized that UEs may be able to transmit and receive on multiple bands simultaneously, but using different radio technologies, for example, high-rate packet data (HRPD)/1x round-trip time (RTT) or Universal Mobile Telecommunications System (UMTS). The introduction of interband carrier aggregation in LTE-A suggests that at least some UEs would be capable of receiving

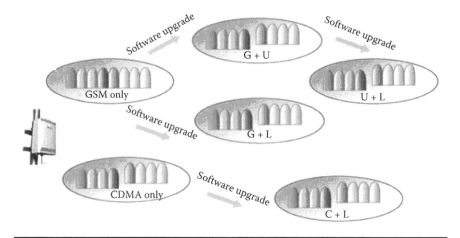

Figure 7.18 GSM/CDMA upgraded to LTE by software. (C + L—CDMA and LTE; U + L—UMTS and LTE; G + U—GSM and UMTS; G + L—GSM and LTE)

and transmitting data simultaneously on different bands, and that this would typically be achieved by the use of multiple independent radio transceivers. A typical UE is likely to have a defined number of multiband transceivers and a defined baseband processing capability that simultaneously supports other radio technologies.

The multistandard radio (MSR) specification establishes the minimum RF characteristics for LTE, Universal Terrestrial Radio Access (UTRA), and the GSM/ Enhanced Data Rates for GSM Evolution (EDGE) MSR base station. Requirements for multiple radio access technology (RAT) and single RAT operation of multistandard radio base stations are met at the present (see Figure 7.18).

7.3.3 Strategy of Cognitive Radio in LTE-A

The future of software-defined radio is cognitive radio, and software radio provides an ideal platform for the realization of cognitive radio. The main components of a cognitive radio transceiver are radio front-end and baseband processing units, which were originally proposed for software-defined radio. A cognitive network is a network composed of elements that, through measuring, sensing, learning, and reasoning, dynamically adapt to varying network conditions in order to optimize end-to-end performance. A cognitive network is shown in Figure 7.19.

In wireless networks, there has been a trend toward increasingly complicated, heterogeneous, and dynamic environments. We have used adaptive radio technologies that adjust themselves to accommodate anticipated events in the live network for 2G and 3G. Beyond adaptive radios, cognitive radio's software-defined radio based platform can further handle unanticipated channels and events. Cognitive radios require sensing, adaptation, and learning. Cognitive radios can sense their environment and learn how to adapt. An enhancement of the traditional software radio concept is that

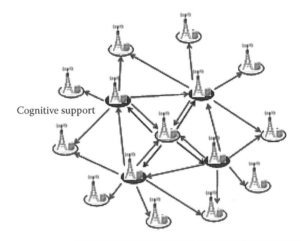

Figure 7.19 Cognitive network definition.

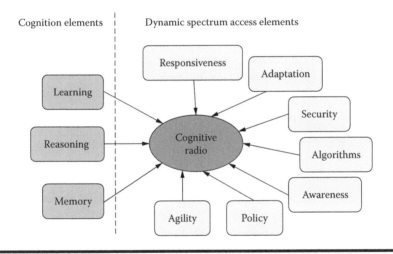

Figure 7.20 CR system function.

the radio is aware of its environment and its capabilities, is able to independently alter its physical layer behavior, and is capable of following complex adaptation strategies. Cognitive radio can improve link performance and spectrum utilization, and we believe it will be widely used in enhanced mobile networks such as LTE-A.

Cognitive radio refers to wireless architectures in which a communication system does not operate in a fixed assigned band, but rather searches and finds an appropriate band in which to operate. The basic idea of the initial cognitive function is shown as Figure 7.20. The receivers obtain channel quality information and interference information from the surrounding radio environment by observing.

After the transmitters receive the necessary feedback information from their corresponding receivers, they determine the strategies that react to the radio environment. For more intelligent function, learning is adopted to estimate the utility of possible strategies to improve system performance.

Cognitive radio is widely used in several areas of wireless communication, and many researchers are trying to put the idea of cognitive radio into the LTE network. The feature of autonomous component carrier selection is a potential simple cognitive method for LTE-A. Each node selects the carrier(s) that are suitable, but only to the extent that excessive interference is not created for neighboring nodes. The decisions are based on collection of local measurements, and each node "learns" the local environment via sensing. In LTE-A, we should move toward a cognitive radio light approach.

The proposed autonomous component carrier selection scheme is further illustrated in Figure 7.21 with a simple example. There are four existing eNBs. A new eNB 5 is being switched on, and hence is ready to first select its primary component carrier (PCC). The current selection of PCC and secondary component carriers (SCCs) is illustrated for each eNB with "P" and "S," respectively. Component carriers not allocated for PCC or SCC are completely muted and not used for carrying any traffic.

As a new eNB is initialized, the new inter-eNB measurements based on reference signal received power levels for the purpose of estimating the path loss between

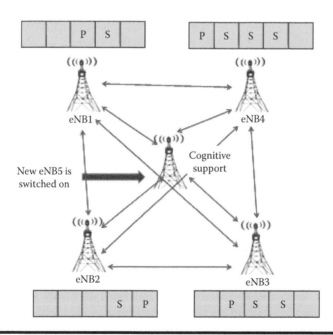

Figure 7.21 **The autonomous component carrier concept.**

neighboring eNBs should be taken. In FDD systems, this implies that eNBs are able to listen to the downlink band as well. It is proposed that the new eNB carries out the measurements on the primary component carrier of the surrounding cells and that knowledge of their corresponding reference signal transmit power is available (signaled between eNBs), so that the inter-eNB path loss can be estimated. So the new eNB is able to select the best component carrier (CC) for primary. After the new eNB has selected its primary component carrier and it is ready to transmit and carry traffic, the eNB will constantly monitor the quality of the PCC to make sure that it continues to have the desired quality and coverage. If poor quality is detected, recovery actions will be triggered to improve the situation.

In this case, the new eNB is not allowed to cause too much interference, which may interrupt communication or decrease the service quality of the four existing eNBs. Hence, the new eNB can opportunistically utilize the spectrum when it is not being occupied by the existing eNBs. The schemes need to achieve a balance of the trade-off between the utility of the new eNB and its influence on existing eNBs.

In LTE-A implementation scenarios, there may be many HeNBs (picocells) in a macrocell. These HeNBs (picos) are located in an ad hoc way and the topology of HeNBs (picos) would even be dynamic since they are placed and turned on and off by customers. Thus, cognitive methods should be used in a deployment with both macrocells and femtocells, which is shown in Figure 7.22. The HeNBs (picos) should take the responsibility to adjust their transmit strategy to minimize the impact on the macroUE (MUE). First, the HeNB should be able to learn its position in the macrocell and interference with a nearby macroUE, and then adjust its subcarriers or transmit power based on local sensing based on its location and interference.

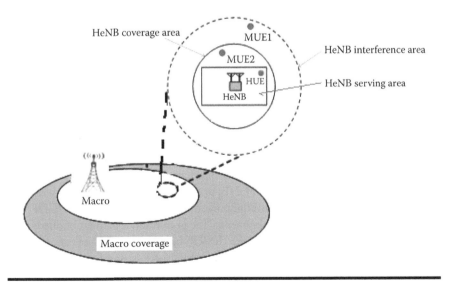

Figure 7.22   HeNB is employed in the center of macrocell.

So, dynamic spectrum management is an efficient method to avoid the interference between eNBs. When a spectrum is idle, the cognitive radio systems choose the spectra that have low interference. If the macrocell comes back to use the spectrum occupied by HeNBs, cognitive radio systems should obtain the information in time. Based on the information, the HeNBs might decrease the transmit power to avoid too much interference to macrocell if there is no other candidate spectra. In such a scenario, by considering the traffic load and the potential to cause interference at each other's UEs, a dynamic resource block (RB) (in terms of time and frequency resource units) allocation and power control scheme of the macro eNB (MeNB) and HeNBs (picos) for interference management can be allocated.

In conclusion, cognitive radio is one of the research frontiers in the wireless communication field and cognitive radio networks will provide high bandwidth to mobile users via heterogeneous networks and cooperative spectrum-sharing techniques in LTE-A. Through this capability, noise and interference estimation and avoidance is easier, and the best spectrum band and the most appropriate operating parameters can be selected and reconfigured. MIMO can decrease the interference by adjusting the signal orthogonal to the interference channel of existing users in cognitive radio networks to increase the throughput of new users and decrease the interference for existing users. However, there are still some research challenges for LTE-A heterogeneous networks. For example, distributed cooperative spectrum sensing needs further research to better balance the trade-off between accuracy and overhead. Considering the error of spectrum sensing, the resource allocation schemes should restrict the outage probability that the new users will interrupt the communication of existing users.

7.4 Interference Analysis of Heterogeneous Network

7.4.1 General Description

The new nodes (picocells, HeNBs, femtocells, or relays) we have talked about change the topology of the system to a much more heterogeneous network with a completely new interference environment in which nodes of multiple classes compete for the same wireless resources. In a heterogeneous network, the co-channel interference would be the interference between the macro eNB and the local eNB, between local eNBs, or between UEs. Co-channel interference management is particularly critical when the micro- or picocell, HeNB, or relay node coverage overlays a macrocell. Figure 7.23 presents the UE throughput cumulative density function (CDF) for macrocell only and co-channel deployments. The performance gain is obtained from using the reference signal received power (RSRP)-based serving cell selection under co-channel deployments. It can be seen that while there is significant throughput improvement for a small percentage of UEs, there is only marginal improvement in tail and median UE throughputs. But Figure 7.24 shows the UE throughput CDF

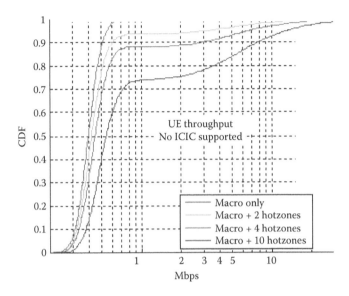

Figure 7.23 UE throughput CDF for macro only and co-channel deployments.

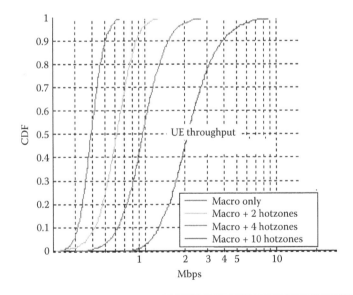

Figure 7.24 UE throughput CDF for range expansion and coordinated interference management.

via enhanced serving cell selection and coordinated interference management. This can provide more substantial gain in tail and median UE throughputs.

Interference problems between different deployment layers in HetNet have traditionally been avoided by using different carriers for the different layers, but what is now being discussed for LTE is the possibility of considering exploiting carrier aggregation and cross-carrier scheduling techniques between different layers. As we know, the interference coordination methods defined in Rel-8 systems aim to solve the interference problem in homogeneous networks. Due to the smaller transmission power and coverage of the lower power nodes, current interference coordination methods cannot be directly applied to handle the interference problem in heterogeneous networks. Thus, an enhanced interference coordination method should be considered for heterogeneous networks.

7.4.1.1 Interference Analysis

Interference management in a heterogeneous network is critical due to strong interference conditions from an unbalanced power profile and co-channel overlaid deployment. In particular, the macro's transmit power is much higher than the low-power nodes. In the case of co-channel deployment of macro eNBs and outdoor pico nodes, pico UEs will suffer from macro interference, and pico nodes will suffer from macro UE interference, while macro UEs will suffer from HeNB interference if macro UEs are close to the CSG HeNB. The DL control channel hearability is one critical issue in heterogeneous network deployment when the macrocells interfere with the local cells. So the LTE-A interference avoidance mechanism is important to efficiently support highly variable traffic loading as well as increasingly complex network deployment scenarios with unbalanced transmit power nodes sharing the same frequency and coordination among macro and local cells autonomously through the X2 interface.

The DL control channel is located in the first one to three symbols of the subframe. The principle of interference mitigation for improving control channel hearability is coordinated muting or softening. The coordinated muting or softening could be realized by interlacing the control channel symbols between the macro and local cells, or interference coordination. The interference coordination could be accomplished through fractional frequency reuse, coordinated component carrier selection, or almost blank subframes.

Interference for data traffic should be managed together with DL control channel interference coordination. The interference management scheme for data traffic also depends on the deployment scenario in a heterogeneous network. The data traffic in the macro and picocells could be easily coordinated through coordinated scheduling, coordinated beamforming, or joint processing to mitigate the interference. The interference management for HeNB deployment would rely on frequency reuse or component carrier selection due to the fact that the macro eNB cannot engage with and control a HeNB in the same way as a picocell.

The most basic mean to operate a HetNet is to apply complete frequency separation between different layers, but the drawback of operating layers on different carrier frequencies is that it may lead to resource utilization inefficiency. Carrier aggregation of component carriers combined with cross-carrier scheduling is supported in LTE-A and multicarrier operations in HetNet appear to be the key for compatibility with legacy UEs, although these UEs cannot access more than one component carrier.

In a heterogeneous system, CA-based solutions are attractive for situations with large availability of spectrum and UEs with CA capability; non-CA-based (i.e., co-channel) solutions are important to enable efficient heterogeneous network deployments with small bandwidth availability and UEs without CA capability. A multicarrier system may be configured so that cells of a certain power class are allocated only a subset of the DL carriers and are not allowed to transmit on the remaining DL carriers. As a result, low-power picocells in the vicinity of a high-power macrocell may use carriers not used by the high-power cell to serve their own UEs, without being interfered with by the DL transmission from the high-power cell. Similarly, open access cells in the vicinity of a CSG cell (e.g., closed HeNB) may use carriers not used by the CSG cell to serve its own users, without being impacted by the interference from the CSG cell, thereby preventing outage to non-allowed users in the RF coverage of the CSG cell.

With UL transmissions in heterogeneous networks, two cases should be distinguished. At the edge of the macrocell, very high interference levels exist (HeNB 2 in Figure 7.25) because macro UEs utilize high transmit power to overcome high path loss toward their serving base stations. In this case, local cell users (femto UEs) are the main victims. On the other hand, close to a macro eNB, femto→macro interference may become a serious threat (HeNB 1 in Figure 7.25).

7.4.1.2 Interference Scenarios

Heterogeneous deployments consist of deployments where low-power nodes are placed throughout a macrocell layout. The interference characteristics in a heterogeneous deployment can be significantly different from those in a homogeneous deployment. The interference scenarios can be classified into the following categories: interference due to HeNBs, interference due to different transmit power (i.e.,

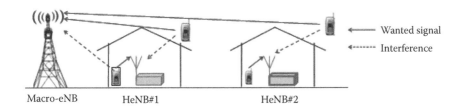

| Macro-eNB | HeNB#1 | HeNB#2 |

—— Wanted signal
◄········· Interference

Figure 7.25 Interference paths in UL.

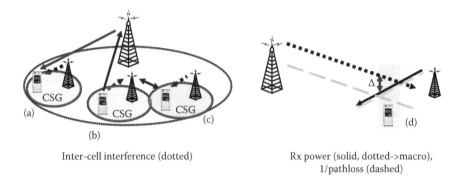

Inter-cell interference (dotted)

Rx power (solid, dotted->macro), 1/pathloss (dashed)

Figure 7.26 Examples of interference scenarios in heterogeneous deployments.

between high-power [e.g., macro eNB] and low-power [e.g., pico eNB]) base stations, interference in multihop deployments (i.e., relays). Examples here are given in Figure 7.26. In case (a), a macro user with no access to the CSG cell will be interfered with by the HeNB; in case (b), a macro user causes severe interference toward the HeNB; and in case (c), a CSG user is interfered with by another CSG HeNB. On the right-hand side, case (d), path loss-based cell association (e.g., by using biased RSRP reports) may improve the uplink but at the cost of increasing the downlink interference (up to Δ in Figure 7.26) to non-macro users at the cell edge.

Case (a) and case (d) are the most important scenarios in LTE-A. In these scenarios, preliminary results indicate that methods for handling the uplink and downlink interference toward data as well as Layer 1 and Layer 2 control signaling, synchronization signals and reference signals are important. Such methods may operate in time, frequency, and/or spatial domains.

Currently, in order to improve network capacity without additional spectrum (single carrier) and costly network planning (deploy underlaid cells instead of splitting overlay cells), the interference avoidance and control method will be adopted and can be categorized into the complete frequency separation scheme, the CA-based scheme, and the non-CA-based scheme.

The complete frequency separation scheme is to operate a HetNet to apply a different frequency between different layers and thereby avoid any interference between layers. This is the conventional approach to HetNets that has been used since the early days of GSM mobile communication. With no macrocell interference toward the underlaid cells, cell splitting gains are achieved when all resources can simultaneously be used by the underlaid cells. The drawback of operating layers on different carrier frequencies is that it may lead to resource utilization inefficiency.

7.4.1.3 Control Channel Interference

Interference between the physical DL shared channel (PDSCH) from the relevant cells can be avoided or mitigated by scheduling, for example, choosing different frequency domain resources. Interference between CRSs from adjacent cells can usually be avoided by cell ID planning to give suitable frequency shifts. The remaining issue is interference between control channels.

Good control channel performance is a precondition for both good uplink and downlink data channel performance. For an LTE-A HetNet system, in order to provide proper network functionality, it is assumed that a block error rate (BLER) of less than 1% for control channels is needed. In Figure 7.27, the approximations of minimum SINR for different control channels are listed.

The two control channels with worse performance are the physical HARQ indicator channel (PHICH) and physical DL control channel (PDCCH). For example, in LTE Rel-8 the main mechanism for mitigating the interference of the PDCCH is randomization of the REs in the frequency domain over the system bandwidth and on the time domain over the orthogonal frequency-division multiplexing (OFDM) symbols reserved for the control channel. In the case of the PDCCH, power boosting (reducing power on some control channel elements [CCEs] to allow for increased transmit power on other CCEs) can be used, but at the cost of reduced capacity for the control channel area. On the other hand, less intercell interference is generated in the PDCCH when fewer CCEs are occupied, the SINR in the PDCCH could be modeled as follows:

$$\mathrm{SINR}_{\mathrm{PDCCH}} = S / \left(\sum_{i=1}^{M} Load * I_i + N \right)$$

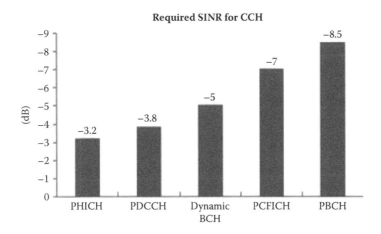

Figure 7.27 SINR thresholds for 1% BLER on different control channel. (BCH—broadcast channel; PCFICH—physical control format indicator channel; PBCH—physical broadcast channel)

where *S* is the average receiving power per subcarrier of the PDCCH; *Load* is the PDCCH load factor, which equals the percent of CCE being occupied for sector *i*; I_i is the average receiving interference power per subcarrier in the control region from sector *i*; and *N* is noise power. From the equation we can see that PDCCH load control is an effective method to mitigate the intercell interference in the PDCCH.

If we adopt the plain co-channel deployment of macrocells and pico/HeNBs, the macrocell UEs would have general downlink control channel reception problems if located close to the pico/HeNB. Although pico/HeNB power control is enabled, it is not completely circumvented, as relatively small macrocell coverage holes still exist.

In a CA-based scenario, two different cells (for example, a macrocell and a picocell) will transmit control channels on two different component carriers, and transmit data channels with co-channel transmission. This means that macro UEs close to pico/HeNBs can always be served with good quality on the carrier free of pico/HeNBs. Macro UEs located farther away from pico/HeNBs can also be served without problems on the co-channel deployed carrier. A cell can also use different transmission power for the control channels on two different component carriers.

In a non-CA-based scenario, ICIC across the cell layers of channels/signals such as the PDCCH, the physical broadcast channel (PBCH), and the primary synchronization signal (PSS) and the secondary synchronization signal (SSS) can be done by the time/frequency domain ICIC schemes in order to avoid collisions across the cell layers. This may imply either new locations of these channels/signals, or introduction of blank subframes and requirements on interference cancellation in the UE.

7.4.2 CA-Based ICIC Technology of the Heterogeneous Network

Carrier aggregation can be very beneficial in heterogeneous network deployments. Multiple carriers enable interference management between different power class cells as well as between open access and CSG cells. Long-term time or frequency domain resource partitioning can be carried out by exclusively dedicating carriers to a certain power class cell (macro/pico/CSG).

Carrier aggregation–based HetNets mean that the available bandwidth will be split into multiple carriers. One of the carriers is configured with a primary serving cell (PCell) for UEs connected to the macro network whereas the PCell for UEs connected to the pico network is configured on the other carrier. The other carrier in each cell area is configured with secondary serving cells (SCells). This is shown in Figure 7.28.

To improve cell-edge user throughput and coverage and deployment flexibility, LTE-A will adopt enhanced intercell interference coordination (eICIC). CA can be used for frequency domain coordination, time domain coordination, power control, and fractional frequency reuse (FFR). Overlaid deployment of cells with different TX power schemes (see Figure 7.29) are also to be introduced.

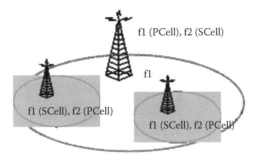

Figure 7.28   PCell and SCell allocation for carrier aggregation–based HetNets.

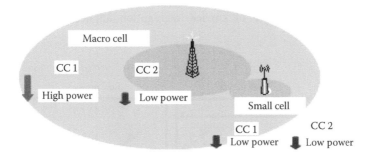

Figure 7.29   Example of cell deployment for heterogeneous network deployment.

For the CA-based solution, frequency-division multiplexing (FDM) of cell layers by means of cross-carrier scheduling (CA with cross-carrier scheduling with a carrier indicator field [CIF]) can be used for heterogeneous deployments. Downlink interference for control signaling can be handled by partitioning component carriers in each cell layer into two sets, one set used for data and control, the other set used mainly for data and possibly control signaling with reduced transmission power. One example is illustrated in Figure 7.30. For the data part, downlink interference coordination techniques can be used. Rel-8 and Rel-9 terminals can be scheduled on one component carrier while an LTE-A terminal capable of carrier aggregation can be scheduled on multiple component carriers. Time synchronization between the cell layers is assumed in this example.

In this case, component carrier–level interference coordination for the PDCCH is useful for a reliable PDCCH transmission in an operation scenario with a heterogeneous network because UE needs to receive the PHICH and PDCCH correctly in order to know the PDSCH resource assigned and subsequently perform PDSCH decoding. Another example is for the scenario where a control channel in a small cell (femto or relay) suffers strong interference from the macrocell and a component carrier with PDCCH-less operation is added to a heterogeneous network.

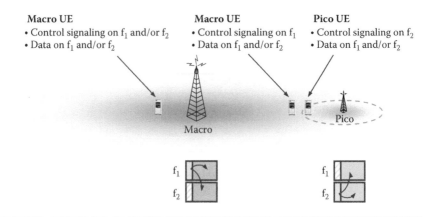

Figure 7.30 One example of carrier aggregation applies to heterogeneous deployments.

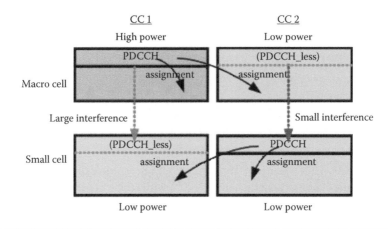

Figure 7.31 Example of PDCCH-less operation for heterogeneous network deployment.

The transmit power in one component carrier in the macrocell is decreased to reduce the intercell interference with a femto/relay cell. In order to allow reliable PDCCH transmission, it is beneficial to transmit all PDCCHs on the component carrier with less intercell interference. In such an operation, the component carrier where PDCCH is not transmitted is PDCCH-less. Figure 7.31 shows an example of PDCCH-less operation in a heterogeneous network. In the macrocell, a higher transmit power is used on DL CC1 while a lower transmit power is used on DL CC2. Since PDCCHs on DL CC1 in the femto/relay cell suffer strong interference from the macrocell, PDCCH-less operation is configured for DL CC1 in the femto/relay cell. The PDSCH resources can be efficiently utilized by an RB-level

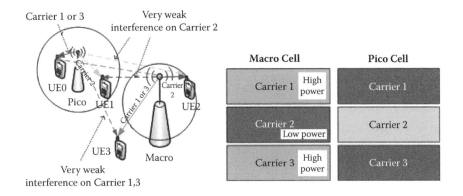

Figure 7.32 A cell can use different transmission powers on different component carriers.

interference coordination between a macro and femto/relay depending on the traffic load of the macro and femto/relay.

In this example, for the macrocell: PDCCH-less operation in CC2 is configured for reliable control channel reception in small cells; PDCCH in CC1 schedules PDSCH in CC1 and CC2. For the small cell: PDCCH-less operation is configured in CC1; all PDSCHs are scheduled from the PDCCH in CC2; and physical resource block (PRB)-level ICIC for the PDSCH (FFR) is performed.

Another example of carrier reuse is shown in Figure 7.32. Carriers 1 and 3 are open-access shared carriers with unrestricted power, which are used by macro- and picocells up to their maximum power. Carrier 2 is the open-access shared carrier with low power that is used by the macrocell with reduced power (relative to its maximum power) and by the picocell with its full power (which is low by configuration).

Picocell-served UEs 0 and 1 can be scheduled on carrier 2. UE 0 can also be scheduled by the picocell on carriers 1 and 3 since the interference from the macrocell seen by that UE on those carriers is very weak compared to the signal power received from the picocell (high SINR). UE 1 experiences strong interference from the macrocell on carriers 1 and 3, so it could be scheduled by the picocell on carrier 2, but only if no other interference coordination between the macrocell and picocells is present. UE 2 can be scheduled by the macrocell on carrier 2, since it is close enough to the macrocell and falls within the range of coverage of low-power carrier 2. UE 3 is outside of the coverage range for carrier 2 (due to low transmit power), so it cannot be scheduled there.

In these examples, allocating only a subset of carriers to the macrocells enables multiple picocells in their vicinity to significantly expand their coverage on carriers not used by the macrocell. UEs that report significant interference from the other macro sites on a certain carrier would not be scheduled on that carrier.

In addition to carrier partitioning, fractional frequency reuse schemes within a carrier can also be used in a heterogeneous network, and this can be much more dynamic than carrier partitioning. Application of fractional frequency reuse to cells in heterogeneous networks allows for a shared carrier to be efficiently used with finer frequency resource partitioning.

7.4.3 Non-CA-Based Interference Technology of the Heterogeneous Network

For non-CA-based deployments of heterogeneous networks, the main motivation of interference coordination is to avoid transmission collision in the same time/frequency resources, so data and control ICIC can be implemented by using either the time or frequency domain. On the other hand, the dominant interference condition has been shown when non-CSG and CSG users are in close proximity to a femtocell. In this case, Rel-8 and Rel-9 ICIC techniques are not fully effective in mitigating control channel interference; therefore, enhanced interference management is needed that should have backward compatibility with Rel-8 and Rel-9 UEs.

Since access to picos is open to all UEs, macro eNBs will cause interference to UEs served by the pico eNB in the downlink. The time domain and the frequency domain partitioning schemes are considered equivalent because both time and frequency can be treated as resources for the data channel. Therefore, it is suggested to adaptively implement the time domain partitions based on the topology and user distribution. In addition, the macro can be silenced in one part of the bandwidth, while the pico can operate in the complete bandwidth. The advantages and disadvantages of time domain and the frequency domain partitioning schemes are described in Table 7.3.

LTE-A introduces a set of new downlink time and frequency domain–enhanced intercell interference coordination concepts for HetNet cases where operators have only one carrier available for LTE deployment. In the enhanced ICIC feature of non-CA (assume time synchronization between cell layers), UEs are configured with measurement (radio link monitoring [RLM], RRM, and CSI*) restriction patterns via dedicated radio resource control (RRC) signaling, and the configured UEs are only allowed to make measurements in certain subsets of subframes. The measurement fluctuation based on the cell-specific reference signal (CRS) among different subframes would occur at the victim UEs. The CRS at the victim cell would still suffer serious interference from non-ABSs (almost blank subframes) in the aggressor cell, but in ABS this kind of interference can be mitigated. RLM, RRM, and CSI measurement accuracy will be impacted by this kind of fluctuation.

---

* RLM: Measurements to determine radio link failure (RLF). RRM: CRS measurements from the UEs for handover decisions including RSRP and reference signal received quality (RSRQ). CSI: Measurements from the UE for packet scheduling and link adaption decisions including CQI, PMI, and resource indicator (RI).

Table 7.3 Time Domain and Frequency Domain Partitioning Schemes

	TDM	*FDM*
Advantages	Since the transmissions of pico- and macrocells are orthogonal to each other, there is no interference from the macro to the pico.	Can be used for asynchronous networks
Disadvantages	This method is not possible in asynchronous networks. The picocell cannot be located very close to the cell edge. Otherwise, all the macros and picos in the network need to decide on a fixed partitioning. Fixed partitioning would reduce performance. Overlap between picos is not possible. Loss in throughput due to inefficient use of resources	Picocells cannot overlap each other. Some of the picos will always have interference from the macro if they are scheduled in the first half of the total bandwidth. At the cell edge, the pico allotted in the first half of the bandwidth will also face interference from the neighboring macro.

7.4.3.1 Time Domain Scheme

7.4.3.1.1 Downlink

By restricting the transmission of DL control signals during part of the time, possibly in combination with time shifting,* at least one cell layer can reduce interference to control signaling on other cell layers. The basic idea of time-division multiplexing (TDM) solutions (shown in Figure 7.33) is to configure some subframes of the aggressor cell as ABSs to avoid interfering with UEs served by victim cells. Almost blank subframes are subframes with reduced transmit power (including no transmission) on some physical channels and/or reduced activity. The eNB ensures backward compatibility toward UEs by transmitting necessary control channels and physical signals as well as system information. The ABS is a fully backward-compatible subframe with reduced transmission, and control signaling such as CRS, PSS, SSS, PBCH, SIB1, paging, and PRS are still transmitted. Multicast-

---

* The interfering cell configures an ABS/MBSFN subframe at subframe 0, 5 of the interfered cell to protect the PBCH/PSS/SSS/SIB1 (system information block 1). The interfering cell configures an ABS/MBSFN subframe at one of the subframes 0, 4, 5, or 9 of the interfered cell to protect paging. (Paging protection or idle-mode eICIC will not be considered in LTE-A).

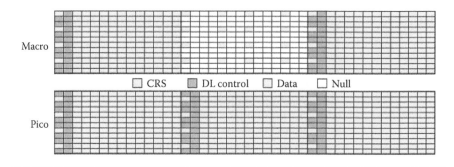

Figure 7.33 Time-domain scheme.

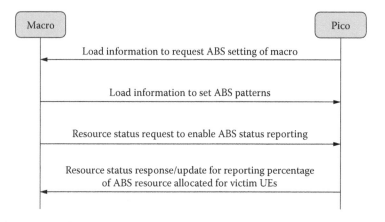

Figure 7.34 X2 interface to exchange the ABS information dynamically (macro–pico configuration).

Broadcast Single-Frequency Network (MBSFN) subframes can be used when they are also included in ABS patterns. MBSFN subframes cause less interference than ABS because their PDSCH region is CRS-free. However, the MBSFN subframe can be configured only on subframe numbers 1, 2, 3, 6, 7, 8, which is less flexible.

For eNBs such as macro, micro, and pico with an established X2 interface, the configuration of an ABS muting pattern can be more dynamically negotiated and optimized. The aggressor eNB notifies its victim eNB of its almost blank subframe pattern. The victim eNB then uses the ABS pattern to serve its UEs (with reduced interference). Note that time synchronization among relevant eNBs across the network is needed to enable this feature. The ABS pattern is set up through either X2 signaling (macro–pico configuration) or O&M (operation and maintenance) configuration (macro–femto configuration).

A typical X2 coordination to apply ABS-based eICIC and detailed X2 signaling between eNBs is shown in Figure 7.34.

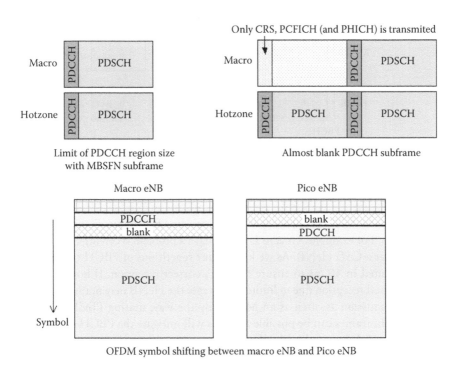

Figure 7.35 Example of increasing sparseness of PDCCH region.

A pico can first request an ABS setting from a macro by sending load information with an invoke indication with only one case, which means "I want ABS." It is possible for the pico to exchange more information, such as ABSs the pico wants, to the macro by vendor private signaling if the macro and pico are from the same vendor. The macro can set ABS patterns according to its own capacity requirement and load. The macro can also trigger ABS status reporting from the pico to learn about the percentage of ABS resource usage in the pico and further update its ABS setting.

In LTE Rel-8, the main mechanism for mitigating the effect of interference on the PDCCH is randomization of the resource elements in the frequency domain over the system bandwidth and on the time domain over the OFDM symbols reserved for the control channel. Interference management for the PDCCH is not supported in Rel-8 or Rel-9. It seems that to significantly mitigate the interference, extreme limitation of the usage ratio in the PDCCH region is necessary. For example, increasing the sparseness of the PDCCH region or almost blank subframe, or symbol shifting by limiting the usage ratio can mitigate interference without any specification change or constraint, but this decreases the number of co-scheduled UEs and may degrade system performance. Figure 7.35 offers some methods of increasing the sparseness of the PDCCH region.

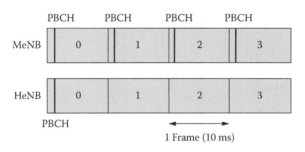

Figure 7.36 Muting PBCH of HeNB.

Actually, the TDM-based method is muting some RE resources and is reserved for pico UEs. For example, a macro UE may have a problem receiving the PBCH if it locates near a CSG HeNB. As we know, four repetitions of PBCH transmission will be executed in 40 ms to ensure the UE's correct reception. If home UEs are capable of good reception due to limited coverage, the HeNB may not have to make PBCH transmission as often as an MeNB. By the way, muting PBCH transmission at certain frames can be possible and this will mitigate the PBCH interference between the HeNB and the MeNB. When muting that region, the HeNB may allocate data on that region with an RB-based inference mitigation scheme or let it empty to fully avoid interference (see Figure 7.36). In reality, the muting method option is dependent on the impact of its PBCH reception quality.

In brief, to make pico UEs work, a time domain scheme where some parts of the time/frequency resources are reserved only for pico UEs is fairly reasonable. Without interference from the MeNB, the pico UEs can work well in the reserved resources and the cell-edge UE performance can be improved.

7.4.3.1.2 Uplink

Similar to the downlink, the interference problems among the uplink channels, such as PUCCH, PUSCH, and physical random access channel (PRACH) in a heterogeneous network can be avoided by coordinated transmission. In the LTE Rel-8 physical layer design, the PUCCH resource is identified by a cyclic shift for a constant amplitude zero autocorrelation (CAZAC) base sequence and an RB index (position in frequency domain). Thus, uplink control information (UCI) from UEs in different RBs is naturally orthogonal in the frequency domain, while UCI from UEs in the same RB is orthogonal in the code domain. In the scenario where a low-power node is deployed far from the MeNB, the PUCCH/PUSCH transmitted by UEs that are served by the MeNB may bring large interference to the PUCCH/PUSCH from UEs served by the low-power node, and impact the detection performance of the PRACH from UEs served by the low-power node. It will thus lead to poor throughput performance, and random access inside the low-power node may suffer continual failure.

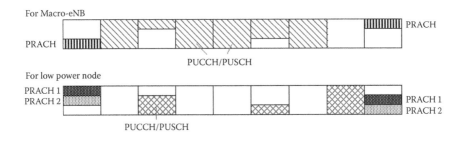

Figure 7.37   TDM-based PUCCH, PUSCH, and PRACH scheduling.

For the PUCCH, because the base sequences and the derived reference signal sequences for the PUCCH that is used in MeNB and low power node are typically not orthogonal, the UCI interference between MeNB UEs and low-power-node UEs is not negligible. The existing intercell interference randomization mechanism is not robust enough in this case.

Because PRACH is used for initial network access, radio link reestablishment, handover, uplink synchronization, and scheduling requests, the detection failure of the PRACH will impact the performance of UE access and uplink synchronization.

For example, since the time domain structure of PRACH is very different from that of PUCCH and PUSCH, the orthogonality is hard to maintain if both of them are allocated in the same subframe. For homogeneous networks, this problem is not significant because the power levels of these two kinds of uplink channels are the same; even if there is leakage from one channel to the other, the leakage power is still relatively small. However, for heterogeneous networks, especially when these two kinds of uplink channels are for eNBs with different required transmission power levels, the intercell interference problem may become significant.

To mitigate the intercell interference between the PRACH in the low-power node and PUCCH/PUSCH in the overlaid macrocell, it is necessary to coordinate the PRACH allocation between the low-power node and the macrocell. The PRACH for the low-power node is only allocated in subframes where there is no PUCCH/PUSCH in the macrocell, as illustrated in Figure 7.37.

7.4.3.2 Frequency Domain Scheme

The frequency domain scheme refers to the bandwidth to be assigned to each hot zone and macro node during deployment. Usually we have three simple static frequency allocation schemes as shown in Figure 7.38. The first is the same bandwidth allocation to both macro nodes and low-power nodes; the interference mitigation will rely on frequency selective scheduling and beamforming. The second is a nonoverlap scheme; this case allows some interference mitigation by assigning nonoverlapping bandwidth to different nodes. The last is an overlap scheme; this case assumes that macro nodes will be assigned to part of the frequency band, but hot zone eNBs can still use the

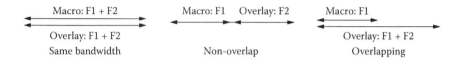

Macro: F1 + F2   Macro: F1   Overlay: F2   Macro: F1

Overlay: F1 + F2   Overlay: F1 + F2
Same bandwidth   Non-overlap   Overlapping

Figure 7.38   Frequency domain scheme.

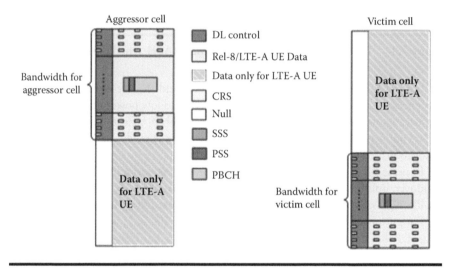

Figure 7.39   Control channel shrinking.

entire band. While assigning nonoverlapping bands is a natural way to avoid interference between the macro and the hot zone, it may not be that straightforward to see the advantage of the overlapping scheme. This particular example of a macro using partial bandwidth and the hot zone taking the entire band reflects the following concept: Hot zone UEs will have a private band on which significant throughput gain can be obtained because of the removal of the macro interference. At the same time, macros UEs may not see much interference from the hot zone eNB. Hence, they may not see a big improvement even if the interference from hot zone eNBs is removed as in the nonoverlapping scheme. The effectiveness of such an overlapping strategy depends on the relative number of UEs attached to the hot zone and macro eNBs.

Using different parts of the carrier bandwidth to transmit DL control signals in different cell layers is another solution that has been considered in LTE-A, which involves reduced bandwidth for control channels (PDCCH, PHICH, PCFICH, etc.) and physical signals (synchronization, CRS), such that the control channels and physical signals can be totally orthogonal to those in another layer. Figure 7.39 shows that Rel-8 and Rel-9 UE accesses and operates on the reduced bandwidth without change, whereas LTE-A UE accesses the reduced bandwidth as a Rel-8/Rel-9 UE, but may be scheduled over the entire bandwidth. Reduced bandwidth

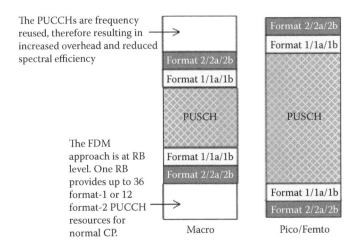

Figure 7.40 Resource by coordinated transmission for PUCCH. (RB—resource block; CP—cyclic prefix)

for the aggressor and victim cells is backward compatible with Rel-8 and Rel-9 UE. Figure 7.39 is an illustration to show the frequency location of PSS/SSS/PBCH and does not imply that they are always present in each subframe.

The DL control channel does not span the whole bandwidth of the carrier. In this way, DL control channels with high power could transmit on the non-overlapped frequency region between eNBs. Obviously, the non-CA-based FDM solution is totally orthogonal, and this will be helpful to protect the victim cell's detection performance. Compared with the CA-based solution, the UE does not need to have CA capability.

For the uplink, the interference to the PUCCH in the heterogeneous network can be avoided by FDM-based coordinated transmission. For example, a macro system and a lower-power node system will use different uplink resources to transmit the PUCCH to avoid interference. The method is to make the UEs that may cause uplink interference to other eNBs transmit the PUCCH in orthogonal resources, as shown in Figure 7.40.

7.4.3.3 Time-Shifting Scheme

Time-shifting solutions, including OFDM symbol shifting and/or subframe shifting, can eliminate control channel interference and interference to other common control channels such as PBCH, primary and secondary synchronization channels (P/S-SCH), but the picocell's control channels and CRS would still suffer interference from the macrocell's PDSCH. Thus, interference management between the control channel and data channel would be needed. For example, macrocell power reduction or muting on the portion of a symbol that overlapped the control region

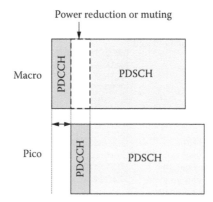

Figure 7.41 Time shifting with power reduction at the symbol level.

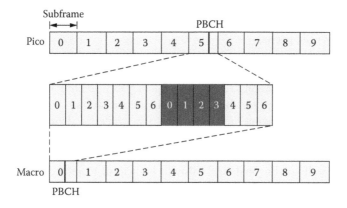

Figure 7.42 Subframe level shifting.

of the picocell could be used to decrease the interference from the macrocell data channel to the picocell control channel (see Figure 7.41).

Another example is PBCH interference. Because the PBCH is different from the PDCCH in transmission characteristics, four repetitions of transmission will be executed during 40 ms to ensure the UE's correct reception and the aforementioned interference mitigation mechanisms designed for PDCCH may not work well. Typically, the PBCH is transmitted at the first subframe in a radio frame; the pico or HeNB may transmit its PBCH at a different DL subframe. In this case (see Figure 7.42), the pico/HeNB still has a synchronized frame structure with the MeNB and only the PBCH is shifted to different subframes (e.g., the sixth subframe in this example; note that the sixth subframe will always be the DL subframe for all configurations).

To support this case, the UE will know whether the cell is a pico/HeNB or not, and then go to the determined subframe for PBCH reception. This can be done by partitioning a synchronization signal to two groups, one for MeNB and another for the pico/HeNB's usage. This will increase the complexity of cell selection and put limitations on cell ID configurations.

In addition to the TDM and FDM schemes, downlink interference could also be mitigated by a power control technique for improving macro UE performance by restricting the expected received power of femto UE. Actually, the interference to the PUCCH in HetNet can also be mitigated by power control in the uplink.

7.4.3.4 Interference Occurs between the Pico and the Macro

In macro–pico co-channel deployment, the introduction of new low-power nodes benefits the system average throughput due to the cell-split gain, while the UEs may suffer from interference caused by another tier and suffer poor edge coverage performance.

7.4.3.4.1 TDM Scheme for Interference Avoidance

A typical interference-limited case when adopting a resource element is that a noticeable fraction of cell-edge pico UEs (PUEs) will suffer from macro cell interference. For those PUEs to work well, it is fairly reasonable that some parts of the time/frequency resources are reserved only for the PUEs and that the MeNBs will be mute or power controlled in a fraction of the subframes. Without interference from the MeNB, the PUEs can work well in the reserved resources and performance can be improved.

In the macro muted subframes, the pico is able to schedule users that would otherwise experience excessive interference from the macro layer. In general, center UEs can be scheduled in any subframe. Muting of more subframes at the MeNB enables higher off-load potential by making it possible for more users to be served by the pico (see Figure 7.43). In this case, measurement restriction is needed to

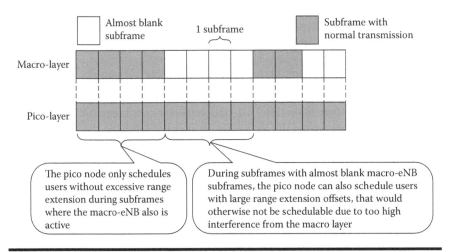

Figure 7.43   Muting some subframes at the macro eNB.

ensure that victim UEs make measurements at subframes only where interfering cells are muted (ABS or MBSFN). However, this is not possible for Rel-8 and Rel-9 UE. Furthermore, muting subframes at the MeNB also means lower macrocell capacity, so the muting pattern at the macrocells needs to be carefully optimized to achieve real gains from TDM-based eICIC in a macro + pico scenario. For the macro + pico case, the muting pattern can be coordinated via the backhaul, such as X2.

7.4.3.4.2 Power Control Scheme of Interference Avoidance

The goal of optimization is to enhance the edge UEs' SINR by increasing the received signal or decreasing the interference from a different tier. The system will calculate the appropriate objective uplink power Po (Po_{PUSCH} or Po_{PUCCH}) for each cell of different tiers in the area for which it has responsibility. The coverage information, power control parameters, and measurements of the neighborhood can also be sent via the X2 interface to assist the local cell power control parameter optimization. The main motivation of this power control scheme is to increase the performance of the edge users of the HetNet system with little degradation to other tiers by uplink transmission power-level cooperative adjustments.

7.4.3.4.3 Picocell Interference toward Fast-Moving Macro UEs

When a fast moving UE passes a small picocell quickly and spends only a short period of time within the pico, it is undesirable to do handovers for this UE in such a short period of time, because it is a waste of system resources. However, interference from a co-channel picocell can be significant such that the UE may lose its connection when passing the picocell. Basically we can consider pedestrian UEs as CSG subscribers to the picocell and fast-moving UEs as non-CSG subscribers. To avoid interfering with passing macro UE, the picocell mutes its transmission over the resource elements that are used by the overlay macrocell for the PBCH, PSS, and SSS.

Upon receiving the measurement reports indicating that a high-speed UE is about to enter a picocell of small size, the macrocell takes at least one of these actions: using downlink control information (DCI) 1C and persistent or semipersistent scheduling for this UE, or commanding the picocell to set PCFICH to 3, but limit its usage of PDCCH to help reduce interference toward the macrocell PDCCH.

The macrocell can command the picocell to constrain the usage of certain PRBs, which can be used by the macrocell to schedule the passing UE, or leverage the picocell FFR pattern and the UE is scheduled on the RBs and the picocell uses less power.

A picocell is typically much smaller than a macrocell; thus, its signal strength decreases/increases faster around its cell boundary. Especially in a co-channel picocell deployment, UE SINR decreases drastically after the UE passes the border between the serving cell and the target cell. If handover cannot be triggered and completed within a very short period of time, RLF can happen. So configuration of handover parameters becomes more complicated. Static configurations of handover parameters (offset, time-to-trigger, etc.) as defined in the standard specifications are not appropriate when picocell deployments become prevalent. It is necessary to have dynamic configuration schemes for specific deployment scenarios:

■ **Scenario 1:** UEs moving from different directions into the picocell need different configuration parameters because the rate of SINR change at the cell boundary is different.

■ **Scenario 2:** UEs moving into pico cells at different locations need different configuration parameters because the rate of SINR change at the cell boundaries is different.

■ **Scenario 3:** UEs moving toward a picocell within a cluster need to trigger and complete handover faster than UEs moving into a standalone picocell because the rate of SINR change at the boundary will be larger.

■ **Scenario 4:** UEs moving toward a picocell with lower transmission power need to trigger and complete handover faster than those moving into a picocell with more transmission power because the rate of SINR change at the boundary will be larger.

■ **Scenario 5:** UEs moving toward a heavily loaded picocell need to trigger and complete handover faster than those moving into a lightly loaded picocell because the probability of RLF will be higher.

7.4.3.5 Interference Occurs between Femtos and between Femtos and Macros

In a heterogeneous network comprising macrocells and HeNBs/CSG cells that have overlapping bandwidth deployments, certain interference problems can arise. In this section, the need for time-frequency resource partitioning and coordination between macrocells and HeNBs to mitigate interference scenarios in HeNB deployments will be discussed. The interference scenarios between femtos and between femtos and macros are shown in Table 7.4.

The severity of this problem can be quite high when the separation between the MeNB and the the HeNB is large because the macro UE (MUE) sets its UL transmit power based on the receiver SINR requirement at the MeNB. For example, the path loss equation for a typical 2-GHz carrier frequency macrocellular environment (from TR 25.814) used in system evaluations is given by PL (dBm) = $128.1 + 37.6\log_{10}(R)$. As we know, the UL power control equation can be approximated as

$$P_{Tx,MUE} = \max\{P_{CMAX}, I_{MeNB} + SNR_{req} + PL\}$$

Table 7.4   Interference Scenarios

Number	Aggressor	Victim
1	UE attached to HeNB	Macro eNB uplink
2	HeNB	Macro eNB downlink
3	UE attached to Macro eNB	HeNB uplink
4	Macro eNB	HeNB downlink
5	UE attached to HeNB	Other HeNB uplink
6	HeNB	Other HeNB downlink

where P_{CMAX} is the maximum allowed MUE transmit power per the power class (23 dBm). I_{MeNB} is the co-channel interference at the MeNB receiver and can be calculated as follows:

$$I_{MeNB} = 10 \log_{10}(10^{I_{Femto}/10} + 10^{N_{background\_noise}/10})$$

SNR_{req} (signal-to-noise ratio) is the required SINR for the MUE UL transmission to support the desired MCS level, and PL is the path loss from the MeNB to the MUE.

Figure 7.44 gives a summary of the dependence on distance of path loss (PL) and MUE transmit power.

7.4.3.5.1 Interference Analysis

The HeNB may use the same spectrum as the eNB. Two factors will help with the interference problem: (1) smaller cell radius and, therefore, much lower transmission power in the HeNB and (2) wall penetration loss isolating indoor HeNBs against the outside.

Now we suppose that all macrocells transmit at the same power level, are in the shape of a hexagon with radius of *Rm*, and cover the entire plane; macro UEs transmit at a same power level and are uniformly distributed in the macrocells but not in the HeNBs. All HeNBs are in a circle with a radius of *Rf* and uniformly distributed on the plane without overlap. Femto UEs are uniformly distributed in the HeNBs but not in the macrocells. All HeNBs transmit at the same power level, and all HeNB UEs transmit at the same power level. All macrocells are synchronized; the HeNBs operate independently and are quasi-synchronized to each other and to the macrocells. Parameters of interference analysis are depicted in Figure 7.45.

Assume the presence of *n* = 1 to a total of *N* HeNBs within a macrocell.

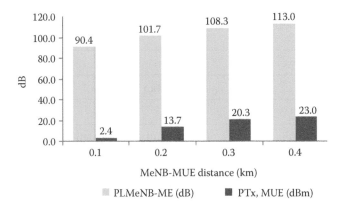

Figure 7.44 Summarize the dependence on distance of PL and MUE transmit power (I_{MeNB} = −98 dBm and SNR_{req} = 10 dB). (PLMeNB-ME—path loss between MeNB and MUE)

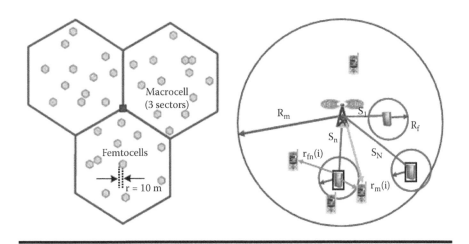

Figure 7.45 Parameters of interference analysis.

Rm: radius of macrocell (0.5 to 2 km)
Rf: radius of the HeNB (10 to 50 m)
Sn: range from the macrocell to the nth HeNB, for $n = 1$ to N HeNB
$r_m(i)$: range from the macrocell to mobile I
$r_{fn}(i)$: range from the nth HeNB to mobile i

For simplified analysis assumptions, refer to Table 7.5.

Table 7.5 Analysis Assumptions

Parameter	Value
Macro cell Tx power	~43 dBm
HeNB Tx power	10–20 dBm
UE Tx power	23 dBm
Propagation model	Exponential with a path loss exponent α: 2–6
Macro-cell coverage	3 sector cellular
HeNB coverage	Omni
In-building penetration loss β	1–30 dB
Noise floor (1.25 MHz BW)	–113 dBm

7.4.3.5.2 DL Interference to HeNB UE

Interference seen by a UE being served by an HeNB will be a combination of the DL TX from the macrocell and the received DL signals from all other HeNBs. Interference from the macrocell could be dominant and the other interference can be mitigated by penetration loss of in-building HeNB coverage and dynamic channel allocation by the HeNB.

$$SIR_{DL}(mobile(i))$$
$$\propto \frac{Tx(femtocell(n)) - 10\alpha \log_{10}(r_{f_n}(i))}{Tx(macrocell) - 10\alpha \log_{10}(r_m(i)) - \beta + \displaystyle\sum_{\substack{p=1 \\ p \neq n}}^{N} \{Tx(femto(p)) - 10\alpha \log_{10}(r_{f_p}(i)) \\ -2\beta[dB]\}}$$

7.4.3.5.3 DL Interference to a Macro UE

DL transmissions to UEs served by a macrocell could be interfered with by HeNB DL transmissions within the macro sector. Interference is dominated by an interference over thermal (IoT) rise from HeNBs and other interference can be mitigated by penetration loss of in-building HeNB coverage and dynamic channel allocation by the HeNB.

$$SIR_{DL}(mobile(i)) \propto \frac{Tx(macro - cell(n)) - 10\alpha \log_{10}(r_m(i))}{\displaystyle\sum_{p=1}^{N}[Tx(femto(p)) - 10\alpha \log_{10}(r_{f_p}(i)) - \beta[dB]]}$$

7.4.3.5.4 UL Interference to a Macrocell

UL interference to a macrocell is the interference due to the sum of the UL HeNB transmissions within the serving macro sector. Interference will be seen as an IoT rise over thermal and can be mitigated by macrocell antenna patterns, penetration loss of in-building HeNB coverage, and dynamic channel allocation by the HeNB.

$$SIR_{UL}(macroBTS) \propto \frac{Tx(mobile(i)) - 10\alpha\log_{10}(r_m(i))}{\frac{1}{3}\sum_{p=1}^{N} k[Tx(femto(p)) - 10\alpha\log_{10}(S_p(femto)) - \beta[dB]]}$$

7.4.3.5.5 UL Interference to an HeNB

Interference seen by an HeNB will be a combination of the UL TX from UEs on the macrocell and the received signals from all other UEs on the HeNBs. Interference can be dominated by the UL MUE transmissions (near–far problem) and the interference from all other UEs on the HeNBs can be mitigated by the penetration loss of in-building HeNB coverage and dynamic channel allocation by the HeNB.

$$SIR_{UL}(femto(n))$$
$$\propto \frac{Tx(mobile(i)) - 10\alpha\log_{10}(r_{f_n}(i))}{\sum_{\substack{N \\ \text{Sum order} \\ \text{macro} \\ \text{modules}}} \{Tx(macrocell) - 10\alpha\log_{10}(r_m(i)) - \beta\} + \sum_{\substack{p=1 \\ p\neq n}}^{N} \begin{array}{l} k\{Tx(femto(p)) \\ -10\alpha\log_{10}(S_p(i)) \\ -2\beta[dB]\} \end{array}}$$

With the previous discussion, we conclude the solutions of interference mitigation.

7.4.3.5.6 Interference: HeNB–Macro Solutions

The level of interference produced by the HeNB on the downlink and uplink transmissions in a macrocell depends on the HeNB position, but more precisely on the path loss between the MeNB and the HeNB. The higher the path loss is, the higher the HeNB impact on MUE performance. In the uplink, the higher this path loss is, the lower the HeNB UE's impact on the MeNB performance. In order to protect downlink control and data channels, the operation of the HeNB is likely to be constrained. Several such classes of constraints that have been considered so far for HetNet include frequency separation, restriction of DL power, restriction of RB usage, restriction of PDCCH usage and aggregation, and transmission of a certain subset of MBSFN and ABSs, as we have discussed previously.

- There are three types of frequency separation (see Figure 7.46): (1) total separation of frequencies between macro and femto, (2) partial separation of frequencies between macro and femto, and (3) dynamic separation of frequencies. (These are the initial options for frequencies downloaded to femtos via configuration management. The femto senses interference and chooses the best alternative by coordination between the femtos and macros. X2 communication is required).

- If there is no an appropriate X2 interface between the HeNB and the eNB for instantaneous information exchanges, the HeNB will be capable of measuring the quantity of received interference power in the DL. The HeNB assumes that a PRB is occupied by the macrocell if this quantity of exceeds a certain value and can make further decisions on whether to occupy the PRB, and the HeNB will then choose a transmission scheme if it decides to occupy the PRB. The received interference power in the UL is also used to identify the UL PRB allocation of the macrocell and vice versa.

- X2 communication is required for power control. Implementation of X2 is expensive and is not preferred by most operators.

- As UE moves into the inner coverage of the CSG cell, the nonallowed UE would generate higher uplink interference to the CSG cell while also suffering from higher downlink interference from the CSG cell. Control signal coverage hole issues can be avoided (see Figure 7.47) via separation in the

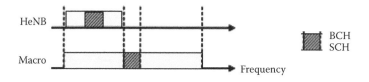

Figure 7.46 Shared HeNB/MeNB frequency deployment. (SCH—synchronization channel.)

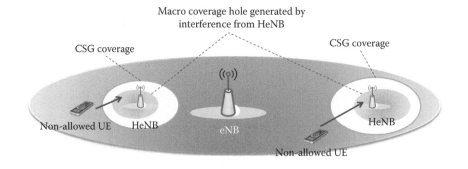

Figure 7.47 Macrocell coverage holes due to the HeNBs.

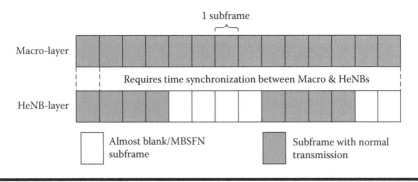

Figure 7.48 Muting some subframes at the HeNB.

time domain. The femtocell aligns its data portion with MeNB control portion and does not schedule any actual data transmission in the overlapping resource blocks; X2 communication is required.

7.4.3.5.7 Interference for HeNB + Macro Case: TDM Scheme

During time periods with almost blank subframes from the HeNBs, the macro users close to the HeNBs will be served (i.e., when there is no excessive HeNB interference). There is no backhaul coordination of the time-domain muting pattern, so configuration of the muting pattern for the HeNBs shown in Figure 7.48 is assumed to come from a centralized entity (such as operation and maintenance [O&M]). The MeNBs are assumed to know the muting pattern used by the HeNBs within their coverage area.

7.4.3.5.8 Dynamic Adaptation of Femtocell Control Channel Transmission Power

Different from macrocells, femtocells are not required to provide continuous coverage over a certain geographical region. They are only required to cover their home users. To minimize femtocell control signal pollution, we need to set the control channel power levels of femtocells to be just enough to meet the needs of their respective users. Thus, instead of a fixed setting, femtocell control channel power needs to vary with respect to the locations, status, and configurations of the corresponding femtocell users.

It is preferable that handover (both femto–macro and macro–femto) takes place in an outdoor area where the curve of the femtocell signal strength is relatively flat. When handover is not needed in the imminent future, the femtocell needs to reduce its transmission power to just maintain service of its indoor users. When no user is at home, the femtocell needs to turn off its transmitter.

UE stores the global cell ID (GCI), tracking area ID (TAI), femto-GW ID, and location of its authorized femtocell. When UE notices that it is close to its femtocell based on the macro fingerprint, global positioning system (GPS), or actual measurement, it will send a message (tracking area update [TAU] message for idle mode and RRC message for connected mode) including the femtocell GCI and femto-GW ID to the network. Based on the received femtocell GCI and femto-GW ID, the network identifies the associated femtocell, adjusts its power level based on the UE location and state, and notifies the UE of the femtocell PCI and other related information to facilitate cell reselection and handover.

The femtocell should know or configure the wake-up occasions of idle- and sleep-mode UEs, and use a high power level only when they are active. The UE that is about to hand-in will be informed of the time interval when the high power level will be used by its femtocell. Neighboring femtocells will use a nonoverlapping time interval for high power transmission to mitigate mutual interference and reduce control signal pollution.

7.4.3.5.9 Femto–Femto Interference

Different from MeNB deployment, femtocell deployment is random and unplanned. In macro networks, a UE can handover to neighboring macrocells when they have a stronger signal. However, a femtocell deployed at home usually has closed access so that handover to neighboring femtocells is not allowed. Note that the percentage of control channel power for femtocells is much larger than that for macrocells since the number of femto UEs is very small. Control channel interference is a much more severe problem than data channel interference.

Each HeNB estimates the fraction of time it needs to transmit, according to the traffic load and channel conditions of its UEs, and reports this ratio to the centralized controller via S1 signaling. For mixed traffic with both delay-sensitive traffic and delay-tolerant traffic, ratios that correspond to both traffic types will be reported. Each HeNB needs to update its report when at least one of the following events occurs: a new traffic session is initiated or the UE channel condition varies over a predefined threshold.

The centralized coordinator determines the sub-bands, carrier frequency, and subframes that each HeNB is allowed to transmit, and notifies each HeNB of its transmission pattern via S1 signaling. An HeNB needs to properly configure the discontinuous reception (DRX) parameters of its UEs according to the transmission pattern of which it is notified by the centralized coordinator.

7.4.3.5.10 Simulation Results for Closed-Access HeNBs

The normalized cell throughput is very close to 1 (macro cell throughput with no HeNB is used as a benchmark reference) with a small number of HeNBs and low HeNB transmission power. The cell throughput variation within this region is quite small with a varying number of HeNBs and their power levels. Beyond this region, however, cell throughput drops very fast when adding more

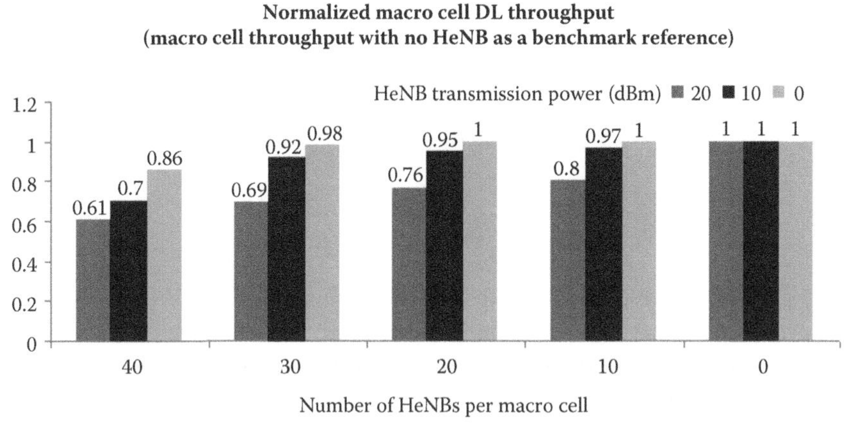

**Normalized macro cell DL throughput
(macro cell throughput with no HeNB as a benchmark reference)**

Figure 7.49   Normalized macrocell DL throughput.

HeNBs, which is shown in Figure 7.49. Similar phenomena can be observed even when power control is applied. It is desirable to set an upper bound for HeNB (closed-access) density, so that macrocell throughput will not be severely impacted.

7.4.3.6 Interference Occurs for Relays

In this section we discuss some of the interference conditions arising from the introduction of relay nodes in a macro network.

7.4.3.6.1 Interference from Relay Node to the UE Being Served by a MeNB

A UE connected to a relay node uses up bandwidth both on the access link as well as on the backhaul link. In this case, an appropriate serving cell selection strategy will have to take the backhaul link of the relay into account; that is, it is not always optimal to connect to the cell with the highest received power. For example, consider a case in which a UE is placed between an MeNB and a relay (i.e., the macro-to-UE link is stronger than the macro-to-relay link). In this case, it is preferable for the UE not to connect to the relay even if the received power from the relay is stronger than that from the MeNB. In such a situation, the UE would see strong interference from the relay node while being served by the MeNB, which is shown in Figure 7.50.

7.4.3.6.2 Interference from the MeNB to UE Being Served by a Relay Node

Since a relay node will typically have significantly lower transmit power as compared to an MeNB, it may be desirable for a UE to connect to the relay even if the received

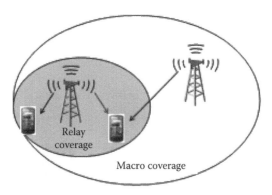

Figure 7.50 Interference from the relay node to the UE being served by the macro eNB.

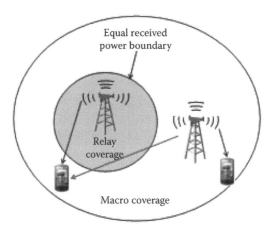

Figure 7.51 Interference from the macro eNB to the UE being served by the relay node.

power of the relay node is lower than that of the MeNB; for example, if the relay node has a very good backhaul link and if the path loss between the UE and the relay node is better than the path loss between the UE and the MeNB. In this case, in the absence of interference from the MeNB, the relay node can serve the UE while causing significantly less interference to the network as compared to the case where the UE is served by the MeNB. Moreover, multiple relay nodes can simultaneously use the bandwidth vacated by the MeNB, thus creating cell-splitting gains on the relay–UE link. The deployment of relay nodes in such a configuration can therefore provide significant capacity benefits to the network. However, this configuration also results in the UE seeing strong interference from the MeNB. This scenario is illustrated in Figure 7.51.

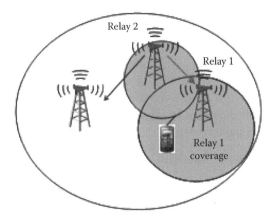

Figure 7.52   Interference from relay node 2 to relay node 1.

7.4.3.6.3 Interference between Two Relay Nodes

The above two scenarios describe strong interference seen at the UE from either the MeNB or the relay node. Other interference conditions are created by the presence of multiple relays close to each other. In this case, one relay may be transmitting on the DL frequency band (to a UE) while another relay eNB may be receiving on the same frequency band (from the MeNB). The transmitting relay may have a much stronger signal than the MeNB at the receiving relay. Similar conditions may occur on the UL frequency band where one relay may be receiving (from a UE) while another is transmitting (to the MeNB). This scenario is illustrated in Figure 7.52.

Operation in the scenarios described above yields significant capacity benefits for the network. However, these scenarios also imply operating conditions where the interference power seen at a UE or a relay node may be substantially larger than that of its serving eNB. In other words, the UE or relay node will see highly negative geometries from its serving eNB. This may necessitate the introduction of acquisition signals and control channel techniques that allow for robust operation in such conditions. Enhanced ICIC techniques will also be needed to provide reliable data reception in such conditions.

7.5 Mobility Management of Heterogeneous Networks

Deployment of picocells and relays would create smaller cells where interference characterizes changes more rapidly than in the case of a network consisting of macrocells only. LTE-A will support seamless mobility, which means users will not suffer packet data losses during handovers. As a consequence, handover frequency would increase, potentially increasing instances of radio link failure and even undesirable transition to the RRC_IDLE state in a heterogeneous network.

Because TDM/FDM interference avoidance schemes have been applied in heterogeneous networks, the RB resources should be coordinated between macro and low-power nodes, and the dynamic configuration of handover parameters should be performed during the handover procedure. An example of a two-step handover and dynamic configuration of handover parameters is described in the following text.

The first measurement report contains the following information elements of each of the neighboring cells in the same frequency to help the macrocell set appropriate handover parameters:

- RSRP samples in the last measurement
- Load
- Transmission power
- FFR pattern
- Picocell location (indoor or outdoor, edge or center)

The last four parameters may be obtained from X2.

Upon receiving the first measurement report, the source cell needs to prepare the target cell and forward buffered user data over X2. The macrocell determines the handover target and commands it to constrain usage (block or reduce power) of certain RBs. The macrocell will then only schedule UE on certain RBs and send an RRC reconfiguration message to the UE to provide a positive transmission power offset and handover parameters derived from the first measurement report. This procedure is depicted in Figure 7.53.

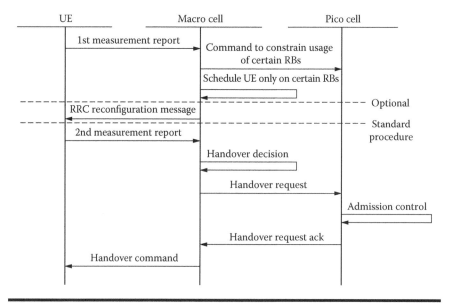

Figure 7.53   Two–step handover in a heterogeneous network. (ACK—acknowledgment)

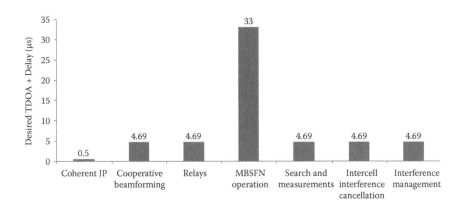

Figure 7.54 Time synchronization accuracy requirements.

7.6 Network Synchronization

Heterogeneous network time synchronization is a practical requirement for TDD systems and also necessary to get single-frequency network (SFN) gains for MBSFN operation. In the LTE-A system, we strongly agree that some advanced techniques require time synchronization even for the FDD network, including: positioning, CoMP, relay, cell search and measurements, intercell interference cancellation, interference management, and so on. If eNBs are not synchronized, they can severely impact the performance of LTE-A eNBs in the vicinity.

Heterogeneous networks require symbol-level synchronization of low-power nodes and MeNBs to protect the control channel and to enhance data channel interference coordination. The synchronization between HeNBs and MeNBs is a problem that needs to be resolved. For efficient intercell coordination, an FDD HeNB with DL sniffing capability seems desirable so that it can easily keep synchronized with the MeNB via wireless channel from the MeNB to the HeNB. Moreover, faster coordination becomes possible via a direct over-the-air channel when compared to the method using synchronized backhaul.

It should be noted that the synchronization requirements to support CoMP, relays, and other advanced features are not very stringent. In general, it should suffice if the time difference of arrival (TDOA) plus the delay spread difference between neighboring cells is within the cyclic prefix. Note that for small cells (e.g., 500 m) and low delay spreads, the TDOA plus the delay spread difference translates to the transmitted timing difference. Further details on the time synchronization accuracy required for the different techniques are shown in Figure 7.54 and explained below.

1. Coherent joint processing needs coherence over the sub-band to enable sub-band-level channel quality indicator (CQI) feedback; achieving very good timing accuracy is necessary.

2. Cooperative beamforming will cause moderate performance degradation with larger delays (e.g., minimal degradation with a 10μs delay).

3. Relay synchronization to the eNB is a requirement for relays because they cannot transmit and receive in the same band at the same time.

4. MBSFN operation requires a long cyclic prefix defined in 3GPP standard.

5. Search and measurements with less than cyclic prefix (CP) timing can acquire all cells with one fast Fourier transform (FFT).

6. Intercell interference cancellation and interference management with less than CP timing can demodulate a neighboring cell's signal using the same FFT. Significantly higher complexity is required in a fully asynchronous system.

Accuracy in the range of a few microseconds is easily achieved by use of an external source of time such as GPS or other equivalent positioning systems. When cells are out of GPS coverage, self-synchronization techniques should be used to achieve synchronization. Since the TDOA between different cells is typically well within the CP length, an asynchronous cell (e.g., HeNB) may set its timing to the earliest arriving path from a synchronized cell. For example, if an HeNB uses a near-macro eNB to achieve its synchronization, the error in timing is only 1–2 μs. Moreover, as the UE and HeNB are likely to be very close to each other, the actual TDOA between the signals of the HeNB and the synchronous eNB at the UE is going to be even smaller.

Chapter 8

Interference Suppression and eICIC Technology

It is well known that the performance of cellular systems is limited due to the presence of intercell interference. The intercell interference problem naturally results in a lowering of the data rates achievable for users at the edges of the cell, especially in an overlaid co-channel deployment between the macro evolved Node-Bs (MeNBs) and low-power nodes. Transmissions to users with low signal-to-interference plus noise ratio (SINR) require information redundancy and hence a low code rate to achieve the desired decoding quality, resulting in a corresponding reduction in the data rate. In a live Long Term Evolution (LTE) network, intercell interference coordination (ICIC) is implemented to improve interference limitations found in cellular systems deployed with universal reuse and also improves user equipment (UE) throughput at the cell edge. Mitigating interference could therefore include interference avoidance, interference suppression, and intercell interference coordination.

Interference avoidance (transmitter processing) includes spatial processing (e.g., beamforming) and interference coordination (scheduling). This method is less complex but has large feedback overhead or limited resource reuse.

Interference suppression (receiver processing) includes linear equalization, interference rejection in the spatial domain (interference rejection combining [IRC]), and interference cancellation. This method is more complex, but needs low feedback overhead and full resource reuse.

Intercell interference coordination is an implicit interference management method via radio resource management (RRM) scheduling. For the uplink, soft fractional frequency reuse and fractional frequency reuse schemes can be deployed; for the downlink, a scheme of power limitation on frequency blocks can be used.

In a reuse 1 deployment, it is critical to manage the uplink (UL) and downlink (DL) interference levels. The eNBs can send UL overload indications to neighbor eNBs via the X2 interface based on frequency and power resource allocations. Power control parameters can also be adapted based on overload indicators.

ICIC can be done in both the uplink and downlink for the physical DL shared channel (PDSCH) and physical uplink shared channel (PUSCH) or the physical uplink control channel (PUCCH) as we discussed in Chapter 7. In LTE Rel-8, ICIC is not possible on the physical downlink control channel (PDCCH) since the control channel elements are spread over the full band and several orthogonal frequency-division multiplexing (OFDM) symbols. For ideal interference cancellation in LTE and LTE Advanced (LTE-A), the following conditions are required:

1. Interference parameter knowledge at the base station
   - Multiple input multiple output (MIMO): all the streams parameters should be known
   - Intracell: all interferers' parameters should be known; joint multiuser detection can be carried out
   - Intercell: all interferer parameters should be unknown; needs blind detection of interference parameters or signaling between eNBs
2. Interference parameter knowledge at the terminal
   - MIMO: all the streams parameters should be known
   - Intracell and intercell: all interfering user parameters should be unknown; needs blind detection or extra signaling

8.1 Interference Cancellation in LTE-A

It is obvious that ICIC techniques in Rel-8 and Rel-9 have some drawbacks and are not always effective in mitigating dominant interference. RRM measurements and channel feedback may not be accurate and dominant interference conditions on the resources with uncoordinated use can lead to the declaration of radio link failure, even though radio conditions may be very good on the resources with coordinated use.

In LTE-A, enhanced ICIC is an intracarrier interference management technique for heterogeneous networks and many schemes can be applied.

Interference cancellation is bound to be applied in the near future. Turbo successive interference cancellation (SIC) receivers have been taken as the baseline for multiple-input, multiple-output (MIMO) single-carrier Frequency-Division Multiple Access (SC-FDMA) performance evaluation in the Third-Generation Partnership Project (3GPP) LTE-A. It is anticipated that the base station processing capabilities will allow this type of receiver to be implemented by the time LTE-A commercial deployments occur. The gains between 2 and 4 dB have been reported for 2x2 MIMO over a conventional minimum mean squared error (MMSE) receiver.

From cell-specific reference signals (CRSs) or channel state information (CSI)-RS of neighbor cells, the UE can roughly estimate the level of the interference. However, there are two problems: (1) It is hard for UE to know the existence of an interfered PDSCH in the neighbor since it is UE specific. (2) If interference estimation is based on the cell-specific RS, there will be bias in estimation results since the PDSCH could be transmitted with non-codebook-based precoding in an LTE-A network. In these issues, the intercell orthogonal demodulation reference signal (DMRS) can benefit the UE-side intercell interference (ICI) cancellation. So intercell orthogonal DMRSs in LTE-A could be considered for the heterogeneous network (HetNet) for which ICI suppression via coordination among cells is difficult.

In Chapter 3, we mentioned that the precoding matrix indicator (PMI) and rank indicator (RI) coordination between eNBs to steer the beam as another ICIC approach; the ICI suppression depends on UE-side processing in case intercell orthogonal DMRSs exists.

In Chapter 7 we mentioned enhanced intercell interference coordination solutions for heterogeneous network deployments. Since a number of low-power nodes are overlaid by a macrocell and the unbalanced transmitting power are provided between different cell layers, the control and data channels suffer more serious interference than in the case of homogeneous networks. While carrier aggregation (CA)-based solutions are considered as promising approaches, non-CA-based solutions also should be specified to support enhanced ICIC for UEs without CA capability. The method of control channel shrinking (orthogonal in the frequency domain) and time-division multiplexing (TDM)-based control/data channel transmission (orthogonal in the time domain, including multicast-broadcast single-frequency network [MBSFN] subframes and almost blank subframes) could be applied.

In addition, self-organizing networks (SONs) can provide a faster way to identify the interference location and root causes in an operational LTE network, and perform interference reduction and energy savings by suggesting configurations based on traffic, capacity, coverage requirements, and interference measurements (e.g., nighttime, daytime, weekend, weekday, no traffic).

8.2 Multiuser Detection

The ever-increasing appetite for capacity and data rates has motivated the emergence of new multiuser detection (MUD) techniques and multiple access schemes in wireless communication. Multiuser detection is one of the receiver design technologies for separating desired signals from interference and noise. While traditional single-user detection schemes assume that all other users' signals are white noise, MUD utilizes some known information from the other users to perform symbol detection for the desired user. In theory, multipath interference or multiaccess interference is not purely useless white noise, but a pseudo-random sequence signal with a strongly structural feature, and the correlation function of different users and paths are known. It is entirely possible to further eliminate negative effects caused by interference and achieve the purpose of improving system performance with the known structural information and statistics of the pseudo-random sequence. MUD addresses the interference problem by cancelling or suppressing interfering users and multipath effects from the desired user's signal. The single-user detection (SUD) and MUD techniques are shown in Figure 8.1.

The basic idea of MUD is to treat signals of all users as useful rather than as interfering. This can take full advantage of user code, amplitude, and timing and delay information of user signals to significantly reduce multipath multiple-access interference. Reducing the complexity of the multiuser interference suppression algorithm to an acceptable level is critical to put MUD technology into practice. MIMO- and OFDM-based LTE communication invoking MUD techniques has recently attracted intensive research interest. In MIMO systems, the transmitted signals of simultaneous uplink mobile users, each equipped with a single transmit antenna, are received by the different receiver antennas of the eNB. At the eNB the individual users' signals are separated with the aid of their unique, user-specific spatial signature composed of their channel transfer functions or channel impulse responses (CIRs).

MUD requires knowledge of the radio equipment (RE) resources of all users and the channel impulse responses of all users. MUD will depend on the data detection scheme, covariance matrix of transmitted data, and covariance matrix of the noise vector, which is shown in Figure 8.2. Figure 8.3 gives a detailed SUD procedure, so we can see the difference between the two detection algorithms.

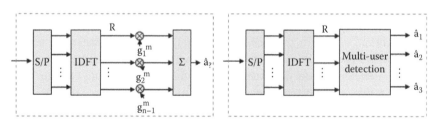

Figure 8.1 SUD and MUD. (S/P—serial/parallel; IDFT—inverse discrete Fourier transform)

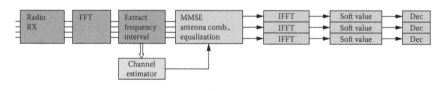

Figure 8.2 Multiuser detection. (FFT—fast Fourier transform; RX—receive antenna; IFFT—inverse fast Fourier transform; Dec—decoding)

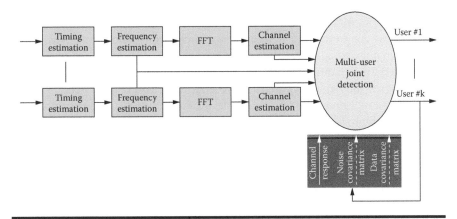

Figure 8.3 **Single-user detection.**

Multiuser detection methods can be separated into linear and nonlinear detection techniques. The group of linear detection methods includes least squares (LS) and MMSE detection, in which no *a priori* knowledge of the remaining users' transmitted symbols is required for the detection of a specific user. In the nonlinear methods, SIC, parallel interference cancellation (PIC), and maximum likelihood detection (MLD), *a priori* knowledge is involved that must be provided by the nonlinear classification operation involved in the demodulation process.

The main disadvantage of multiuser detection is that it increases the complexity of the system equipment and system processing delay, especially for the adaptive algorithms and the system using longer spreading code. Much additional information, which includes the spreading codes of all users and the main statistical parameters of the fading channel, such as amplitude, phase, and delay, must be collected before using the multiuser detection algorithm. For a time-varying channel, this is achieved by ceaselessly estimating each user channel. In general it is very difficult, and the precision of parameter estimation will directly affect the performance of MUD. The varieties of MUD schemes, such as the LS and MMSE detectors, or SIC, PIC, and MLD schemes may be invoked for the sake of separating the different users at the base station on a per-subcarrier basis.

8.3 Receiver Technology and Interference Suppression

8.3.1 Interference Cancellation Receiver

The UE- and eNB-side interference cancellation includes an advanced detection structure, such as interference rejection combining via receiver beamforming if an interference covariance matrix is available for the UE. These techniques are generally implementation issues, except for the aspects that enable the receiver to obtain

ICI information. Although blind detection might be possible as an alternative, performance loss due to the lack of interference information and the increased complexity from more complicated algorithms are the main concerns. Good estimation for the interfering channel is necessary for receiver-side ICI cancellation. The basic principle of the interference cancellation receiver is shown in Figure 8.4.

MMSE, a well-known practical channel estimator with the lowest complexity, describes an approach that minimizes the mean square error (MSE) and can be efficiently implemented in LTE-A; it is a common measure of estimator quality. The MMSE receiver is an optimum linear receiver and is designed by minimizing the MSE between the receiver output and the training signal (see Figure 8.5). The linear detector is chosen such that the mean square error between the soft output of the linear transform of the matched filter output and the actual data is minimized. MMSE is one common criterion to generate a receiver detection algorithm. When a neighbor cell interference correlation matrix is considered by incorporating the inverse of the estimated received signal covariance matrix to the MMSE equalizer, MMSE evolves to a interference cancellation receiver. The received signal covariance matrix usually contains components of the actual desired signal, intercell and intracell interference, and background noise. The interference cancellation mechanism includes the necessary parameter valuation, signal regeneration, and useless signal counteraction, which requires a high-precision channel phase, amplitude, delay parameters, and the valuation of the interfered signal.

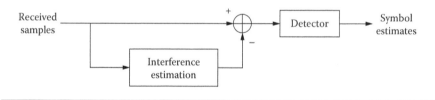

Figure 8.4   The basic principle of the interference cancellation receiver.

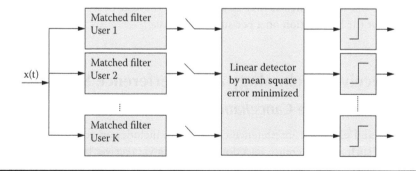

Figure 8.5   MMSE detector.

MMSE is a minimum mean square error estimation, which means that the mean square error of receiving data is derived from CIRs and practical minimum data. The MSE of the channel estimate is defined as

$$\Omega_{MSE} = \frac{1}{N_c} \sum_{m=0}^{N_c-1} \left| \hat{H}(m) - H(m) \right|^2, \text{ and can be approximated as } \Omega_{MSE} \approx \frac{L}{N_c} \hat{\sigma}_r^2$$

The advanced receiver may deploy collection of methods using recursive decoding (e.g., turbo equalization and interference cancellation). Interference cancellation is a nonlinear detection method within the more general class of multiuser detection algorithms. These techniques will increase performance up to 2 dB in some scenarios compared to a linear MMSE receiver because they cancel interstream crosstalk. The complexity and delay issues of these algorithms and possible simplifications need to be studied before future implementation.

Linear multiuser detection algorithms are usually much more demanding in terms of processing power compared with the chosen nonlinear interference cancellation algorithms. The reason is that they require, as part of the detection algorithm, the inversion of a matrix the size of which is proportional to the number of users in the cell.

Basic nonlinear interference cancellation algorithms include interference rejection combining (IRC), SIC, PIC, and the decision-feedback interference cancellation algorithm. Interference elimination detector is generally formed by multi-stage operations (because the advanced receiver is realized by recursive decoding, so multi-stage operations are needed). The basic idea is to estimate multiple-access interference for each user at the receiver and then eliminate some or all of the multiple-access interference from the received signal. This kind of eliminator is similar to the anti-intersymbol interference (ISI) feedback equalizer and the so-called decision-feedback detector. Soft or hard decisions can be used to estimate multiple-access interference. Hard decisions require reliable estimation of signal amplitude.

Interference cancellation (IC) is a method used to improve radio performance of an LTE receiver that cancels intercell interference. The IC receiver improves cell-edge performance in interference-limited cases. It is considered in the UL baseband receiver. In short, the reception of some users (called *cancellees*) is improved by removing the interference caused by other users (called *cancellers*). To do this, the interfering user (a canceller) needs to be decoded and its radio channel estimated. The gamma data corresponding to this user is then reconstructed and subtracted from the total gamma data in the cells where the user is present, before the other users (the cancellees) are decoded. Typically IC receivers are vendor specific, not standard. Interference cancellation receiver alternatives are described as follows:

- **IRC:** The central idea of interference rejection is to remove colored noise as much as possible from the received signal using noise whitening. IRC can utilize correlation in the spatial domain (between antennas) to suppress intercell

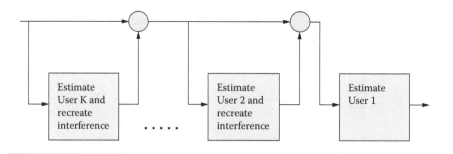

Figure 8.6 Successive interference cancellation scheme.

interference. IRC is much less complex than interference cancellation because in IC the signals from neighboring cells have to be detected.

■ **SIC:** In SIC (see Figure 8.6), a user is only decoded once. The strongest signal is estimated, and its contribution to the received signal is removed before estimating the second-strongest signal. This procedure continues until all desired users are estimated. SIC will introduce 2–3 heavier processing load. Here the users in a cell are divided into two groups: the strong users and the weak users. The strongest users are decoded and removed from the gamma stream before the weak users are processed.

■ **PIC:** In PIC, all users are decoded in a first step, and removed from the gamma data in a suitable way before all the users are decoded in a second step. The procedure could be iterated several times and requires relatively higher complexity. When all users are received with equal strength, PIC outperforms SIC. When they are of different strengths, the performance of SIC is superior.

8.3.2 Interference Rejection Combining

In LTE, the maximum ratio combining (MRC) algorithm is employed by the LTE PUSCH receiver to combine the received antenna signals into one signal before linear equalization and decoding. MRC means to combine the multiple received signals in such a way that the wanted signal's power is maximized, compared with the interference and the noise power, and the SINR is enhanced. When the MRC receiver is used, the receiver is only aware of interlayer interference by the user's own transmitted layers. However, interference suppression is not done for interuser interference. Actually, the LTE UL PUSCH/PUCCH/physical random access channel (PRACH) receiver that employs MRC can maximize the signal-to-noise ratio (SNR) after the antennas are combined, but the spatial dimension of the receiver is not fully exploited as only the diagonal elements of the spatial interference-plus-noise covariance matrix are utilized. In the ideal case, the output of the MRC maximizes the SNR and will therefore work satisfactorily provided there is no strong cell interference (i.e., no spatially colored noise). MRC is used to combine the signals received from both the serve link and selected macro links. The MRC's model is as follows:

$$SINR_{MRC} = SINR_{ServeLink} + \sum_{n=0}^{N} SINR_{MacroLink\_n}$$

where *SINR* is the signal-to-interference plus noise ratio and *N* is the number of macro diversity links for a given UE. The SINR value in the serve link and macro links is calculated using this equation:

$$SINR = \sum_{Ant=1}^{Nrx} \frac{\sum\limits_{PRB=1}^{K_{PRB}} S_{Nrx,K_{PRB}}}{\sum\limits_{PRB=1}^{K_{PRB}} \left(\sum\limits_{Interf=1}^{N_I} I_{Nrx,K_{PRB},N_I}^{InterCell} + n \right)}$$

where *S* is the signal strength per physical resource block (PRB) per antenna; $I^{InterCell}$ is the interference strength per PRB per antenna, assuming the intracell interference is ideally zero; *Nrx* is the number of RX antennas per cell; K_{PRB} is the number of transmission PRBs assigned to a certain UE; N_I is the number of intercell interferences; and *n* is the noise spectral density.

However, in a typical LTE network there will be cases of strong cell interference, for both FDD and TDD systems, which could reduce the UL transmission capacity significantly. IRC is a method to enhance the transmission capacity by mitigating the undesirable intercell interference.

IRC works independently of the UE and eNB, so it is not standardized by 3GPP. Using IRC with an antenna configuration with *N* receivers (N = 1, 2, 4) makes it possible to cancel out X – 1 interferers with the best result. This means that for single-input multiple-output (SIMO) 1 × 2, only the interference from one disturber can be cancelled out completely. If there are more interferers present, the IRC algorithm seeks a best-effort solution.

From the technology point of view, IRC is based on the covariance matrices* estimation of the noise of various diversity branches. After a matrix is inverted, it is used to calculate the needed matrix. IRC eliminates coherent noise, which it receives from different antennas. Due to the technical characteristics, IRC technology can merge and eliminate correlative interference. If noise is irrelevant, the IRC performance is similar to ordinary MRC.

---

* In statistics, each element of the covariance matrix is the variance between the various elements of the vector, which is an extension from scalar random variables to a high-dimensional random vector. Suppose *x* is a column vector consisting of *n* scalar random variables, and μk is the expected value of the *k*th element, that is, $\mu k = E(Xk)$. The covariance matrix is then defined as: $\Sigma = E\{(X - E[X])(X - E[X])T\}$.

Channel estimation and disturbance covariance matrix calculation, including signals from all other users, are executed for each user. The disturbance covariance matrix is estimated by averaging the autocorrelation d•d^H of the disturbance over a number of subcarriers. The estimation of the disturbance covariance matrix is on a slot basis (e.g., no averaging over multiple transmission time intervals [TTIs]).

IRC is achieved by incorporating the spatial intercell interference covariance matrix estimate in the linear equalizer. When neighbor cell interference correlation matrix is considered, MMSE evolves to IRC.

Ideal IRC with an 8-element x-polar antenna array can provide up to 10-dB UE SINR gain so that MCS is enhanced from 16QAM (quadrature amplitude modulation) to 64QAM, shown in Figure 8.7.

It is worth mentioning that in some scenarios, for example, 16QAM or no interference, the performance of MRC could be slightly better than IRC. So an internal IRC–MRC switch is implemented (every slot).

IRC utilizes correlation in the spatial domain of the interfering signals between multiple receiving antenna elements of the victim receiver to suppress intercell interferences. In doing so, the received interference is suppressed by spatial whitening before receiver equalization and decoding. So it can improve UL capacity in case of high intercell interference. An IRC receiver calculates and applies a set of antenna weights to maximize the SINR of the signal after combination, taking into account the direction of arrival of the desired and interfering signals. An estimated interfering channel is used to construct interference covariance for the MMSE–IRC receiver. IRC processing is shown in Figure 8.8.

For IRC, the eNB may perform interference estimation, essentially constructing the interference covariance matrix needed in the IRC. In a most trivial way, the eNB may estimate its own desired signal, subtract its own cell contribution from

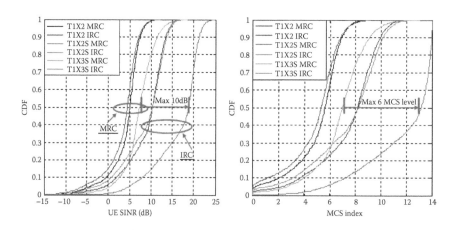

Figure 8.7   Ideal IRC gain compared to MRC.

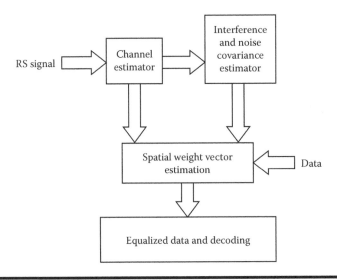

Figure 8.8   IRC processing.

the received signal, and calculate a sample average over one PRB to estimate the covariance matrix of the remaining signal. The interference rejection capability of the receiver depends also on the number of receive antennas and streams of interest. Performance gains are increased with a larger number of receive antennas. For example, in a situation of two receive antennas present at the eNB, if one desired stream is received, there is only one degree of freedom left for efficient interference suppression.

In general, the MMSE/MRC type of receiver does not make any assumptions about or use of the other cell's interference. The more advanced techniques of IRC will assume a larger degree of interference knowledge, which is shown in Figure 8.9. IRC will be aided with network information in order to estimate the interference in a realistic manner; hence, this type of receiver is more appealing for future implementation in UE.

The IRC gain compared to MRC depends strongly on the radio channel conditions, the spatial correlation, the number of disturbers, and the number of receiver antennas. In beneficial conditions, an IRC gain of more than 10–20 dB is expected; for a typical case, the gain is expected to be 0–5 dB.

The interference rejection is based on an estimation of the covariance matrix of the noise from the diversity branches. This covariance matrix is then inverted and used to calculate the needed metric.

Let us define the received frequency domain spatial vector of size $N_r x1$ for the considered user at frequency k by $\bar{Y}(k) = \left[Y_1(k)Y_2(k)\cdots Y_{N_r}(k) \right]^T$, as shown in Figure 8.10. Furthermore, assume the corresponding spatial channel vector is $\bar{H}(k) = \left[H_1(k)H_2(k)\cdots H_{N_r}(k) \right]^T$. With this notation, the received

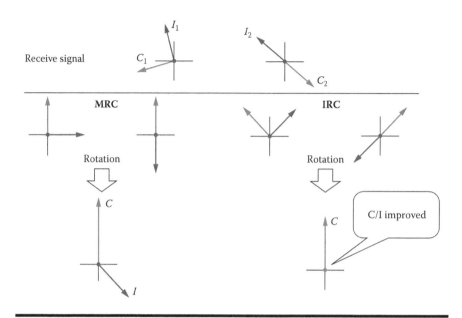

Figure 8.9 MRC and IRC model.

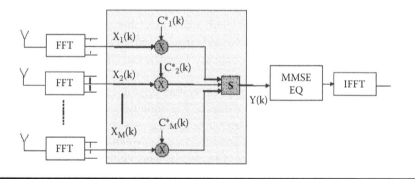

Figure 8.10 Proposed IRC solution for PUSCH/PUCCH.

frequency domain spatial vector can be expressed as $\bar{Y}(k) = \bar{H}(k)X(k) + \bar{D}(k)$, where $X(k)$ denotes the transmitted signal component for subcarrier k and $\bar{D}(k) = \left[D_1(k)D_2(k)\cdots D_{N_r}(k) \right]^T$ is the received spatial disturbance vector containing the interference plus additive white Gaussian noise (AWGN) at all Nr receiver antennas. Figure 8.10 also shows that the IRC coefficient vector for subcarrier k is denoted by $\bar{C}(k) = \left[C_1(k)C_2(k)\cdots C_{N_r}(k) \right]^T$, which we assume is calculated according to the optimum combining method. Thus, the total received signal after IRC processing can be written as $Z(k) = \bar{C}^H(k)\bar{Y}(k) = \bar{C}^H(k)\bar{H}(k)X(k) + \bar{C}^H(k)\bar{D}(k)$.

8.3.2.1 IRC Gain

IRC is an optimization feature for high-load LTE networks in interference-limited scenarios. The main drawback of IRC is the added baseband computational complexity, which grows strongly with the number of receiver antennas. The number of interferers the IRC processing can suppress depends on the number of receiver antennas. The basic rule is that the maximum number of interferers that can be suppressed is the number of receiver antennas − 1, through complex demodulation algorithms. Some suppression of interference is achieved even with a higher number of interferers, but the suppression gain decreases very rapidly. Within the radio network scenario project, a new model has been introduced in order to calculate the gain from using IRC:

$$\text{IRC gain} = \text{nominal gain} + \text{fast fading gain [dB]},$$

Here the nominal gain is expressed as: nominal gain $= 10 \cdot \log (N_{Rx} - N_{stream} + 1) - 10 \cdot \log (N_{stream})$, where N_{Rx} is the number of receivers and N_{stream} is the number of MIMO streams for the UE–eNB connection considered. For SIMO 1×2, $N_{Rx} = 2$ and $N_{stream} = 1$; hence nominal gain = 3 dB.

If the number of interferers is large, the interference starts to look like white Gaussian noise. IRC can do nothing to improve performance in the white noise–limited case. IRC processing actually slightly decreases performance against white noise. So IRC gain on the link level is directly related to interferer characteristics. The more spatially colored the interference, the higher the IRC gain. UL IRC improves cell-edge performance better in the high- than in low-load case. In the low-load case, UL IRC provides little gain in average cell throughput.

Finally, it should be noted that IRC can be seen as a way of achieving more robust gains and also exploiting the network information in a potentially lighter way than transmit-centric schemes, which are typically investigated in collaborative multipoint (CoMP) studies. Indeed, interference-aware receivers rely mostly on downlink signaling and network coordination of resources. On the opposite side, transmit-centric CoMP relies on heavy UE feedback, UL resource utilization, and low-latency information exchange between transmission points.

8.3.3 Successive Interference Cancellation

SIC is one of the simplest and the most intuitive methods to eliminate multiple-access interference.

The first step of SIC is to sort the user signals according to the received power. Each time only one user is selected, the strongest user in a cell is first decoded and then removed from the gamma stream before the second-strongest user is decoded and removed, and so on. The order of cancellation is typically aligned to the order of the users regarding receive power and cancellation is starting from the user(s)

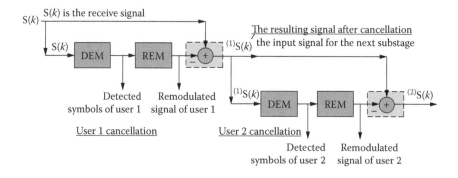

Figure 8.11   Serial interference cancellation. (DEM—demodulate; REM—remodulate)

with the highest receive power. Users are detected and subtracted from the received signal, one after the other (serially/sequentially), in multiple stages. Its working principle is detailed in Figure 8.11. When the number of users is increased, the SIC will produce a great time delay. If each stage delays 1 bit, K users will achieve $K-1$ bit delay. SIC may produce an aberration transmission effect because of imprecise evaluation. The operation sequence is decided by the signal power; the higher the power, the earlier the operation. Therefore, the weakest signal gets the most benefit. The performance of the serial interference canceller is decided to a great extent by the received user signal power distribution. If the distribution varies greatly, the performance's improvement is obvious. The performance of SIC is a great improvement over conventional detectors; there is little change in the hardware and it is easy to implement. The receiver uses soft outputs from the turbo decoder to regenerate a time domain signal that is converted to frequency domain and used to subtract intersymbol interference in case of turbo equalization and other users in case of SIC or PIC. In a live network, we can reduce ICI at the receiver side using the SIC technique in the uplink. SIC-capable eNB decodes the interference data coming from the UE. If the data is decoded without error, it subtracts the re-encoded interference data from the received signal so that eNB decodes the desired data without interference from the macro UE.

SIC detection has a significant flaw. Its performance depends on the reliability of the initial data estimation. Unreliable initial data detection will produce great interference in the second stage. In the SIC, demodulation of every user introduces a certain degree of processing delay. When the number of users is great, delay will accumulate to a level that the system cannot endure. Moreover, when the signal power strength of any user changes, the user processing sequence has to be reordered. Therefore, in the SIC scheme, a groupwise successive interference cancellation (GSIC) is proposed. This scheme divides users into groups and performs SIC. Each group of users should not be too large, and generally four users are suitable.

Advanced receivers based on iterative soft interference cancellation receivers (e.g., turbo SIC) are nonlinear receivers whose performance improves with the reliability of the interference reconstruction as the number of iterations increases. The turbo equalizer/SIC performs multiuser interference cancellation by utilizing an iterative loop between decoding and equalization. The complexity of the turbo SIC receiver depends on the number of iterations in the IC stage as well as on the number of turbo decoding iterations. When looking at the different contributors to the complexity in the receiver, turbo decoding plays the major part. Generally, turbo decoding is a many times more complex operation than equalization. By reducing the number of turbo decoding iterations for each turbo equalizer iteration, the overall increase in receiver complexity can be kept very low.

8.3.3.1 SIC-Receiver Modeling

Advanced receivers for multiuser MIMO (MU-MIMO) or single-user MIMO (SU-MIMO) can reduce multiuser or interstream interference. The turbo SIC detection strategy in a multiuser context consists of an iterative linear filtering with successive interference cancellation of the users, as depicted in Figure 8.12. The example here is for multiuser interference cancellation (same for MIMO); other functions can benefit from the iterative process (channel estimation, demapping, synchronization). Computational complexity can be kept reasonable by splitting the number of turbo code iterations among the interference cancellation iterations. Nonlinear receiver performance improves with the reliability of the interference reconstruction as the number of iterations increases.

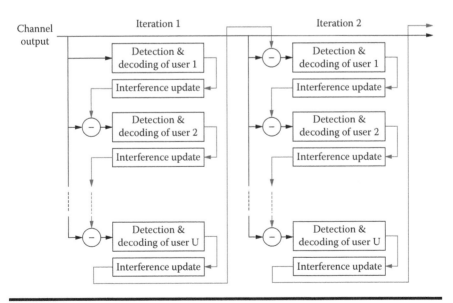

Figure 8.12 SIC process assuming natural ordering of the users.

For SIC receiver in a coordinated cell, the correctly received serving signal can be deducted from the total interference in the macro link of the macro diversity UE. So the $SINR_{MacroLink}$ (in the equation below) can be enhanced and the total combining gain can be further improved. If an ideal interference cancellation scheme is assumed, the $S_{MacroLink}$ can always be successfully cancelled out and the upper bound can be studied by using the IC.

The SIC SINR calculation can be according to this equation:

$$SINR_{MacroLink} = \sum_{Ant=1}^{Nrx} \frac{\sum_{PRB=1}^{K_{PRB}} S_{Nrx, K_{PRB}}}{\sum_{PRB=1}^{K_{PRB}} \left(\sum_{Interf=1}^{N_I} I_{Nrx, K_{PRB}, N_I}^{InterCell} - S_{Nrx, K_{PRB}}^{MacroLink} + n \right)}$$

where $S_{MacroLink}$ is the signal of the UE served by the coordination cell, S is the signal strength per PRB per antenna, $I^{InterCell}$ is the interference strength per PRB per antenna, Nrx is the number of Rx antennas per cell, K_{PRB} is the number of transmission PRBs assigned to a certain UE, N_I is the number of intercell interferences, and n is the noise spectral density.

8.4 ICIC Technology

Multicell wireless systems are inherently exposed to intercell interference originating from the spatial reuse of the same radio resources in neighboring cells. Intercell interference is a clear limiting factor in the performance achievable with LTE Rel-8, especially for cell-edge UEs. Accordingly, many of these systems, including the Global System for Mobile Communication (GSM), Universal Terrestrial Radio Access (UTRA), and LTE have employed some typically static, quasi-static, or interference randomization–based ICIC technology. One of the routinely employed static coordination solutions used in GSM is frequency reuse n, where each cell is assigned one out of the available n number of frequency bands (e.g., $n = 12$) such that no two directly neighboring cells get the same frequency. An obvious drawback of reuse n solutions is that only a fraction of the bandwidth is available in one cell and that may become the bottleneck in achieving high data rates. Mainly because of the bandwidth waste and the planning effort associated with reuse n solutions, the LTE Rel-8 system has been designed with the assumption of reuse 1 as the primary deployment case. Intercell interference in LTE appears in the form of "collisions" when the schedulers of neighboring cells assign the same resource block (RB) to concurrent transmissions that interfere with each other.

Normally, the cell-edge data rate is just about 10% of the peak rate, which needs to be improved. The cell-edge data rate is limited by intercell interference in the reuse 1 system in the LTE system. Reuse 1 is most efficient for overall efficiency, but it is bad for the cell edge. The target of ICIC is to push the cell-edge data rate

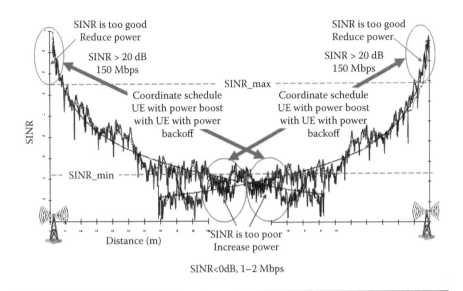

Figure 8.13 Cell-edge data rate and the peak rate.

up, potentially by cooperation and coordination between the cells. A comparison of the cell-edge data rate and the peak rate is shown in Figure 8.13.

Since the scheduler selects users for transmission on a RB (TTI = 1 ms) basis, the set of colliding transmissions will change rapidly from one TTI to the other, making it particularly challenging to follow this fast fluctuation in the interference coordination. During the standardization process, the 3GPP has extensively studied a range of intuitively appealing and feasible ICIC techniques, most of which center on lower-complexity heuristics with or without intercell communication.

8.4.1 ICIC Technology and Evolution

In conventional single-cell transmission/reception in LTE, the interference from neighboring cells is avoided or mitigated through coordination among neighboring cells. The Rel-8 interference management scheme relies on average value interference measurement. This is typically achieved by estimating the signal power from the sounding reference signal (SRS) and subtracting it from the total received power. It is difficult to estimate from which cell the interference is coming and to separate intracell, noise and jamming, and intercell interference. The various interference management techniques used in LTE can be classified as follows:

1. Interference randomization includes frequency selective scheduling (FSS), frequency hopping, soft frequency reuse, cell planning of sequences for sync channels, random access channels (RACHs), and pilot channels.

2. Interference control includes power control (fractional path loss–dependent power control), and interference over thermal (IoT) control.
3. Interference cancellation and reduction:
   - To aid intercell coordination, LTE Rel-8 defines two indicators that are exchanged between base stations: the high-interference indicator (HI) and the overload indicator (OI). HI provides information to neighboring cells about the part of the cell bandwidth upon which the cell intends to schedule its cell-edge users. OI provides information on the uplink interference level experienced in each part of the cell bandwidth. For the downlink, intercell interference coordination can be realized using a relative narrowband transmit power (RNTP) indicator. On receiving those parameters from neighboring eNBs, the eNB analyzes the intercell interference pattern, determines its own values of the ICIC parameters, and finally sends these values of the parameters to neighboring eNBs.
   - Coding and signal processing at the transmitter or receiver (e.g., IRC receivers, beamforming) to suppress interference.

However, these techniques are all implemented at a single cell without intercell cooperation on the physical layer. The multisite CoMP is a candidate interference cancellation technology in LTE-A for further coordinating intelligent physical resource allocation to reduce interference and improving the system performance.

On the other hand, uplink interference control can be achieved through a combination of power control and IoT control. For the LTE uplink, the standardized power control performs fractional path loss compensation (FPC), which allows a trade-off between near-cell user throughput and cell-edge user throughput. IoT control for the uplink adjusts the uplink SINR target based on interference reports from neighboring cells. These reports are the OIs and the HIs, the indicators that the eNBs exchange with each other via the standards defined for the X2 interface. In a live network, if the X2 interface is not available, information on transmit power can also be transmitted on the broadcast channel. As an example, other types of information such as traffic load information can be broadcast together. Neighboring Home eNB (HeNB) DL receiver is capable to capture the information periodically. UE relaying can be used to share coordination information too. The HeNB may request that its UE relay its coordination information or gather neighboring HeNB coordination information. The shared information enables the neighboring HeNBs to coordinate interference in each RB. Thus the HeNB can assign high transmit power on the RBs where neighboring HeNBs are not assigning high power. By exchanging coordination information, equilibrium of power assignment can be established between HeNBs.

8.4.2 ICIC Consideration of LTE-A

Enhanced intercell interference coordination (eICIC) should be considered in LTE-A. The eICIC mechanism is important to fully exploit the potential benefit of heterogeneous networks, and hence, we have concluded that some of the new eICIC techniques should be used for heterogeneous scenarios for LTE-A. UL CoMP, power control, time/frequency resource partitioning, and even coordinated beamforming are approaches of ICIC enhancement that exploit interference measurement (see Table 8.1).

We have known that CoMP is another approach to mitigate interference between cells and improve system performance through intra-eNB and inter-eNB dynamic coordination in the scheduling/transmission and/or joint transmission between/from multiple cell sites. Interference cancellation by way of CoMP is a mechanism to deal with intersite interference in a coordinated manner such as coordinated scheduling (CS) and coordinated beamforming (CB), and joint processing (JP). As a basic design principle of CoMP, the UEs feed back channel state information to the eNB for selected interfering cells in addition to the serving cell CSI. The CSI is exchanged between the cells, and the eNBs coordinate the scheduling decision and corresponding precoder on each time frequency resource to reduce the UEs' observed intercell interference. According to the level of coordination, CoMP can be categorized as scheduling-level coordination, beam-level coordination, and precoder-level coordination.

In Chapter 7, we gave many approaches of RB-level eICIC for data channel and time and frequency domain eICIC schemes for the control channel. For example, MBSFN subframes and almost blank subframes that we mentioned previously

Table 8.1 ICIC Technology Evolution

	ICIC	*eICIC*	*CoMP*
Control domain	Frequency domain control	Time domain control	Spatial domain
Principle	Channel quality indication (CQI) feedback with packet scheduler (or X2)	Time domain resource sharing	Coordination from multiple cells
Base transceiver station (BTS) synchronization required	No	Yes	Yes
Transport requirements	No	Only control plane	High-capacity transport for full gains (fiber)
3GPP standard release	Rel-8	LTE-A also works with Rel-8 UEs	Rel-10/11 study item

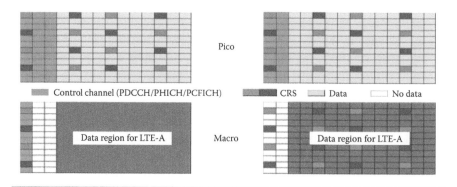

Figure 8.14 MBSFN subframes and almost blank subframes for eICIC. (PHICH—physical hybrid ARQ indicator channel; PCFICH—physical control format indicator channel)

could be used to reduce interference to the co-channel PDCCH in the other cell layer, which is shown in Figure 8.14. The macrocell and picocell can configure their MBSFN and almost blank subframes in a cooperative manner in the time domain. For MBSFN, the macrocell configures subframe N as an MBSFN subframe with only one OFDM symbol used for the control region, while the picocell configures this subframe as a normal subframe. Thus, only one OFDM symbol in the control region at the picocell would suffer the interference from the macro, and the PDCCH performance at the picocell could be improved compared with the situation where all of the three OFDM symbols are the victims. The almost blank subframe (ABS) refers to the subframe that carries only the CRS in the whole subframe region. Similar to the previous MBSFN, the interference of the control channel could be alleviated.

For example, we assume that the eNB1 sends the ABS pattern and the RNTP, UL HII (high interference indicator), and UL OI (overload indicator) to the eNB2. We also assume that the DL subframe n is configured as an ABS. The ABS may have no other signals that are transmitted in ABSs except the CRS. If the primary synchronization signal (PSS), secondary synchronization signal (SSS), physical broadcast channel (PBCH), system information block (SIB)1, paging, and the positioning reference signal (PRS) coincide with an ABS, they will be transmitted in the ABS. In CoMP scenarios, an aggressor cell informs a victim cell of certain ABSs in a time-domain manner, and based on this information the victim cell can perform appropriate user scheduling with RRC signaling to UEs for related RRM and CSI measurements (see Figure 8.15).

In the DL subframe, which is configured as an ABS by eNB1, eNB2 can do PDSCH scheduling without being concerned with the interference from eNB1 regardless of the RNTP message.

It is worth mentioning that the performance of the victim PDCCH and PDSCH could be improved further if the UE is enabled with the capability to detect and cancel the neighbor cell's CRS.

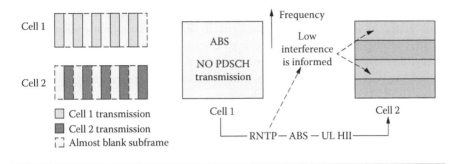

Figure 8.15 ABS. (HII—high interference indicator)

Another method of decreasing the interference is power reduction or muting on the portion of symbols of the macrocell that overlapped the control region of the picocell by scheduling and power control.

From the previous discussion we can see that Rel-8 and Rel-9 frequency domain ICIC and LTE-A time domain ICIC can coexist. Each eNB can send both Rel-8 and Rel-9 and LTE-A ICIC messages to its neighboring cell. Further, possible LTE-A eICIC methods include space domain coordination, network power control, network interference cancellation, network coding, network MIMO, and network beamforming, which means an aggressor cell informs a victim cell of certain kinds of network information in a space, time, code, or frequency domain manner, so as to be utilized at a victim cell for appropriate avoidance schemes.

8.4.3 CoMP for ICIC

A simple CoMP transmission scheme that relies on resource management cooperation among eNBs for controlling intercell interference is an efficient way to improve the cell-edge spectral efficiency. The scheduler, which is distributed across eNBs, can be used to determine the channel quality of a radio channel between an eNB and a UE and then adjust the number of data streams used for communication with a UE. The ICIC enhancement currently being studied for LTE-A can be classified into dynamic interference coordination and static interference coordination. In dynamic interference coordination, the utilization of frequency resources, spatial resources (beam pattern), or power resources is exchanged dynamically among eNBs. This scheme is flexible and adaptive to implement resource balancing in unequal load situations. For static interference coordination, both static and semi-static spatial resource coordination among eNBs are being considered. Possible ways for coordination among multiple points include those discussed in the following text.

Interference control among multiple points can be implemented by adjusting the transmit parameters of mutually interfering points such that their transmit power and interference at a target UE can be reduced to optimize the overall system throughout. Adjusting the power of each point includes the special case of assigning nonoverlapping RBs to different UEs. Basically, it is a fractional reuse scheme coordinated among multiple points. In this case, each individual point is not required to have any knowledge of the transmit symbols. At a UE, a typical requirement for supporting ICIC is the measurement and reporting of the reference signal received power (RSRP) corresponding to each point.

Interference control among multiple points can also be implemented by a TDM-based cooperative method (silencing, puncturing, symbol-level time offset, etc.) where the node causing severe interference to other nodes stops its transmission in a coordinated subframe as a time domain extension of Rel-8/9 ICIC techniques. In such cooperation, victimized nodes gain benefit from the "protected" subframes in which significant interference from aggressor nodes exists.

Interference control among multipoints can also be implemented by interference randomization with hopping among multipoints. Another way to achieve noncolliding resource allocation is to make the information packet to be sent to each UE available to all points. So each point transmits a fully or partially redundant subpacket to allow increment redundancy (IR) or chase combining at the UE. The potential gain seems to come from interference randomization. Interference control only mitigates the cross-interference as seen by UEs, but cannot turn the destructive interference into something useful. So if synchronization of the content at the symbol level among multiple points is possible (i.e., all multipoints are equipped with the same symbol content via exchanging the content in a synchronous network), multipoint simultaneous transmission can then occur to deliver more total transmit power to each UE. At the same time, site diversity gain can also be achieved. In addition to symbol-level content synchronization, we can further categorize constructive transmission according to whether the amplitude and phase of the transmitted signal is coherent or not across all antennas among multiple points.

In the case of coherent transmission across multipoints, signal waveforms sent from all antennas can be coherently amplitude/phase adjusted to allow better delivery of signal power to the UEs and often at the same time mitigate the interference leaked to other UEs that are served simultaneously in a multiuser fashion. The complex-valued weights for all antennas across multiple points are derived in a centralized fashion based on a single optimization problem from channel information to all antennas. After the global weights are derived at the single point, the weights for the antennas belonging to other points are conveyed from that point. Obviously, this kind of coordination has stringent requirements on the exchange of information via backhaul. Alternatively, the weight derivation can be done at each point according to the same rule, based on the same set of global information so that the weights calculated in a distributed manner are essentially the same as if they were calculated at a single point.

In the case of noncoherent transmission across multipoints, coherent transmission among antennas within a single point is still possible, but only local weights are derived in a distributed manner, typically based on nonglobal information on channels. In the case of no channel information at a participating point for a target UE, open loop rank 1 or 2 transmissions can still be helpful because constructive noninterfering signals are still sent from each participating point. A single frequency network (SFN) type of multipoint coordination is a typical example of noncoherent, but still constructive, transmission.

8.4.4 SON for ICIC

As we have mentioned previously, automatic interference control is a key piece of functionality in eNB self-optimization. For example, the load balancing shown in Figure 8.16 that makes a UE tune/attach to a cell other than the RF-wise most favorable one (strongest), will contribute to decreased interference. Interference reduction via eNB or cell switch off will utilize the SON functionality as another ICIC functionality. Again focus will be on switching off those cells that do not support traffic for a defined period of time. With energy reduction, the functionality will be used predominantly in hierarchical cell structures for reducing the interference from femto- or picocell systems used in homes or businesses. The added benefit to this will be that because of the nature of femto/pico networks, they will tend to be more densely deployed and therefore more prone to interference.

In this section, we focus on the approaches to manage interference, intercell interference coordination, and UL interference control, and how they can be combined with SON principles to result in automated and robust techniques to manage intercell interference in LTE networks.

8.4.4.1 ICIC as SON Functionality

In a reuse 1 deployment, it is critical to manage the uplink and downlink interference level. The eNBs can send uplink overload indications to neighbor eNBs via the X2 interface based on frequency and power resource allocations. Power control parameters can be adapted based on overload indicators, which allows control of the interference levels to ensure coverage and system stability.

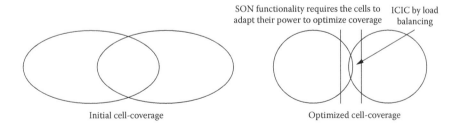

Figure 8.16   ICIC by load balancing.

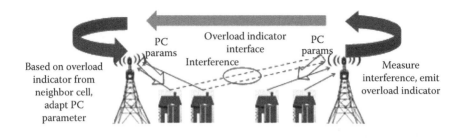

Figure 8.17 **Uplink and downlink interference between eNBs intelligently managed by SON feature. (PC—power control)**

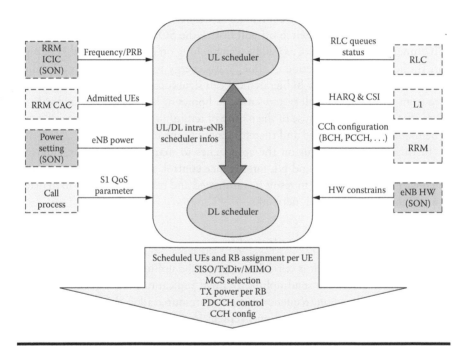

Figure 8.18 **SON impact on EICIC. (CAC—call admission control; QoS—quality of service; RLC—radio link control; HARQ—hybrid automatic repeat request; CCH—control channel; BCH—broadcast channel; PCCH—paging control channel; HW—hardware; SISO—single input, single output; TxDiv—transmit diversity)**

Honestly speaking, ICIC does not have the same impact on the uplink and the downlink. The uplink transmission of a UE at the cell border impacts all the UE of the neighboring cells that are scheduled on the same resource blocks. The downlink transmission to any UE in the cell impacts the UE from the neighboring cells that are standing at the cell border and using the same frequency resources.

Nevertheless, these (few) cell-edge UEs can escape the interference thanks to frequency selective scheduling. If the cells are not loaded, the cell-edge UE can find interference-free frequency blocks and get scheduled on them. Thus, the uplink is considered with higher priority. Uplink and downlink interference between eNBs intelligently managed by the SON feature is shown in Figure 8.17.

8.4.4.2 SON Impact on eICIC

The SON-based interference control functionality will be based on semistatic and dynamic interference coordination, which is shown in Figure 8.18. The semistatic scheme has been used in Rel-8 by automatically adjusting its power and scheduling of physical resource block settings to avoid intercell interference. The enhancements will include the additional ability to adjust network parameters, enabling self-configuration and self-optimization of downlink and uplink resources in a coordinated way between cells. By applying these restrictions or preferences to the downlink and uplink resources in a coordinated way between cells, the signal-to-interference ratio and cell-edge data rates/coverage on the corresponding time/ frequency resources in a neighbor cell are improved.

For example, self-configuration of the power setting per PRB for cells with strong neighborhood relationships in a multicell area get different restricted or preferred frequencies in an optimized way. This is based on the analysis of the neighbor cell list function of the SON; that is, the information regarding which neighbor cells will generate interference for the downlink traffic or which cells are interfered with by uplink traffic. On the other hand, the interference on the uplink can vary greatly between different resources used by macro- or picocells. Therefore, the power control operation needs to be resource specific in order to maintain the rise over thermal (RoT)* at reasonable levels.

---

* Take CDMA for example, rise over thermal tells how much stronger signal and interference are compared to noise. It is known in CDMA as a double-edged sword: high RoT gives high throughput but also makes CDMA systems unstable. In perfect joint detection, RoT directly determines system capacity: the higher the better.

Glossary of Acronyms

AAA: authentication, authorization, and accounting
ABS: almost blank subframes
ACK/NACK: acknowledgment/negative acknowledgment
AF: application function
AKA: authentication and key agreement
AMBR: aggregate maximum bit rate
AMC: adaptive modulation and coding
ARP: allocation and retention priority
AS: access stratum or application server
AuC: authorization center
BCH: broadcast channel
BF: beamforming
BLER: block error rate
BS: base station
BTS: base transceiver station
CA: carrier aggregation
CAPEX: capital expenditure
CB: coordinated beamforming
CBS: coordinated beam switching
CC: component carrier
CCPCH: common control physical channel
CDF: cumulative density function
CDM: code division multiplexing
CDMA: code-division multiple access
CGF: charging gateway function
CIO: cell individual offset
CM: cubic metric
CN: core network
CP: cyclic prefix
CQI: channel quality indication
C-RNTI: cell radio network temporary identifier
CRS: cell-specific reference signal

CS: circuit switched
CS/CBF: coordinated scheduling/coordinated beamforming
CSCF: call setup control function
CW: code word
DB: database
DCI: downlink control information
DeNB: donor eNB
DFT-S-OFDM: discrete Fourier transform spread OFDM
DL: downlink
DMRS: demodulation reference signal
DRA: dynamic resource allocation
DRB: data radio bearer
DRX: discontinuous reception
DSCP: differentiated services code point
DTX: discontinuous transmission
DwPTS: downlink pilot time slot
ECM: EPS connection management
EMM: EPS mobility management
eNB: evolved Node-B (eNode-B)
EPC: Evolved Packet Core
ePDG: evolved packet data gateway
EPS: evolved packet system
E-UTRAN: Evolved Universal Terrestrial Radio Access Network
EV-DO: Evolution-Data Optimized
FA: foreign agent
FDD: frequency-division duplex
FDMA: frequency division multiple access
FFT: fast Fourier transform
GBR: guaranteed bit rate
GDAS: generalized distributed antenna system
GERAN: SM/EDGE Radio Access Network
GGSN: Gateway GPRS Support Node
GLP: Grassmannian line packing
GPRS: general packet radio service
GPS: global positioning system
GSM: Global System for Mobile Communication
GTP: GPRS Tunneling Protocol
GUMMEI: globally unique MME ID
GUTI: globally unique temporary ID
HA: home agent
HARQ: hybrid automatic repeat request
HO: handover
hPCRF: home PCRF

HRPD: high-rate packet data
HSPA: High-Speed Packet Access
HSS: Home Subscriber Server
HW: hardware
ID: identifier
IE: information element
IETF: Internet Engineering Task Force
IFFT: inverse fast Fourier transform
IMS: IP multimedia subsystem
IMSI: International Mobile Subscriber Identity
I/Q: in-phase/quadrature
IR: increment redundancy
ISI: intersymbol interference
L2: Layer 2
L3: Layer 3
LTE: Long Term Evolution
MAC: Medium Access Control
MAG: Mobile Access Gateway
MAP: Mobile Application Part
MBMS: Multimedia Broadcast/Multicast Service
MBR: maximum bit rate
MBSFN: multimedia broadcast/multicast service single frequecy network
MCC: Mobile Country Code
MCCH: Multicast Control Channel
MCE: MBMS Coordination Entity
MCS: modulation and coding scheme
MGW: media gateway
MIMO: multiple-input, multiple-output
MIP: mobile IP
MM: mobility management
MME: mobility management entity
MNC: Mobile Network Code
MNO: mobile network operator
MOD: modulation
MRC: maximum ratio combining
MS: mobile station
MTCH: MBMS traffic channel
M-TMSI: M-Temporary Mobile Subscriber Identity
NAS: non-access stratum
NE: network element
NRT: non-real-time
NTP: Network Time Protocol
Nx: N times

OC: orthogonal code
O&M: operation and maintenance
OFDM: orthogonal frequency-division multiplexing
OFDMA: orthogonal frequency-division multiple access
OL-SM: open loop spatial multiplexing
OPEX: operating expenses
PA: power amplifier
PARA: peak to average power rate
PAPR: peak-to-average power ratio
PBCH: physical broadcast channel
PCC: policy and charging control
PCEF: policy and charging enforcement function
PCFICH: physical control format indicator channel
PCH: paging channel
PCRF: policy and charging rules function
PDCCH: physical DL control channel
PDCP: packet data convergence protocol
PDG: packet data gateway
PDN: packet data network
PDP: packet data protocol
PDSCH: physical DL shared channel
PDSN: Packet Data Serving Node
PDU: protocol data unit
PGW: packet data network (PDN) gateway
PHICH: physical hybrid ARQ indicator channel
PLMN: public land mobile network
PMCH: physical multicast channel
PMIP: proxy mobile IP
PMI/RI: precoding matrix indicator/rank indicator
PPS: pulse per second
PRACH: physical random access channel
PRB: physical resource block
PS: packet switched
P/S: parallel/serial
PSS: primary synchronization signals
PUCCH: physical uplink control channel
PUSCH: physical uplink shared channel
QAM: quadrature amplitude modulation
QCI: QoS class identifier
QoS: quality of service
QPSK: quadrature phase-shift keying
RA: random access
RAC: routing area code

RACH: random access channel
RAT: radio access technology
RAU: radio access update
RB: resource block
RBC: radio bearer control
RC: radio configuration
RE: resource element
RF: radio frequency
RLC: radio link control
RN: relay node
RNC: radio network controller
RNL: radio network layer
RNS: radio network subsystem
RNTI: radio network temporary identifier
RRC: radio resource control
RRM: radio resource management
RS: reference signals
RT: real time
RV: redundancy version
SAE: system architecture evolution
SC-FDMA: single carrier frequency-division multiple access
SCH: synchronization channel
SCID: scrambling identity
SC-OFDMA: single-carrier OFDMA
SCTP: Stream Control Transmission Protocol
SDF: service data flow
SDU: service data unit
SFN: single-frequency network
SGSN: serving GPRS support node
SGW: serving gateway
SIM: subscriber identity module
SNR: signal-to-noise ratio
SON: self-organizing network
S/P: serial/parallel
SR: scheduling request
SRS: sounding reference signal
SSS: secondary synchronization signals
S-TMSI: SAE temporary mobile subscriber identifier
SVD: single value decomposition
SW: software
TA: tracking area
TB: transport block
TAC: tracking area code

TAI: tracking area identity
TAU: tracking area update
TDD: time-division duplexing
TDM: time-division multiplexing
TEID: tunnel endpoint identifier
TFT: traffic flow template
3GPP: Third Generation Partnership Project
TNL: transport network layer
TPC: transmit power control
TTI: transmission time interval
TX: transmission
UE: user equipment
UICC: Universal Integrated Circuit Card
UL: uplink
UMTS: universal mobile telecommunications system
UN: ratio interface between DeNB and the RN
USIM: Universal Subscriber Identity Module
UTC: Coordinated Universal Time
UU: air interface between the UE and the RN
VCC: voice call continuity
VCC AS: voice call continuity application server
VLAN: virtual local area network
VoIP: Voice over Internet Protocol
VPN: virtual private network
WCDMA: wideband code-division multiple access
WiMAX: Worldwide Interoperability for Microwave Access
WLAN: wireless local area network

Bibliography

3rd Generation Partnership Project (3GPP)

3rd Generation Partnership Project (3GPP). TR 21.905 v11.0.1 (2011–12): "Vocabulary for 3GPP Specifications."

3rd Generation Partnership Project (3GPP). TS 36.211 v10.4.0 (2011–12): "Evolved Universal Terrestrial Radio Access (E-UTRA); Physical Channels and Modulation."

3rd Generation Partnership Project (3GPP). TS 36.212 v10.4.0 (2011–12): "Evolved Universal Terrestrial Radio Access (E-UTRA); Multiplexing and Channel Coding."

3rd Generation Partnership Project (3GPP). TS 36.213 v10.4.0 (2011–12): "Evolved Universal Terrestrial Radio Access (E-UTRA); Physical Layer Procedures."

3rd Generation Partnership Project (3GPP). TS 36.214 v10.1.0 (2011–03): "Evolved Universal Terrestrial Radio Access (E-UTRA); Physical Layer—Measurements."

3rd Generation Partnership Project (3GPP). TS 36.216 v10.3.1 (2011–09): "Evolved Universal Terrestrial Radio Access (E-UTRA); Physical Layer for Relaying Operation."

3rd Generation Partnership Project (3GPP). TR 36.814 v. 9.0.0 (2010–03): "Further Advancements for E-UTRA Physical Layer Aspects (Release 9)."

3rd Generation Partnership Project (3GPP). TS 36.300 v11.0.0 (2011–12): "E-UTRAN Overall Description: Stage 2."

3rd Generation Partnership Project (3GPP). TR 36.913 v10.0.0 (2011–03): "Requirements for Evolved UTRA (E-UTRA) and Evolved UTRAN (E-UTRAN)."

3rd Generation Partnership Project (3GPP). TR 36.912 v10.0.0 (2011–03): "Feasibility Study for Further Advancements for E-UTRA (LTE-Advanced)."

3rd Generation Partnership Project (3GPP). TR 36.819 v11.1.0 (2011–12): "Coordinated Multi-Point Operation for LTE Physical Layer Aspects (Release 11)."

3rd Generation Partnership Project (3GPP). R1-091660 (2009–03): "ITU-R Submission Template."

3rd Generation Partnership Project (3GPP). TS 36.413 v10.4.0 (2011–12): "Evolved Universal Terrestrial Radio Access Network (E-UTRAN); S1 Application Protocol (S1AP)."

3rd Generation Partnership Project (3GPP). TS 36.423 v10.4.0 (2011–12): "Evolved Universal Terrestrial Radio Access Network (E-UTRAN); X2 Application Protocol (X2AP)."

3rd Generation Partnership Project (3GPP). TR 25 996 v10.0.0 (2011–03): "Spatial Channel Model for Multiple Input Multiple Output (MIMO) Simulations."

Carrier Aggregation (CA)

3rd Generation Partnership Project (3GPP). TR 36.808 v. 0.3.0: "Carrier Aggregation: BS Radio Transmission and Reception (Release 10)."

IST-1999-11571 EMBRACE D8, 2001, "TDD versus FDD Access Schemes [R]."

Next Generation Mobile Networks. "Next Generation Mobile Networks Spectrum Requirements Update [R]." A White Paper Update by the NGMN, 2009.

3rd Generation Partnership Project (3GPP). TR 36.913 v. 8.0.0: "Requirements for Further Advancements for E-UTRA," 2008.

REV-080030: "Technology Components," Ericsson, April 2008.

R1-082569: "Consideration on Technologies for LTE-A," CATT, June 2008.

R1-082448: "Carrier Aggregation in Advanced E-UTRA," Huawei, June 2008.

R1-082468: "Carrier Aggregation in LTE-A," Ericsson, June 2008.

RP-100661: "Revised Carrier Aggregation for LTE WID—Core Part."

R1-084319: "Considerations on SC-FDMA and OFDMA for LTE-Advanced Uplink," Nokia Siemens Networks, Nokia.

R1-094415: "Concept of Carrier Segment for LTE-A." 3GPP TSG-WG1 #58 bis

R1-084403: "Multi-Antenna Uplink Transmission for LTE-A," Motorola.

R1-082945: "Uplink Multiple Access Schemes for LTE-A," LG Electronics.

R1-082609: "Uplink Multiple Access for LTE-Advanced," Nokia Siemens Networks, Nokia.

R1-091878: "Concurrent PUSCH and PUCCH Transmissions," Samsung.

R1-083232: "Carrier Aggregation for LTE-A: E-NodeB Issues," Motorola.

R1-083820: "Uplink Access for LTE-A: Non-aggregated and Aggregated Scenarios," Motorola.

R1-092173: "PDCCH Beamforming for LTE-A," Motorola.

R1-093206: "Designing Issues for Carrier Aggregation," ZTE.

R1-110857: "Remaining Issue Regarding Resource Allocation for Channel Selection," NTT DOCOMO.

R1-100849: "Discussion on DM-RS Power Boosting," Ericsson, ST-Ericsson.

R1-102946: "Power Headroom Reporting for Uplink Carrier Aggregation," Nokia Siemens Networks, Nokia Corporation.

R1-090266: "Non-Contiguous Resource Allocation in Uplink LTE-A," Motorola.

Collaborative Multipoint (CoMP)

3rd Generation Partnership Project (3GPP). TR 36.814 v1.0.1: "Further Advancements for E-UTRA Physical Layer Aspects (Release 9)[S]," 2009.

R1-094279: "Extended ICIC-A Rel-10 CoMP Scheme," Ericsson, ST-Ericsson, October 2009.

R1-092147: "A Progressive Multi-Cell MIMO Transmission with Sequential Linear Precoding Design in DL TDD Systems," Alcatel-Lucent Shanghai Bell, Alcatel-Lucent, May 2009.

R1-082886: "Inter-cell Interference Mitigation through Limited Coordination," Samsung, August 2008.

R1-091936: "Spatial Correlation Feedback to Support LTE-A MU-MIMO and CoMP: System Operation and Performance Results," Motorola, May 2009.

R1-100331: "Coordinated Beamforming/Scheduling Performance Evaluation," Nokia Siemens Networks, Nokia, January 2010.

R1-101431: "CoMP Performance Evaluation," Nokia Siemens Networks, Nokia, February 2010.

R1-092691: "Preliminary CoMP Gains for ITU Micro Scenario," Qualcomm Europe, July 2009.

R1-101354: "System Level Performance with CoMP CB," LG Electronics Inc., February 2010.

R1-101355: "System Level Performance with CoMP JT," LG Electronics Inc., February 2010.

R1-093016: "Consideration on Performance of Coordinated Beamforming with PMI Feedback," Alcatel-Lucent, August 2009.

R1-101173: "Performance Evaluation of CoMP CS/CB," Samsung, February 2010.

3rd Generation Partnership Project (3GPP). R1-090821: "Solutions for DL CoMP Transmission: For Issues on Control Zone, CRS and DRS," Huawei, Qualcomm Europe, RITT, and CMCC.

3rd Generation Partnership Project (3GPP). R1-092822: "Views on the Relationship among CoMP Sets," CMCC.

R1-100820: "Evaluation Scenarios and Assumptions for Intra-Site CoMP," NTT DOCOMO.

R1-090100: "SRS Transmission Issues for LTE-A," Samsung.

R1-084336: "Analysis on Uplink/Downlink Time Delay Issue for Distributed Antenna System." Huawei, CMCC, RITT, CATT.

R1-091618: "System Performance Evaluation for Uplink CoMP," Huawei.

R1-091664: "CoMP TP for TR," Qualcomm Europe.

R1-092363: "LTE-A Downlink DM-RS Pattern Design," Huawei.

R1-093030: "DMRS Design Considerations for LTE-A," Huawei.

R1-093833: "System Performance Comparisons of Several DL CoMP Schemes," Huawei.

R1-093846: "Common Feedback Design for CoMP and Single Cell MIMO," Huawei.

R1-084115: "Downlink CoMP Transmitting Scheme Based on Beamforming," ZTE.

R1-082886: "Inter-Cell Interference Mitigation through Limited Coordination," Samsung.

R1-091688: "Potential Gain of DL CoMP with Joint Transmission," NEC Group.

R1-090193: "Aspects of Joint Processing in Downlink CoMP," CATT.

R1-091835: "Consideration on UE Feedback in Support of CoMP," Texas Instruments.

R1-091520: "Analysis of Feedback Signalling for Downlink CoMP," CATT.

R1-090745: "Cell Clustering for CoMP Transmission/Reception," Nortel.

R1-083569: "Further Discussion on Inter-Cell Interference Mitigation through Limited Coordination," Samsung.

R1-090325: "Coordinated Beamforming/Precoding and Some Performance Results," Motorola.

R1-091969: "Considerations on the Selection Method for CoMP Cells," Potevio.

R1-091976: "Multi-Cell PMI Coordination for Downlink CoMP," ETRI.

R1-084444: "Aspects of Coordinated Multi-Point Transmission for Advanced E-UTRA," Texas Instruments.

R1-093036: "Practical Analysis of CoMP Coordinated Beamforming," Huawei.

R1-093152: "CSI-RS Design for Virtualized LTE Antenna in LTE-A System," Fujitsu.

R1-094234: "Remaining Issues for Rel. 9 Downlink DM-RS Design," NTT DOCOMO.

R1-094467: "DM RS Sequence Design for Dual Layer Beamforming," LG Electronics.

R1-093841: "Further Design and Evaluation on CSI-RS for CoMP," Huawei.

R1-093865: "Mapping of UL RS Sequence for Clustered DFT-S-OFDM," NEC Group, NTT DOCOMO.

R1-093506: "UL RS Enhancement for LTE-Advanced," NTT DOCOMO.

R1-094784: "DM-RS Design for Rank 5–8," LG Electronics.

R1-092651: "Discussions on CSI-RS for LTE-Advanced," Samsung.

R1-094867: "Details of CSI-RS," Qualcomm Europe.

R1-094869: "UE-RS Patterns for Ranks 5 to 8," Qualcomm Europe.

R1-093862: "Rel-8 Cell-Specific RS as CSI-RS for LTE-A," NEC Group, NTT DOCOMO.

R1-094909: "Views on CSI-RS Design Issues for LTE-Advanced," NTT DOCOMO.

R1-100884: "Discussion on PRB Bundling for Rank 1–8," CATT.

R1-100745: "Increasing Sounding Capacity for LTE-A," Texas Instruments.

R1-101175: "Interference Mitigation Based on Rank Restriction and Recommendation," Samsung.

3rd Generation Partnership Project (3GPP). R1-100616, Potevio, "Proposal for an Enhanced SRS Scheme for CoMP," January 2010.

R1-101224: "Views on SRS Enhancement for LTE-Advanced," NTT DOCOMO.

R1-100451: "Required CSI-RS Density for Rel-10 SU-MIMO Transmission," Texas Instruments.

R1-103016: "PRB Bundling for Rel-10," Samsung.

Cover, T.M. and A. A. El Gamal. 1979. "Capacity Theorems for the Relay Channel," *IEEE Trans. Inform. Theory*, vol. 25, no. 5, pp. 572–584, Sept.

Zheng, Naizheng; Malek Boussif; Claudio Rosa; Istvan Z. Kovacs; Klaus I. Pedersen; Jeroen Wigard; and Preben E. Mogensen. "Uplink Coordinated Multi-Point for LTE-A in the Form of Macro-Scopic Combining."

Relay

3rd Generation Partnership Project (3GPP). TR 36.806 v0.1.1: "Relay Architectures for E-UTRA (LTE-A) (Release 9)."

TS 36.216: "Physical Layer for Relaying Operation."

R1-100940: "Timing Synchronisation for TDD Type 1 Relay," Alcatel-Lucent Shanghai Bell, Alcatel-Lucent.

R1-101384: "Considerations on Backhaul Interference and Synchronization for Relay," CMCC.

R1-100976: "Synchronization in Backhaul Link," ZTE.

R1-100752: Support of Synchronization between eNB-UE and RN-UE Link," LG Electronics.

R1-100435: "Type 1 Relay Timing and Node Synchronization," Alcatel-Lucent Shanghai Bell, Alcatel-Lucent.

R1-100558: "Relay Node Synchronization and DL/UL Subframe Timing," CMCC.

R1-094040: "Considerations on the Synchronization of Relay Nodes," CMCC.

R1-091389: "Synchronization Channel for LTE-A System with Relay," Nortel.

R1-090593: "On the Design of Relay Node for LTE-Advanced," Texas Instruments.

R1-091762: "Cell Edge Performance for Amplify and Forward vs. Decode and Forward Relays," Nokia Siemens Networks, Nokia.

R1-091384: "Discussion Paper on the Control Channel and Data Channel Optimization for Relay Link," Nortel.

R1-090753: "Control Channel and Data Channel Design for Relay Link in LTE-Advanced," Nortel.

R1-093145: "DL Carrier Aggregation Performance in Heterogeneous Networks," Qualcomm Europe.

R1-093169: "Design of the UL Backhaul for a Type I Relay," Texas Instruments.

RP-110398: "New Study Item Proposal: Mobile Relay for EUTRA," CATT, CMCC, CATR.

R1-094824: "Relay Performance Evaluation," CMCC.

R1-102918: "Simulation Study on Downlink RN to RN Interference," ZTE.

Multiple-Input, Multiple-Output (MIMO)

R1-084376: "Uplink SU-MIMO in LTE-Advanced," Ericsson.

R1-101071: "Codebook for Uplink Rank 3 Precoding," Huawei.

R1-092100: "UL Transmit Diversity Schemes in LTE-Advanced," NTT DOCOMO.

R1-091818: "Cubic Metric Friendly Precoding for UL 4Tx MIMO," Huawei.

R1-082812: "Collaborative MIMO for LTE-Advanced Downlink," Alcatel Shanghai Bell, Alcatel-Lucent.

R1-091402: "Way Forward on Channel Dependent Precoding for UL SU-MIMO."

R1-091795: "Considerations on Downlink Antenna Mapping," Huawei.

R1-092274: "MIMO AH Summary," Samsung.

R1-072843: "Way Forward on 4-Tx Antenna Codebook for SU-MIMO."

R1-092389: "Adaptive Codebook Designs for DL MU-MIMO," Huawei.

R1-084201: "Consideration on DL-MIMO in LTE-Advanced," LG Electronics.

R1-083570: "Codebook-Based Precoding for 8 Tx Transmission in LTE-A," Samsung.

R1-083567: "Discussions on 8-TX Diversity Schemes for LTE-A Downlink," Samsung.

R1-083830: "A Structured Approach for Studying DL-MIMO Enhancements for LTE-A," Motorola.

R1-091353: "On CSI Feedback Signalling in LTE-Advanced Uplink," Nokia Siemens Networks, Nokia.

R1-082501: "Collaborative MIMO for LTE-A Downlink," Alcatel Shanghai Bell, Alcatel Lucent.

R1-083239: "ICIC with Multi-Site Collaborative MIMO," Mitsubishi Electric.

R1-084350: "Beamforming Enhancement in LTE-Advanced," Huawei, CMCC, CATT.

R1-084466: "Design Aspect for Higher-Order MIMO in LTE-Advanced," Nortel.

R1-092608: "4Tx UL Codebook: Antenna Turn-Off Elements and Unitary Structure/CM-Preserving," Motorola.

R1-092339: "The Benefits of One PA Mode for UEs Supporting Multiple Pas," Sharp.

R1-092130: "Codebook Design for 4Tx Uplink SU-MIMO," LG Electronics.

R1-091752: "Performance Study on Tx/Rx Mismatch in LTE TDD Dual-Layer Beamforming," Nokia, Nokia Siemens Networks, CATT, ZTE.

R1-090289: "Supporting 8Tx Downlink SU-MIMO for Advanced E-UTRA," Texas Instruments.

R1-091842: "Progressing on 4Tx Codebook Design for Uplink SU-MIMO," Texas Instruments.

R1-091888: "Codebook Design for 8 Tx Transmission in LTE-A," Samsung.

R1-071510: "Details of Zero-Forcing MU-MIMO for DL EUTRA," Freescale Semiconductor Inc.

R1-091976: "Multi-cell PMI Coordination for Downlink CoMP," ETRI.

R1-091515: "Beamforming Based MU-MIMO for LTE-TDD," CATT, CMCC.

R1-092000: "Discussion on Non-Codebook Based Precoding," CATT, RITT, Potevio.

R1-092027: "Uplink SU-MIMO in LTE-Advanced," Ericsson.

R1-093474: "Coordinated Beamforming with DL MU-MIMO," Texas Instruments.

Sadek, M., A. Tarighat, and A. H. Sayed, "A Leakage Based Precoding Scheme for Downlink Multiuser MIMO Channels," *IEEE Trans. Wireless Commun.* 6, no. 5 (2007): 1711–1721.

Spencer, Q. H., A. L. Swindlehurst, and M. Haardt, "Zero-Forcing Methods for Downlink Spatial Multiplexing in Multiuser MIMO Channels," *IEEE Trans. on Sig. Proc.* 52, no. 2 (2004): 461–471.

R1-093331: "Grid of Beams: A Realization for Downloadable Codebooks," Alcatel-Lucent.

R1-093996: "Consideration on MU-MIMO and Related Signaling Support in LTE-A," Texas Instruments.

R1-094710: "Transparency of the MU-MIMO," Huawei.

R1-094943: "Discussion on DL MU-MIMO in LTE-A," Fujitsu.

R1-102791: "Further Development of Two-Stage Feedback Framework for Rel-10," Alcatel-Lucent, Alcatel-Lucent Shanghai Bell.

R1-103026: "Views on the Feedback Framework for Rel. 10," Samsung.

R1-100051: "A Flexible Feedback Concept," Ericsson, ST-Ericsson.

R1-101219: "Views on Codebook Design for Downlink 8Tx MIMO," NTT DOCOMO.

R1-103106: "Double Codebook Based Differential Feedback," Huawei.

HetNet

R1-101594: "TP on Heterogeneous Networks," Ericsson.

R1-105081: "Summary of the Description of Candidate eICIC Solutions," CMCC.

R1-093466: "Component Carrier Operation without PDCCH," Panasonic.

R1-093788: "Technology Issues for Heterogeneous Network for LTE-A," Alcatel-Lucent, Alcatel-Lucent Shanghai Bell.

R4-091976: "LTE-FDD HeNB Interference Scenarios," Vodafone, AT&T, Alcatel Lucent, picoChip Designs, Qualcomm Europe.

R2-093853: "HeNB Issue," LG Electronics Inc.

R3-090700: "Self-Synchronization," Qualcomm Europe.

R1-092722: "Time Synchronization Requirements for Different LTE-A Techniques," Qualcomm Europe.

R1-100903: "Uplink Interference Mitigation via Power Control," CATT.

R1-100701: "Importance of Serving Cell Selection in Heterogeneous Networks," Qualcomm Incorporated.

R1-100350: "Downlink CCH Performance Aspects for Co-Channel Deployed Macro and HeNBs," Nokia Siemens Networks, Nokia.

R1-101982: "LTE Non-CA Based HetNet Support," Huawei.

R1-102678: "Interference Management for Control Channels in Outdoor Hotzone Scenario," Kyocera.

R1-102150: "On Range Extension in Open-Access Heterogeneous Networks," Motorola.

R1-103626: "Considerations on PBCH eICIC for CSG HeNB," ITRI.

R1-103873: "Performance Analysis of PDCCH Interference Mitigation Techniques in Outdoor Hotzone Het-Net," ZTE.

Institute of Electrical and Electronics Engineers (IEEE). "IEEE Standard Definitions and Concepts for Dynamic Spectrum Access: Terminology Relating to Emerging Wireless Networks, System Functionality, and Spectrum Management." IEEE Communications Society, IEEE Std. 1900.1 TM, September 2008.

Wang, Wei. "A Brief Survey on Cognitive Radio." Institute of Information and Communication Engineering Zhejiang University, P.R. China.

Self-Organizing Network (SON)

3rd Generation Partnership Project (3GPP). TR 32.500 v8.0.0 (2008-12): "3rd Generation Partnership Project; Technical Specification Group Services and System Aspects; Telecommunication Management; Self-Organizing Networks (SON); Concepts and Requirements (Release 8)."

3rd Generation Partnership Project (3GPP). TR 32.501 v8.0.0 (2008-12): "3rd Generation Partnership Project; Technical Specification Group Services and System Aspects; Telecommunication Management; Self Configuration of Network Elements; Concepts and Requirements (Release 8)."

3rd Generation Partnership Project (3GPP). TR 32.511 v8.1.0 (2009-03): "3rd Generation Partnership Project; Technical Specification Group Services and System Aspects; Telecommunication Management; Automatic Neighbour Relation (ANR) Management; Concepts and Requirements (Release 8)."

3rd Generation Partnership Project (3GPP). TR 32.762 v8.0.0 (2009-03): "3rd Generation Partnership Project; Technical Specification Group Services and System Aspects; Telecommunications Management; Evolved Universal Terrestrial Radio Access Network (E-UTRAN) Network Resource Model (NRM) Integration Reference Point (IRP): Information Service (IS) (Release 8)."

3rd Generation Partnership Project (3GPP). TR 36.300 v8.8.0 (2009-03): "3rd Generation Partnership Project; Technical Specification Group Radio Access Network; Evolved Universal Terrestrial Radio Access (E-UTRA) and Evolved Universal Terrestrial Radio Access Network(E-UTRAN); Overall Description; Stage 2 (Release 8)."

TS 37.320: "Radio Measurement Collection for Minimization of Drive Tests (MDT); Overall Description; Stage 2."

Next Generation Mobile Networks (NGMN) Alliance. "Next Generation Mobile Networks Recommendation on SON and O&M Requirements," December 5, 2008.

Next Generation Mobile Networks (NGMN) Alliance. "Informative List of SON Use Cases, Annex A (Informative) of Use Cases Related to Self-Organising," April 2007.

Nomor 3GPP Newsletter "Self-Organizing Networks in LTE," May 2008.

Next Generation Mobile Networks (NGMN) Alliance. "Self-Organizing Networks (SON) for LTE Network. Overall Description," Version 1.53, April 17, 2007.

Next Generation Mobile Networks (NGMN) Alliance. "Use Cases Related to Self-Organising Network. Overall Description." Version 2.02, April 16, 2007.

3rd Generation Partnership Project (3GPP). TS 36.902: "Evolved Universal Terrestrial Radio Access Network (E-UTRAN); Self-Configuring and Self-Optimizing Network (SON) Use Cases and Solutions."

3rd Generation Partnership Project (3GPP). TS 32.502: "Telecommunication Management; Self-Configuration of Network Elements Integration Reference Point (IRP); Information Service (IS)."

3rd Generation Partnership Project (3GPP). "Performance Results for Cell Global Identity Detection in E-UTRAN," 3GPP TSG RAN, WG4 #48bis, R4-082493, Ericsson, Edinburgh, UK, September 2008.

3rd Generation Partnership Project (3GPP). "Location Based Cell Border Information Share during Handover," 3GPP TSG RAN, WG3 #64, R3-091295, Alcatel-Lucent, San Francisco, CA, May 2009.

R1-091678: "Performance Prediction of Turbo-SIC Receivers for System-Level Simulations," Orange, Nokia, Nokia Siemens Networks, Texas Instruments.

R1-084321: "Algorithms and Results for Autonomous Component Carrier Selection for LTE-Advanced," Nokia Siemens Networks, Nokia.

Quality of Service (QoS)

TS 22.278: "Service Requirements for Evolved Packet System (EPS) (Release 8)."

TS 23.107: "Quality of Service (QoS) Concept and Architecture (Release 7)."

TS 23.203: "Policy and Charging Control Architecture (Release 8)."

TS 23.401: "GPRS Enhancements for E-UTRAN Access (Release 8)."

TS 23.402: "Architecture Enhancements for Non-3GPP Accesses (Release 8)."

TS 24.008: "Mobile Radio Interface Layer 3 Specification: Core Network Protocols; Stage 3 (Rel-8)."

TS 29.212: "Policy and Charging Control over Gx Reference Point (Release 8)."

TS 36.300: "Evolved Universal Terrestrial Radio Access (E-UTRA) and Evolved Universal Terrestrial Radio Access Network (E-UTRAN); Overall Description; Stage 2 (Release 8)."

TS 36.321: "Medium Access Control (MAC) Protocol Specification (Release 8)."

TS 36.322: "Radio Link Control (RLC) Protocol Specification (Release 8)."

TS 36.331: "Radio Resource Control (RRC) Protocol Specification (Release 8)."

TS 36.413: "S1 Application Protocol (S1 AP) (Release 8)."

TS 36.423: X2 Application Protocol (X2AP) (Release 8Multiuser Detection (MUD)

Alexander, P. D., M. C. Reed, et al. "Iterative Multiuser Interference Reduction: Turbo CDMA." *IEEE Trans. Commun.* 47, no. 7 (1999): 1008–1014.

Boutros, J., and G. Caire. "Iterative Multiuser Joint Decoding: Unified Framework and Asymptotic Analysis." *IEEE Trans. Inform. Theory* 48, no. 7 (2002): 1772–1793.

Divsalar, D., M. Simon, and D. Raphaeli. "Improved Parallel Interference Cancellation for CDMA" *IEEE Trans. Commun.* 46 (1998): 258–268.

Duel-Hallen, A., J. Holtzman, and Z. Zvonar. "Multiuser Detection for CDMA Systems." *IEEE Person. Commun.* 2, no. 2 (1995): 46–58.

El Gamal, H., and E. Geraniotis. "Iterative Multiuser Detection for Coded CDMA Signals in AWGN and Fading Channels." *IEEE J. Select. Areas Commun.* 18, no. 1 (2000): 30–41.

Lu, B., and X. Wang. "Iterative Receivers for Multiuser Space-Time Coding Systems." *IEEE J. Select. Areas Commun.* 18, no. 11 (2000): 2322–2335.

Moher, M. "An Iterative Multiuser Decoder for Near-Capacity Communications." *IEEE Trans. Commun.* 46, no. 7 (1998): 870–880.

Enhanced Intercell Interference (EICIC)

R1-094659: "Autonomous CC Selection for Heterogeneous Environments," Nokia Siemens Networks, Nokia.

R1-102885: "Possibility of UE-Side ICI Cancellation in HetNet," Panasonic.

R1-103126: "Enhanced ICIC for Control Channels to Support HetNet," Huawei.

R1-103458: "Analysis on the eICIC Schemes for the Control Channels in HetNet," Huawei.

R1-103498: "Evaluation of R8/9 Techniques and Enhancements for PUCCH Interference Coordination in Macro-Pico," CATT.

R1-084319: "Considerations on SC-FDMA and OFDMA for LTE-Advanced Uplink," Nokia Siemens Networks, Nokia.

Dominique, Francis, Christian G. Gerlach, Nandu Gopalakrishnan, Anil Rao, James P. Seymour, Robert Soni, Aleksandr Stolyar, Harish Viswanathan, Carl Weaver, and Andreas Weber. "Self-Organizing Interference Management for LTE." *Bell Labs Technical Journal* 15, no. 3 (2010): 19–42.

Index

P